George Whitehead

# Understanding
# Behavioral
# Research

# Understanding Behavioral Research

Nancy S. Harrison
California State University, Hayward

Wadsworth Publishing Company, Inc.
Belmont, California

Psychology Editor: Kenneth King
Production Editor: Kathie Head
Designer: Dare Porter
Copy Editor: Claire Annchild Connelly

© 1979 by Wadsworth Publishing Company, Inc., Belmont, California 94002. All rights reserved. No part of this book may be reproduced, stored in a retrieval system, or transcribed, in any form or by any means, electronic, mechanical, photocopying, recording, or otherwise, without the prior written permission of the publisher.

Printed in the United States of America
4  5  6  7  8  9  10

**Library of Congress Cataloging in Publication Data**
Harrison, Nancy S
 Understanding behavioral research.

 Bibliography: p.
 Includes index.
 1. Psychology, Experimental.  2. Psychological research.  3. Experimental design.  4. Psychometrics.
5. Report writing.  I. Title.
BF181.H34                               150'.7'2                          78-17306
ISBN 0-534-00597-7

ISBN 0-534-00597-7

**Acknowledgments**

The Cohen and Smith Case Study is from Ronald Jay Cohen and Frederick J. Smith, "Socially reinforced obsessing: Etiology of a disorder in a Christian Scientist," *Journal of Consulting and Clinical Psychology*, 1976, 44, 142–144. Copyright 1976 by the American Psychological Association. Used by permission of Ronald Jay Cohen and the publisher.

The Cross and Davis Marijuana Study is from Herbert J. Cross and Gary L. Davis, "College students' adjustment and frequency of marijuana use," *Journal of Counseling Psychology*, 1972, 19, 65–67. Copyright 1972 by the American Psychological Association. Used by permission of Herbert J. Cross and the publisher.

The Dugan and Sheridan Hand Temperature Experiment is from Michelle Dugan and Charles Sheridan, "Effects of instructed imagery on temperature of hands," *Perceptual and Motor Skills*, 1976, 42, 14. Used by permission of the authors and the publisher.

The Loftus and Zanni Question-Wording Study is from Elizabeth F. Loftus and Guido Zanni, "Eyewitness testimony: The influence of the wording of a question," *Bulletin of the Psychonomic Society*, 1975, 5, 86–88. Used by permission of Elizabeth F. Loftus and the publisher.

The Lutkus Mirror Image Drawing Study is from Anthony D. Lutkus, "The effect of 'imaging' on mirror image drawing," *Bulletin of the Psychonomic Society*, 1975, 5, 389–390. Used by permission of the author and the publisher.

The McDaniel and Vestal Source Credibility Study is from William F. McDaniel and Leslie C. Vestal, "Issue relevance and source credibility as a determinant of retention," *Bulletin of the Psychonomic Society*, 1976, 5, 481–482. Used by permission of William F. McDaniel and the publisher.

The Melamed Survey is from Leslie Melamed, "Therapeutic abortion in a midwestern city," *Psychological Reports*, 1975, 37, 1143–1146. Used by permission of the author and the publisher.

The Schutz and Keislar Word-Class Study is from Samuel R. Schutz and Evan R. Keislar, "Young children's immediate memory of word classes in relation to social class," *Journal of Verbal Learning and Verbal Behavior*, 1972, 11, 13–17. Used by permission of Samuel R. Schutz and the publisher.

The Sullivan and Skanes Teacher Evaluation Study is from Arthur M. Sullivan and Graham R. Skanes, "Validity of student evaluation of teaching and the characteristics of successful instructors," *Journal of Educational Psychology*, 1974, 66, 584–590. Copyright 1974 by the American Psychological Association. Used by permission of the authors and the publisher.

The Wallace et al. Teaching-Method Study is from Warren G. Wallace, John J. Horan, Stanley B. Baker, and George R. Hudson, "Incremental effects of modeling and performance feedback in teaching decision-making counseling," *Journal of Counseling Psychology*, 1975, 22, 570–572. Copyright 1975 by the American Psychological Association. Used by permission of the authors and the publisher.

The Yeatts and Brantley Typing Study is from Linda M. Yeatts and John C. Brantley, "Improving a cerebral palsied child's typing with operant techniques," *Perceptual and Motor Skills*, 1976, 42, 197–198. Used by permission of Linda M. Yeatts and the publisher.

# Contents

| | |
|---|---:|
| **Preface** | xi |
| **Chapter One  Getting Started** | 2 |
| Section A  Learning from Others' Observations | 3 |
| The Nature of Scientific Knowledge | 3 |
| Types of Observational Strategies in the Behavioral Sciences | 8 |
| *The Cohen and Smith Case Study* | 9 |
| *The Melamed Survey* | 14 |
| Reading Experimental Reports | 19 |
| Overall Plan of Reports of Experiments | 19 |
| Sources of Reports of Experiments | 20 |
| How to Find a Specific Journal Article | 21 |
| How to Find Experimental Reports on Particular Topics | 22 |
| Choose Practice Articles | 23 |
| Section B  Learning from Your Own Observations | 25 |
| Necessary Skills | 25 |
| Sources of Ideas for Experiments | 26 |
| **Chapter Two  What's the Research About?** | 28 |
| Section A  Reading Introduction Sections | 29 |
| *The Introduction Section of the Cross and Davis Marijuana Study* | 29 |
| Variables | 31 |
| Association between Variables | 32 |
| Statistical Expressions Related to the Concept of Correlation | 33 |
| *The Introduction Section of the Sullivan and Skanes Teacher Evaluation Study* | 36 |
| *The Introduction Section of the Wallace et al. Teaching-Method Study* | 38 |
| Association versus Causality | 39 |
| Drawing Causal Conclusions | 40 |
| Manipulated and Nonmanipulated Variables | 42 |
| Independent and Dependent Variables | 42 |
| True Experiments, Pseudo-Experiments, Correlational Studies, and Observational Studies | 43 |

| | |
|---|---|
| *The Introduction Section of the McDaniel and Vestal Source Credibility Study* | 46 |
| Factorial Designs | 47 |
| *Excerpts from the Introduction Section of the Schutz and Keislar Word-Class Study* | 49 |
| Section B  Beginning Your Experimental Design | 53 |
| Select Your Variables | 53 |
| Library Research | 57 |
| Decide Whether Your Variables Are Nominal Variables | 58 |
| Designate Independent and Dependent Variables | 59 |
| Manipulation versus Nonmanipulation | 60 |
| Selection versus Measurement | 63 |
| State the Purpose of Your Experiment | 64 |
| State Your Experimental Hypotheses | 64 |
| Some Preliminary Statistical Considerations | 65 |
| **Chapter Three  How's the Research Done?** | **68** |
| Section A  Reading Method Sections | 69 |
| *The Method Section of the Cross and Davis Marijuana Study* | 70 |
| Operational Definitions | 71 |
| Extending Conclusions beyond the Experimental Method (Generalization) | 74 |
| *Most of the Method Section of the Lutkus Mirror Image Drawing Study* | 80 |
| Some Elements of Experimental Design | 81 |
| *The Method Section of the Loftus and Zanni Question-Wording Study* | 85 |
| *The Method Section of the Yeatts and Brantley Typing Study* | 88 |
| Description of Experimental Designs | 89 |
| *Excerpts from the Method Section of the Sullivan and Skanes Teacher Evaluation Study* | 90 |
| Section B  Making Your Experiment More Definite | 93 |
| Decide your Experimental Situation | 93 |
| Decide What Kind of Subjects to Use | 96 |
| Estimate the Number of Subjects | 99 |
| Decide on Between-Subject, Within-Subject, or Before-After Independent Variables | 100 |
| Operationally Define Your Variables | 103 |
| **Chapter Four  Determining the Causes of Relationships** | **110** |
| Section A  Reading Method Sections for Confounds | 111 |
| Confounded Variables and Control Techniques | 112 |
| Possible Sources of Confounding Common to All Independent Variables | 112 |
| *The Method Section of the Wallace et al. Teaching-Method Study* | 116 |
| *The Method Section of the McDaniel and Vestal Source Credibility Study* | 118 |
| Sources of Confounding with Nonmanipulated Variables | 119 |

| | |
|---|---|
| Sources of Confounding with Manipulated Variables | 121 |
| *The Method Section of the Schutz and Keislar Word-Class Study* | 128 |
| **Section B  Completing Your Experiment** | **134** |
| **Part 1  Completing Your Experimental Design** | **134** |
| Dealing with Subject Bias | 134 |
| Dealing with Experimenter Bias | 135 |
| Dealing with Variable-Connected Confounds | 136 |
| Between-Subject Variables Only: Avoiding Extra Differences between Groups | 140 |
| Within-Subject Variables Only: Avoiding Extra Differences between Levels | 143 |
| Before-After Variables Only: Making Causal Conclusions Appear More Reasonable | 147 |
| Selecting the Number of Subjects Required for Your Experiment | 148 |
| Summarizing Your Experimental Design | 150 |
| **Part 2  Performing Your Experiment** | **151** |
| Prepare for the Physical and Mental Comfort of the Subjects | 151 |
| Prepare for Obtaining Informed Consent | 151 |
| Prepare for Completely Explaining the Experiment As Soon As Possible | 152 |
| Prepare Any Needed Instructions | 152 |
| Prepare for Data Collection | 153 |
| Prepare Specialized Materials | 153 |
| Prepare the Experimental Situation Carefully to Avoid Unnecessary Confounds | 154 |
| Practice the Experiment | 154 |
| Now Do It! | 154 |
| **Chapter Five  What Happens?** | **156** |
| Section A  Reading Results Sections | 157 |
| The Minimal-Statistics Approach | 158 |
| *The Results Section of the Yeatts and Brantley Typing Study* | 158 |
| The Correlational Approach | 160 |
| *The Results Section of the Sullivan and Skanes Teacher Evaluation Study* | 164 |
| Deciding Whether Results Are Accidental (Inferential Statistics) | 165 |
| The Differences-Between-Frequencies Approach | 167 |
| *The Results Section of the Loftus and Zanni Question-Wording Study* | 168 |
| The Differences-Between-Means Approach | 170 |
| *The Results Section of the Cross and Davis Marijuana Study* | 170 |
| Post Hoc Comparisons Require Extra Caution | 172 |
| Variability | 172 |
| *The Results Section of the Wallace et al. Teaching-Method Study* | 176 |
| Main Effects | 177 |
| *Excerpts from the Results Section of the Lutkus Mirror Image Drawing Study* | 178 |
| Interactions | 181 |

| | |
|---|---:|
| *The Results Section of the McDaniel and Vestal Source Credibility Study* | 185 |
| Interactions Occurring When There Are More Than Two Independent Variables | 187 |
| *The Results Section of the Schutz and Keislar Word-Class Study* | 190 |
| Interpreting Main Effects and Interactions | 194 |
| Interpreting the Dependent Variable | 195 |
| A Word on Factor Analysis | 197 |
| Summary | 197 |
| **Section B  Organizing and Analyzing Your Results** | 199 |
| Decide Which Statistical Approaches to Use | 200 |
| Organize Your Results | 200 |
| Look at Your Results | 204 |
| Determine the Scale of Measurement of Your Variables | 209 |
| One New Skill: How to Read Flow Charts | 212 |
| Summarize Your Data | 213 |
| Display Your Results | 216 |
| Think Again about Your Results | 221 |
| Use Inferential Statistics | 221 |
| **Chapter Six   What Do the Results Mean?** | **226** |
| Section A   Reading Discussion Sections | 227 |
| Recognize the Conclusions Drawn by the Authors | 228 |
| Evaluate the Conclusions | 228 |
| *The Discussion Section of the Cross and Davis Marijuana Study* | 233 |
| Practical Implications of Experimental Results | 236 |
| *The Discussion Section of the Wallace et al. Teaching-Method Study* | 237 |
| *The Discussion Section of the Loftus and Zanni Question-Wording Study* | 239 |
| Theoretical Implications of Experimental Results | 241 |
| *Study II of the Schutz and Keislar Word-Class Study* | 246 |
| Combined Results and Discussion Sections | 248 |
| *The Remainder of the Results and Discussion Section of the Sullivan and Skanes Teacher Evaluation Study* | 249 |
| Section B   Interpreting Your Results | 257 |
| Drawing Conclusions | 257 |
| Events That Can Make It Difficult to Draw Conclusions | 259 |
| Deriving Theoretical Implications | 264 |
| Practical Implications | 265 |
| **Chapter Seven   Putting It Together** | **266** |
| Section A   Reading the Whole Report | 267 |
| Titles | 267 |
| Abstracts and Summaries | 269 |
| *The Abstract from the Yeatts and Brantley Typing Study* | 269 |
| *The Abstract from the Cross and Davis Marijuana Study* | 269 |

|   |   |
|---|---|
| *The Abstract from the McDaniel and Vestal Source Credibility Study* | 270 |
| *The Abstract from the Loftus and Zanni Question-Wording Study* | 270 |
| *The Abstract from the Schutz and Keislar Word-Class Study* | 270 |
| Read the Whole Report | 271 |
| *The Dugan and Sheridan Hand Temperature Experiment* | 272 |
| Keep Records of What You've Read | 275 |
| Section B   Publicizing Your Findings | 276 |
| Write a Formal Report | 276 |
| Type a Formal Report | 282 |
| Report Your Findings to Your Local Colleagues | 286 |
| Report Your Experiment Orally | 287 |
| Decide Whether to Report Your Results to a Wider Audience | 287 |
| Perform Follow-up Studies | 289 |
| Report Your Experiments to the Larger Scientific Community | 290 |
| What Next? | 290 |

## Appendix A   Some Information about Journals, with Library of Congress Call Numbers — 293

|   |   |
|---|---|
| Sources of Information about Journals Relevant to Psychology and Allied Fields | 293 |
| Some Journals with Articles Summarizing Original Research in Selected Areas | 294 |
| Some Journals Presenting at Least Some Original Empirical Research in English | 294 |

## Appendix B   Summary of Ethical Principles in the Conduct of Research with Human Participants — 301

## Appendix C   Calculational Procedures for a Few Very Common Statistics — 305

|   |   |
|---|---|
| Descriptive Statistics | 306 |
| Differences-Between-Frequencies Approach: Mode, Median, Range | 306 |
| Differences-Between-Means Approach: Mean, Variance, Standard Deviation | 307 |
| Correlational Approach: Pearson Product-Moment Correlation Coefficient ($r$) | 310 |
| Inferential Statistics | 312 |
| Chi-Square, or $\chi^2$ | 312 |
| Independent-Measures $t$ Test | 315 |
| Related-Measures $t$ Test | 318 |
| Analyses of Variance | 320 |
| One-Way Between-Groups Analysis of Variance | 321 |
| One-Way Within-Subjects Analysis of Variance | 325 |
| Two-Way Between-Groups Analysis of Variance | 329 |
| Two-Way Mixed Analysis of Variance | 334 |
| Newman-Keuls Test | 341 |

| | |
|---|---:|
| Evaluating the Statistical Significance of $r$ | **345** |
| Evaluating the Statistical Significance of $X^2$ | **350** |
| Evaluating the Statistical Significance of $t$ | **352** |
| Evaluating the Statistical Significance of $F$ | **353** |
| **Appendix D  Sample Report Typed in APA Format** | **357** |
| **Glossary** | **371** |
| **References** | **379** |
| **Index** | **383** |

# Preface

This book is most directly involved with experimentation in psychology, and the articles in it are from that area. The examples come more generally from the behavioral and social sciences. However, almost all the skills taught in this book apply to all areas of science.

This book has two purposes. One is to teach people how to be intelligent consumers of scientific information, that is, how to analyze research reports critically. This skill is desirable for anyone having to base decisions on "scientifically demonstrated facts." This includes people in many areas: business, politics, medicine, social work, education, scientific research (of course), and even voters and buyers of advertised goods. The second purpose is to teach people how to design, perform, analyze, and report experiments of their own. They need all the skills required for critically analyzing other peoples' reports, but they also must learn other things.

This book attempts to serve both would-be consumers and producers of scientific research by separating the information relevant for each group. Each chapter is divided into two sections. Section A is devoted to skills necessary for understanding the research of others. Section B concentrates on skills necessary for performing experiments. Readers wanting only to become intelligent consumers of scientific information can read only the first sections without losing continuity. Readers also wanting to learn to perform experiments must read the whole book.

It can be read either in the order given, alternating the first and second sections, or by first reading all the first sections and then all the second sections. If there's enough time, a good sequence for thorough understanding is to read through in the order given, but just to skim the second sections. Then repeat only the second sections while designing and performing original experiments.

The first sections contain reproductions of several published articles, presented one section at a time. These articles were chosen for their relative clarity and simplicity of design. They don't represent a sampling of all available areas of research, nor were they chosen because they're more important than other papers. In addition to reading these articles the student is instructed to find and read one or two additional articles.

The second sections contain instructions allowing the reader to perform an original experiment. A person following all the instructions will independently design, perform, analyze, interpret, and write up a report for an experiment or correlational study.

Many questions are scattered throughout the book, and answers are given to most of them. Some questions are practice problems, appearing in a question-and-answer format. They're designed to ensure that the reader understands each point after it's been introduced in the text. In the first sections there also are questions about the sample articles in the text and about the reader's privately selected articles. The answers about the sample articles are given in the text following the articles. Obviously I couldn't provide answers to the questions about the outside articles, but a reader who has answered all the other questions correctly should be able to evaluate his or her own answers. In the second sections are questions requiring readers to make decisions about the design, performance, analysis, and interpretation of their original experiment. Again, I couldn't give answers, but I hope I've given enough guidance for readers to answer these questions with some assurance.

**A Note to Instructors**

Besides the unusual features of this book I've already mentioned—the separation of "understanding" and "doing," the inclusion of real articles, and the question-and-answer features—there are a few other characteristics you should know in order to use this book.

One is that it's organized according to the four main sections of an experimental report, with Chapter 2 devoted to the introduction, Chapters 3 and 4 covering the method section, Chapter 5 describing results, Chapter 6 involving the discussion section, and Chapter 7 outlining the remainder and bringing everything together. Chapter 1 discusses scientific methods and library research. This organization enabled me to emphasize points relevant to each section, but does fragment coverage a bit. Because of that you might expose students early in the course to a number of complete reports. They needn't understand them in detail, but they should see some whole articles put together.

Another unusual feature is that the first sections require the students to visit a library and actually locate some articles on their own. I've found this to be a valuable experience for the students. However, if it's inconvenient for you, you can circumvent this requirement by providing additional articles for the students in class.

A third unusual feature is that I've tried to include material on how to evaluate and perform both "true experiments" and "correlational (ex post facto) studies" in an integrated manner rather than concentrating on laboratory experiments with manipulated variables to the exclusion of other forms of research or relegating the other forms to a subsidiary level. However, I've avoided using the terms true experiment, correlational study, ex post facto study, observational study, and the like, instead emphasizing the necessity of evaluating whether *each variable* in an experiment was "manipulated" or "nonmanipulated" in drawing conclusions from the study. The reasons I've avoided the usual didactic terms are that (1) they almost never appear in professional publications and (2) there are very few pure cases of either type. Everything, it seems, is a "mixed design"! By limiting myself to talking about one variable at a time, I believe I have made a more useful distinction. I mention the other terms in the text, so you can use them if you prefer, but I don't use them regularly.

Yet another unusual feature of this book is that I've integrated some material that's often covered in separate chapters or sections of existing texts, including ethical considerations, field studies, and small-group (that is, 1-to-3-subject) designs. The ethical considerations are summarized in Appendix B, which is useful in synthesizing the material.

The last aspect I want to mention concerns my coverage of statistics: It's not complete. I've designed this book so that it can be used by students having no statistical background. In Chapter 5, Section A, I've tried to present only those statistical concepts needed for an intuitive understanding of most results sections. In Section B of that chapter there's some instruction on summarizing and displaying data and some instruction on deciding what statistics are appropriate for the student's data. In Appendix C are "cookbook" calculational techniques for a few of the most common parametric statistics, including two-way analyses of variance and the Newman-Keuls test. These aren't intended to substitute for proper instruction in the theory and application of statistics, but rather to allow students to analyze their experimental data.

I've used this text in various forms in several of my classes and find that most students can complete it nearly unaided. Typically I use class time for library visits, for group completion of the assignments at the end of each chapter, and for performance of experiments illustrating the points made in each chapter. Tests of the first sections can consist of questions such as those in the text about experiments in the text, in-class experiments, fake experiments, or published experiments handed out with the test. A better test, though one more difficult to grade, is to have the students abstract or orally describe published experiments and tell what conclusions are or aren't appropriate. Tests of the second sections can consist of requests for experimental designs meeting various requirements. However, the best test of the second sections remains the degree to which the student can complete an original research project.

**A Note to Students**

To get the most out of this book you should try to answer all the questions *in writing*, making sure to write enough so that it is clear. When you see problems, cover the answers and try to write your own *before* reading the ones in the book. Then you'll know whether you understand that topic well enough to continue or whether you should reread that section.

You might be wondering why you should learn this material. Some of you are reading this book because you really are actively interested in learning how to answer questions scientifically. It can be a lot of fun and is deeply satisfying to many people. But some of you are probably interested in the *answers* given by science, not in how those answers were found. Perhaps you just like knowing "scientific facts" or you plan to enter a profession in which you'll apply scientific facts as part of your job. You may not be particularly interested in learning how experiments work, even if you're required to read or perform them.

If you do feel that your interest is really only in the answers found by scientists, you should recognize that you must understand the source of those answers if you're to understand them clearly, much less to apply them appropriately. People who attempt to apply scientific knowledge without understanding its limitations are usually disappointed and confused by the consequences. It's rather like trying to use an electronic calculator to solve problems without understanding what it means to add or multiply numbers. Reading the first sections of this book should help you learn how to understand scientific facts and to recognize their limitations so that you can apply them more intelligently.

You may see the value in learning to evaluate the research of others but feel there's little value in learning how to perform experiments of your own. The first sections were written specifically for such people. However, your understanding and appreciation of the problems faced by researchers, and the limits on the conclusions they can draw, can be complete only if you've tried some experiments of your own. Students often find that after they've performed some original research they like it more than they anticipated. But even if you never learn to love the problem-solving challenges of experimentation, the knowledge you acquire will help you to understand the research performed by others in a way that passive learning can never duplicate.

I'd like to express my appreciation to the professionals at Wadsworth—Ken King, Kathie Head, Dare Porter, and Claire Annchild Connelly—for their help in shaping this book into an attractive, readable package. I must also thank the publisher's readers—S. Joyce Brotsky, California State University, Northridge; Ted Panzer, Chaffey College; Mark Sanders, California State University, Northridge; Paul Stegner, Cañada College; and Albert M. Swanson, California State University, Sacramento. I would especially like to thank the detailed reviewers—Diana Henschel, California State University, Dominguez Hills; Mike Knight, Central State University; H. Kent Merrill, Mesa College; Robert M. Stern, Pennsylvania State University; and Jean Volckman, Pasadena City College—for their extremely helpful and sometimes even encouraging comments. Special thanks are due to Dave Tieman for letting me use his class handouts on how to write a research paper, and to an anonymous teacher at Northwestern University who was the source of the material for the sample report in Appendix D. However, my biggest thanks must go to my students for putting up with ungainly manuscripts as textbooks and letting me know what they couldn't understand, and to Dick for his suggestions about problem topics and his constant support and tolerance over two long years.

# Understanding
# Behavioral
# Research

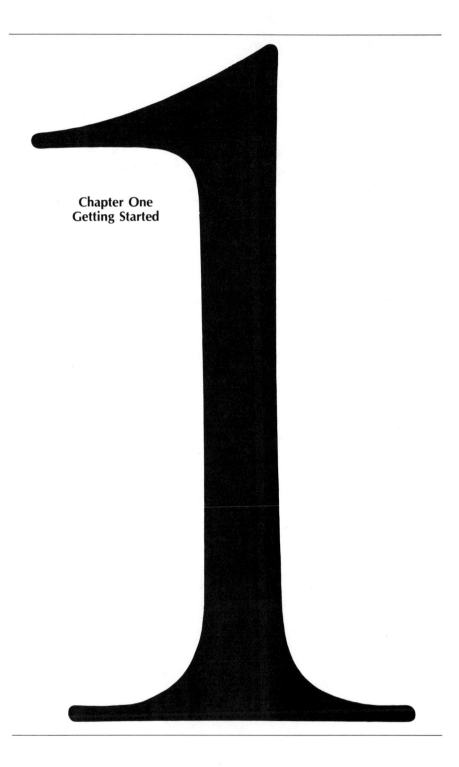

**Chapter One
Getting Started**

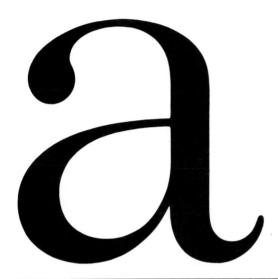

**Section A
Learning from
Others'
Observations**

A sentence that begins with the phrase "It's been scientifically demonstrated that . . ." causes an almost emotional response in many Americans. Some people blindly accept whatever follows such an introduction; others just as blindly reject it. A more useful response than either of these is to ask, "*How* was it demonstrated?" Then one should be able to evaluate the scientific demonstration on its merits.

This book is about understanding and performing scientific demonstrations. Scientists in all areas perform demonstrations and publicize their findings. Students and other researchers study the reports and sometimes use them as a basis for further research.

Why are these demonstrations performed, publicized, and studied? Many scientists perform research simply because it's fun; the problem-solving mental exercise involved has much in common with games and puzzles. Some research, publication, and study occurs because of outside pressures from teachers, promotion committees, and other inglorious causes. But the ultimate reason for these activities is to increase and use scientific knowledge.

### The Nature of Scientific Knowledge

For many people "science" is a group of academic and technical subjects, usually involving complicated instruments and lots of math and summarized as a list of facts, theories, and inventions. In a more general sense, however, science isn't a list of facts or subjects at all, but a way of thinking. Science is one means of obtaining "knowledge" that many people believe to be philosophically valid. There are means other than science, including knowledge obtained from self-insight, religion, or pure logic. Science is one approach, applicable to only some problems.

Each approach to attaining knowledge makes certain assumptions and uses special techniques to do so. People searching for self-insight generally assume that there's some core truth within themselves to be discovered beneath layers of obscuring thoughts or emotions. Some use special techniques such as meditation, mental discipline exercises, drugs, and the like to increase self-insight. People searching for religious knowledge often assume that the truth has been given by some deity and can be discovered through various means. Techniques such as prayer, ritual, and study of religious writings are used to increase religious knowledge. People searching for truth through the use of pure logic must find some starting point they assume to be true, like Descartes's famous observation, "I think, therefore I am." Starting from that point, they use the special techniques of logic to reach other truths. People searching for scientific knowledge also make assumptions and use special techniques.

*Scientific Observation*

Scientists assume that there is a physical world that can be observed, either directly through our physiological senses or by a machine or other device, which in turn is observed by our senses. Philosophers have pointed out that there's no way to prove even the existence of a physical world; it could all be an illusion. Assuming that a physical world exists, we can never know for certain that our perceptions are correct. In fact, we know that our senses sometimes distort information about the world, as in optical illusions. We can use measuring devices like rulers and scales to aid our senses, but we can't be sure that we can perceive *them* accurately. Nevertheless, people searching for scientific knowledge have always made these assumptions and have found them to be satisfactory for many purposes.

Thus science can be used only to investigate events that can be observed. The movements of stars or molecules, the growth of leaves or blood vessels, and the activities of people or animals can all be investigated using the techniques of science. The formation of planets, the development of earthquakes, and the thoughts and wishes of people and animals also can be investigated scientifically insofar as we can discover ways to observe relevant events. If we suppose that there's some event that can't be observed in principle, that event can't be analyzed by scientists. For example, some workers in the area of extrasensory perception suggest that people's ability to use such skills disappears under observation. If that suggestion is true, then those skills can't be studied scientifically. To make sure you understand the kind of question that can be studied scientifically, try these problems.

## Section A  Learning from Others' Observations

**Problems**

**Q.** Some of these questions involve events that can be observed, while others involve events that can't be observed. Pick out the questions you think might be answered by observing events.

1. Who says more often that they're happy, rich people or poor people?
2. Why is the universe here?
3. What happens when an irresistible force meets an immovable object?
4. What did the chicken do after it crossed the road?
5. Do your beliefs or actions determine whether you go to heaven?
6. What happens if you divide a molecule into smaller units?
7. Who gets talked to more at parties, a person who pitches in with the work and looks busy or one who stands quietly and looks alert and contented?
8. Who is better, a person who works hard and provides for others or one dedicated to self-fulfillment?
9. Is there a life form from outer space which we can't see, hear, smell, taste, or feel and which can't be detected by any machines or animals?
10. Do colors *really* look the same to all people with normal vision?
11. Do cows act differently just before earthquakes occur?
12. Can plants learn?

**A.** I can think of ways to observe events related to numbers 1, 4, 6, 7, 11, and 12, although in the case of 12, I had some trouble. I think I'd look at certain plants like Venus's-flytraps and see, perhaps, if they could "learn" not to try to digest plastic flies. You might have been able to think of ways to observe events related to the other questions, but I couldn't.

We've learned that our observations may be influenced by our physiological structure, past learning, expectations, emotions, and desires. However, scientists attempt to observe events in the world objectively, without letting internal factors intrude. In watching, say, a dripping faucet, you can observe the change in the size of the drop, its color, and the sound it makes. You could use a film to slow down the motion visually, a microscope to increase the amount you can see, a scale to be deflected by some amount by each drop, and so on. However, you can't directly observe the "surface tension" or the cohesion of the atoms of hydrogen and oxygen. Those are interpretations of observations rather than basic observations in themselves.

Observations that are as nearly objective as possible form the bases of all sciences. However, it's unavoidable that only certain characteristics of the events are actually observed. A botanist studying a mushroom will observe its appearance, smell, taste, and habitat but is not likely to use a microphone to listen to it. Thus scientific observation is selective. We can only try to select events objectively and not to be influenced by our expectations or wishes.

*Scientific Prediction*

Besides assuming that knowledge can be gained by observing events in the physical world, scientists make an additional assumption: that patterns of events recur, that is, that future events will be much the same as past ones. If the same conditions are repeated, the same outcomes should follow. If *all* the weather conditions on May 12, 1955, are repeated, for example, scientists would expect the same events to follow and would make that prediction.

However, it's obvious that *exact* conditions never recur. In order to make predictions, scientists actually say something like "If the conditions don't change *in any important way*, the same outcome will occur as in the past." Thus they must discover which changes are important.

Groups of observations are needed to clarify which factors are important and which are irrelevant in making predictions. That's one reason why scientists publicize their findings: so other scientists can add many observations together and see the overall pattern. After many observations of rainstorms, for example, a pattern might appear suggesting that changes in air pressure are important factors in predicting rain while changes in air temperature are irrelevant.

*Scientific Theories*

Summarized groups of observations that tell which factors are important in making predictions are sometimes called scientific **theories.**[1] We can use theories to make predictions. For example, we could predict on the basis of a theory that whenever the air pressure drops more than 2 points within a specified time interval, rain will occur within 12 hours.

Theories can differ in their degree of support and acceptance. When the theory is supported by relatively few observations, or when people are uncertain about the pattern of events, they may call the possible pattern a **hypothesis** rather than a theory. As more observations are added and the pattern becomes more clear, the term *theory* is more applicable. When acceptance is very wide, the pattern is sometimes called a **law.** We'll just call them all theories.

Theories can also differ in their degree of sophistication and formality. Some theories can be expressed clearly only by using mathematical equations to describe the patterns of events. Others can be adequately expressed in verbal phrases, such as "the more anxious a person feels, the less well that person is likely to perform on a test." Regardless of the degree of sophistication, theories summarize information and allow predictions to be made.

In addition to these functions, theories interpret or explain the observations

---

[1] Words defined in the Glossary appear in boldface type the first time they occur in this book.

by suggesting some underlying reason for the patterns that have been observed. If a scientist observed that taking an atom apart into smaller units resulted in an explosion, he might interpret the event to suggest that there's a strong force holding the parts of the atom together and that taking the atom apart releases the force. Other scientists may disagree with that interpretation. They believe the observation that the explosion occurred as reported by the first scientist. They can even use that information to predict what will happen if they take atoms apart. But they may interpret the observations differently; that is, they might have different theories.

The possibility that there's more than one interpretation of an event is another reason scientists publicize their observations along with their interpretations. Anyone hearing or reading the report is free to reinterpret the observations.

*Testing Theories*

Good theories summarize a lot of observations, offer satisfying explanations for patterns of events, and allow one to make accurate predictions.

A theory can be tested by checking out the accuracy of predictions based on it. In that case the prediction might be called an **experimental hypothesis.** For example, the "released force" theory about the explosion that occurs when atoms are taken apart would lead us to predict that some kind of force must be released regardless of how the atom is taken apart. It might not be in the form of an explosion, but there must be some heat or movement or something. That prediction could be thought of as an experimental hypothesis, leading to new observations. If we observe events that agree with the prediction, then the theory is supported. However, if we observe events that disagree with the prediction, then the theory is necessarily wrong or more limited than previously thought. For example, observations might show that there are some ways to take atoms apart that don't result in any discoverable force—no explosion, no heat, no anything. Such observations would suggest that the released force theory was incorrect. The predictions, or hypotheses, based on the theory weren't supported by the observations.

Thus science progresses through a combination of observations and theories. The observations form the bases of the theories. The theories summarize the observations, and direct researchers' attention to new observations by making predictions. The new observations test the accuracy of the theories and may require new theories if they disagree with the predictions.

At the early stages of investigating an area the theories may be no more than guesses or hunches based on informal observations. By testing hypotheses based on the theories, scientists can discard useless theories and modify the more useful ones to make them either encompass more observations or be more accurate. As observations are continuously added the theories become more widely accepted until they reach the point where most researchers in the field regard them as "true" in at least a limited sense.

*Practical Applications
of Scientific Knowledge*

Besides increasing scientific knowledge, observations and theories often have practical applications. If we've observed that people tend to buy more when goods are attractively displayed, merchants can apply that information directly. If we have a theory listing factors that can be used to predict which high school graduates could become successful salespersons, employers could certainly use that information. The possible practical applications of observations and theories are a third reason why scientists publicize their findings. Anyone reading a scientific report can apply the information in it to other situations.

Many observations are made and theories developed because people have practical problems that need solutions. Scientists trying to solve practical problems sometimes say they're doing "applied research," while scientists who are less interested in practical applications say they're doing "pure research," but there's no real way to tell the difference. Theories from pure research often have practical applications, and applied research serves quite well to increase scientific knowledge.

Theories and their practical applications are discussed further in Chapter 6.

## Types of Observational Strategies in the Behavioral Sciences

As you've seen, scientists report their original observations so that others can use them to make predictions about future events, perhaps adding other observations to clarify the overall pattern. Anyone reading the original observations can formulate a satisfying theory, or modify an existing one, to account for the new finding. In addition, the observations or theories might have valuable practical applications.

There are different kinds of scientific observations published as scientific reports. Some of them allow clearer predictions and interpretations than others. Next we'll discuss two kinds of studies that are often performed at the very early stages of investigations of an area, before much is known about what's important to observe and record. These are the **case study** and the **survey.** Case studies most typically gather information about a single individual, while surveys gather information from many individuals. Both strategies involve gathering a wide range of information about the area under study. However, as we'll discuss soon, it's difficult to make predictions based on studies of this type. They are better thought of as sources of hunches or hypotheses that can be tested using other kinds of strategies.

To make predictions more easily, more directed observation strategies have been developed, at least some of which can be called experiments. All these directed observations are limited to a relatively few types of events, and they gather a narrow range of information. They're the primary concern of this book and will be discussed later in this chapter and in all the subsequent chapters.

*Case Studies*

In a case study, the goal is to observe and record accurately as many aspects surrounding a particular subject as possible. The subject is most often a single person, but it might be an animal, a group of people, or even a whole society. Case studies are often relatively easy to read, because they're free of much experimental jargon. Researchers performing case studies try to observe events objectively, but they must be selective in what they choose to observe and record.

Read now a case study written by Ronald Jay Cohen and Frederick J. Smith called "Socially reinforced obsessing: Etiology of a disorder in a Christian Scientist." It appeared in the *Journal of Consulting and Clinical Psychology* in 1976, Volume 44, on pages 142 to 144. Part of the report deals with events the authors observed directly, such as what the patient said or did in their presence. Some of the material couldn't have been observed by the authors but must have been reconstructed from what the client told them, such as what happened when she was 15. And some of the comments are clearly interpretations by the authors, as when they point out that the client's miscarriage was the "most crucial" event. As you read this case study, try to distinguish these three kinds of comment.

## The Cohen and Smith Case Study

There is no provision for any form of physical therapy or psychotherapy within the framework of Christian Science because it is this group's belief that

> *drugs are stupid substitutes for Divine Mind; they have no power in themselves; they operate by the "laws of a general belief," . . . a disease, in the terminology of the sect, is a "belief," or a "claim," and the proper treatment is to "deny" it, or to "demonstrate" over it (Podmore, 1963, p. 284).*[2]

Ambivalence with respect to the efficacy of Christian Science treatment is not tolerated because "persistence in the face of alarming physical evidence to the contrary may be what is needed to bring about a healing" (Christian Science Publishing Company, 1966, p. 239). In the present study, the case of a former Christian Scientist who exhibited pronounced obsessive behavior and fears is reviewed and examined in the context of her theological training. It is suggested that the client's faith and subsequent conflict regarding her religious beliefs were instrumental in the initiation of obsessive behavior.

*Presenting Problem*

Mary, an attractive 28-year-old mother of two, was referred to an outpatient mental health center for treatment by a hospital psychiatrist who interviewed her

---
[2] This is a reference to the original article. The use of such references will be discussed shortly.

when she came to the emergency room at 3:00 A.M. asking that her 4-year-old son be treated for rabies. He had been knocked to the ground by a stray dog the previous afternoon. Although she knew it was impossible for him to contract rabies from the dog's paws, she said that thinking about the incident disturbed her so much that she felt impelled to wake her son and bring him to the hospital in the middle of the night. The next day, in the initial interview conducted at the center by the second author, Mary complained of being "afraid of everything." A partial list of things that she said were anxiety arousing included dogs, cats, rabies, botulism, "childhood diseases," and tetanus. Mary said she sought help because the fears began dominating her daily thoughts and consequently her behavior. Her fear of botulism led to her adopting a severely restricted and costly diet for her family in which only fresh meat and vegetables were prepared; jars were held suspect, and all canned food was outlawed. An additional consequence of her fear of botulism was that her children were prohibited from eating at friends' houses, and it was only with a substantial degree of anxiety that she allowed them to purchase school lunches. So great was her fear of contracting tetanus, she would rush to her doctor's office demanding a tetanus shot if she as much as pinpricked a finger. Mary also reported sleeping difficulties which she attributed to her concern over her fears. Overconcern for the welfare of her children prompted her to quit a full-time job she had held for over 1 year.

*Faith, Ambivalence, and the*
*Etiology of the Obsessive Behavior*

Mary was introduced to Christian Science at the age of 15 by Paul, the man she later married. Mary was raised as a Methodist (though religion was not emphasized in her home) but began attending weekly Christian Science services to please Paul. She described her attendance as 80% for the first two years and noted that her faith in Christian Science principles grew over this period to the extent that at age 17 when she was 8 months pregnant with her first child she refused medical consultation in favor of treatment by a Christian Science practitioner. When she went to the hospital to deliver the child she did so convinced, as the practitioner had taught her, that it would be painless. However, she experienced an extremely painful labor. Because the pain did not yield to the practitioner's ministrations and because she wished to "do the best thing for the child," she sought the help of a physician.

This childbirth, as well as subsequent experiences, served to disillusion Mary with respect to the power of Christian Science. One such event occurred during a conversation with a friend who was a Christian Science practitioner. Following the death of a mutually known devout Scientist who had sought treatment at a Christian Science sanitorium, the practitioner made the remark that "it never should have been allowed to happen," implying to her that some form of traditional medical intervention should have been attempted. The suggestion that medical help should be sought when Christian Science appears to be failing borders on heresy and perhaps particularly so when the suggestion is made by a

Christian Science practitioner. Understandably, Mary's questioning of the efficacy of Christian Science healing was increasing but not without a concomitant increase in guilt. According to her, part of her guilt arose from a conflict within herself regarding the nature of God. She struggled to reconcile teachings concerning the basic goodness of God with the evils she saw in the world. She gave considerable thought to questions like, Is God good or is He only half good? Did God make a good earth or is He sadistic? If I'm any kind of human being, why don't I believe in God?

Harshly disillusioned, Mary disavowed Christian Science teachings and for a period of approximately 10 years did not attend services. However, just prior to the onset of the severely maladaptive obsessive behavior which brought her to the clinic, Mary suffered a miscarriage. In addition, she reported three experiences that occurred during her pregnancy that served to reawaken her Christian Science beliefs. The first incident involved a friend who had suddenly become stricken with multiple sclerosis. Mary said that "besides just feeling sorry for her," she found that she could give comfort to her friend by offering prayer. Dissonance theorists, as well as self-perception theorists, might well argue that Mary's calling upon and verbalizing her previously acquired Christian Science skills may in some way have reinforced her belief in the efficacy of the procedures.

Second, during this same period of time, Mary's 3½-year-old son John was complaining of pains in his feet at night. The pain was extreme enough to awaken the boy. Fresh from the experience of having had a close friend contract a major disease, Mary was afraid to even consult a medical doctor concerning her son's condition. Instead, Mary called a practitioner and asked the practitioner to pray for her son. Though the practitioner never saw the child, there was a complete cessation of his symptoms after the second treatment (prayer). This relief at the hands of the practitioner served to rekindle and reinforce Mary's faith in Christian Science teachings.

The third and most crucial event was the miscarriage itself. Mary cited at least two reasons for feeling guilty over the loss of the baby. First, during the pregnancy the family was involved in moving and, in order to economize, Mary helped with the moving of furniture. It is possible that this heavy labor precipitated the miscarriage. Second, and more crucial to the present analysis, neither Mary nor her husband wanted her to become pregnant, and they both openly expressed this. Mary even reported casually "wishing she was not pregnant." When informed by her doctor of the miscarriage, it occurred to her that the baby may have actually been "killed by thought."

It seems reasonable to infer that events surrounding Mary's third and most recent pregnancy served to renew her interest in previously disavowed Christian Science belief. The traumatic experience of losing the baby linked to the notion that the child may actually have been done away with by thought made the power of thought more salient and served to make Mary acutely cognizant of, and cautious with, her own thoughts. In her own words, Mary began "tuning in on disaster." For example, whereas another mother would not become overly concerned if her child complained of a cold, such a complaint would elicit in Mary a variety of thoughts such as "God never created a cold, so how could it exist; How can God's perfect child have anything not harmonious in him?" Indeed, consistent with Christian Science beliefs, disease and pain do not exist and, therefore, should be denied:

> *Bodily conditions they view as effect rather than cause—the outward expression of conscious and unconscious thoughts. On this premise what needs to be healed is always a false concept of being, not a material condition. (Christian Science Publishing Company, 1966.)*
>
> *You say a boil is painful; but that is impossible for matter without mind is not painful. The boil simply manifests, through inflammation and swelling, a belief in pain; and this belief is called a boil. Now administer mentally to your patient a high attention of truth, and it will soon cure the boil. The fact that pain cannot exist where there is no mortal mind to feel it is a proof that this so-called mind makes its own—that is, its own belief in pain. (Eddy, 1875, p. 284.)*

Imbedded in Mary's past were reports of several experiences where this philosophy had apparently been validated. One such event occurred when Mary decided to use the techniques she had been taught in order to quit smoking. She had been smoking up to three packs a day for 3 years. One day before boarding an interstate bus she resolved to use thought to cure herself of her habit. Mary reported that by the time she was ready to get off the bus, she felt as if she had never smoked at all. She has not smoked since. But although the power of concentration has long been known to be instrumental in bringing about positive behavioral consequences (Bain, 1928), a religious reliance on the use of such techniques may be psychologically harmful. As in this case, previous minor concerns such as pinpricks or reports of botulism poisoning in the news media may have inadvertently and paradoxically been made more salient to a person who devotes her energies to denying all but spiritual entities. Actively trying to deny the existence of something may actually increase the amount of time one spends thinking about it; as Mary once so aptly put it, "when all you think about is dieting, all you see is food."

As time went by, Mary became locked into a spiral of ever-increasing ambivalence. If, for example, her son was running a high temperature, her initial reaction would be to deny the existence of sickness. But in spending the day and night trying to mentally treat the problem, her guilt feelings for depriving her son of medical attention increased. Additionally, she would imagine grave consequences (e.g., blindness, paralysis) that might result from her failure to medically treat the problem. The net result of the constellation of factors that appeared to be operating was that the slightest provocation (e.g., the dog knocking her son down with his paws) became a stimulus for panic and frantic seeking of medical attention.

*Note on Treatment*

The primary focus of the present study has been the etiology of an obsessive disorder, and only cursory attention will be given to the treatment procedure undertaken. Mary was seen weekly for a total of 24 sessions. Approximately the first half of the total number of sessions was taken up with a discussion and clarification of her various problems in living, and no attempt was made to deal directly with her relatively stable pattern of obsessiveness. Around the 15th session, Mary reluctantly brought up her involvement with Christian Science (which she had previously avoided mentioning), and a probable basis for her obsessive behavior became clear. The therapist (the first author), wishing to remain neutral with respect to any resolution of Mary's division of loyalty, neither encouraged her to continue therapy or discouraged her from adhering to the tenets of the Christian Science philosophy. Mary elected to continue therapy and was given relaxation training to help reduce anxiety during the day and to help her sleep at night. In addition, she was taught a thought-stopping procedure (described by Lazarus, 1971) to deal with the obsessions directly. From about the 16th session to the termination of treatment, Mary was manifesting none of the urgent and irrational fears that she had experienced when referred for therapy. Additionally, she once again sought and obtained employment. Three months later, she spontaneously called to say that she was continuing to do quite well.

---

What did Cohen and Smith really observe? What they saw and heard was a woman talking. The observations they reported consist largely of rephrasings of what she said. If they had reported only direct observations, they would have limited the report to her words and to descriptions of her behavior as she spoke. However, they chose to interpret her behavior, organizing what she said into what seemed to them to be a meaningful, coherent pattern emphasizing Mary's religious background. Had different therapists reported Mary's case history, presumably they would have observed many of the same behaviors as Cohen and Smith. However, they might have chosen to interpret these observations somewhat differently, perhaps emphasizing what she said about her parents or her early school years.

*Surveys*

Case studies typically concern only one individual; surveys concern large numbers of individuals. Both involve gathering a large amount of information, but researchers performing a survey must limit their observations to fewer events than with a case study. Instead of selecting the observations to report after performing a case study, researchers doing a survey must select the events to observe before they begin the survey.

Once the information has been gathered, the researchers try to organize and interpret it as clearly as they can. Because the information is more limited than in case studies, there's less flexibility in interpretation. However, room still exists for a variety of viewpoints.

The paper you'll read next is a survey performed by Leslie Melamed called "Therapeutic abortion in a midwestern city." It was published in *Psychological Reports* in 1975, Volume 37, on pages 1143 to 1146. As you read it you'll notice that it uses a number of technical terms that may be unfamiliar to you. For now, just ignore them; we'll discuss them in later chapters. Notice how the events described by Melamed are quite limited as compared to the variety of events in the case study by Cohen and Smith.

## The Melamed Survey

The legalization of abortion in the United States has led to the opening of abortion clinics in a large number of cities, and a removal of some of the secrecy surrounding therapeutic abortions. The opening of one such clinic in a midwestern city provided an opportunity to investigate some demographic and behavioral characteristics of those who undergo abortions.

A number of recent studies (Bracken et al., 1973; Harris et al., 1974; Pakter et al., 1973; Kaltreider, 1973; Smith, 1973; Osofsky & Osofsky, 1972; Athenasiou et al., 1973; Martin, 1973) have investigated, among others, factors such as the use of birth control devices, the relationship with the man responsible, reasons for having the abortion, the number of previous pregnancies, and the demographic characteristics of those having abortions. Recent reviews of some of the results may be found in Osofsky and Osofsky (1972) and in David (1973).

No theoretical underpinning was assumed for the present study in that the efforts simply provided empirical data about an important social phenomenon in a new geographical area.

*Method*

The **subjects** were 188 women who had had a therapeutic abortion in an abortion clinic in a midwestern city. The questionnaire used in the study was given to the patients by a nurse between 1 and 2 hr. after they had had the abortion and while they were resting in the recovery room. Respondents were asked to fill out the questionnaires anonymously and to place the completed questionnaires in a receptacle provided. Only one woman who was given the questionnaire refused to answer it, so that responses were available from 187 respondents.

The questionnaire contained both forced-choice and open-ended questions. Included were questions designed to gather demographic information, open-ended questions about menstrual cycle length, times of maximum and minimum fertility, and questions about feelings about the abortion and the man responsible.

An additional variable was generated by combining and recoding responses to a few questions, an estimate of the interval of maximum fertility in the cycle, from respondents' answers to the question about cycle length (Katchadourian & Lunde, 1972). This variable provided an index of the respondent's knowledge of her own fertility cycle. In cases where no cycle length, or a cycle length of fewer than 12 days, was given, the cycle length was assumed to be 28 days.

If a respondent's answer to the question of maximum fertility lay in the estimated fertile interval, and if her answer to the question about minimum fertility lay outside of this interval, then she was scored as having some knowledge of her own fertility cycle, otherwise as not having knowledge.

*Results and Discussion*

The answers to the questions were coded and the results tabulated. The $X^2$ (Hayes, 1973) was the primary statistical test used, both for comparison of frequencies with theoretical data and for tests of the relationship between variables. No $X^2$ values are presented or considered as significant unless they are significant at the .05 level. All missing responses were excluded from consideration, so that the total number of observations varied slightly from question to question. To facilitate comprehension, all scores are presented as percentages although the statistics were calculated from raw frequencies.

**Demographic Characteristics.** Table 1 presents the demographic characteristics of the sample. From these one can see that the modal patient is white, presently unmarried, under 25 yr. of age, Protestant, whose mother is a housewife and father is of a low socio-economic status. However, this modal person is a misleading concept in that there is great variability in all these demographic characteristics.

**Fertility Knowledge.** Of respondents 19.3% who were asked the length of their menstrual cycle either gave answers of 11 days or fewer, or gave no answer at all. This apparent lack of knowledge of their own functioning is even more vividly shown by the fact that only 18.7% were correct in stating both when they were maximally and minimally fertile.

Knowledge of fertility was unrelated to variables such as age, marital status, number of prior pregnancies, or whether or not respondents had been using birth control devices when they became pregnant. It was also independent of the number of men respondents had had intercourse with, and independent of the frequency of intercourse with the same man. Respondents' knowledge of their own fertility, in fact, appears to be independent of a number of "experience" variables that, one would imagine, would give them greater access, and exposure, to information about their own functioning.

**Birth Control Devices.** Of respondents 73.3% were not using any form of birth control when they became pregnant. Among those using contraceptive devices, major contraceptive failures occurred with the use of foam and condoms.

The presence or absence of a birth control device was unrelated either to

## Table 1
### Demographic Characteristics of Those Undergoing Abortions

| Age (yr.) | | Race (%) | |
|---|---|---|---|
| Under 17 | 10.7 | White | 84.2 |
| 17 to 20 | 32.6 | Black | 15.8 |
| 21 to 25 | 26.2 | | |
| 26 to 30 | 13.4 | | |
| 31 to 35 | 10.7 | | |
| Over 35 | 6.4 | | |

| Marital Status (%) | | Religion | |
|---|---|---|---|
| Single | 54.1 | Catholic | 18.8 |
| Married | 20.0 | Protestant | 76.4 |
| Divorced | 15.7 | Jewish | 1.4 |
| Widowed | 4.3 | Moslem | 0.7 |
| Separated | 5.9 | Other | 2.8 |

| Occupation[a] | | | |
|---|---|---|---|
| Level | Own | Mother's | Father's |
| Low status | 32.6 | 26.0 | 53.4 |
| High status | 25.6 | 20.1 | 46.6 |
| Housewife | 18.0 | 46.1 | |
| Student | 23.8 | 7.8 | |

[a] Socio-economic status was based on Reis's (1961) classification. All occupations above 50 were treated as high socio-economic status and the others as low.

the number of men, or to the frequency of intercourse with the man responsible. This suggests that the absence of a birth control device is not related to the single occasion of intercourse but persists as a continuing behavioral pattern. This evidence, tied in with the lack of knowledge of fertility, again suggests that the major problem is one of ignorance of reproductive functioning, independent of those factors which one would expect to influence the availability of information about birth control.

**Relationship with the Man.** Reporting that they had had intercourse with only one man during the previous 2 mo. were 89.8% of respondents; only 1.6% of respondents reported 3 or more men. In addition 70.1% had had intercourse more than 5 times with the man responsible for the pregnancy; 13% had had intercourse only once. Information from these two questions suggests that a fairly stable sexual relationship existed between the respondents and those responsible for the pregnancy. This contention is further supported by the findings that 85.2% of the men were aware of the pregnancy and 82.5% aware of the abortion.

Social support for having the abortion appears to have come from the responsible men; of those aware of the abortion, 85.5% approved of the respondents having the abortion. The approval of the man was not related to whether respondents were single, separated, or widowed on the one hand, or married on the other. It was also not related to the number of men with whom respondents had had intercourse during the previous 2 mo.

The major findings in the present study are concerned with the apparent ignorance about their own reproductive functions, and the absence of use of birth control devices, among those having abortions. The independence of these factors from variables which should, theoretically at least, provide different degrees of access to birth control information, suggests that theoretical access to birth control information has, in itself, no effect on use of knowledge of birth control. There is, however, no suggestion that those who have abortions are any less informed than others about techniques of birth control. Ryder (1973), for example, reports that 26% of couples who use contraceptives fail to delay or prevent a pregnancy within 1 yr. of exposure to risk of pregnancy.

The relatively stable relationship with the man as observed in the present study is similar to the situation reported by Smith (1973) and Bracken et al. (1973). The man involved appears to have provided the major social support for the decision to have the abortion. It is interesting to speculate whether the differences between those who do and those who do not choose to have an abortion could be partially explained by differences in the man's behavior.

---

What was actually observed by Melamed? She observed only the answers to the questions on a printed, anonymous questionnaire. The percentages of respondents giving different answers are the basic observations gathered in this survey. They summarize many characteristics of this group of individuals.

From only these percentages some predictions of a limited nature could be made. If all the conditions observed by Melamed occurred in some other town, scientists would expect the women to respond the same way; that is, they would expect the same proportion of women having abortions to be married, to report stable relations with the man responsible, and so on. However, it's obvious that the exact conditions aren't likely to occur again anywhere else. Some differences will always exist. Scientists would like to be able to predict the conditions under which women can be expected to have abortions, and they must be able to decide which factors in the situation are important in order to make this prediction. Melamed seems to suggest that one important factor is ignorance of reproductive functioning and that perhaps another is social support for abortion from the man. However, it's easy to disagree with these interpretations based on the kind of data collected in an undirected survey such as this one.

*Directed Observations; Experiments*

Case studies and surveys are basic scientific observations. Different people reading them may interpret them as suggesting that one factor is important in making a prediction and that others are less important. However, it's difficult to use such relatively undirected observations to make specific predictions. In order to determine which factors are important in making predictions, more directed observation techniques have been developed. Rather than attempting to observe and record a large number of events surrounding a situation as is done in case studies and surveys, these techniques formally limit the kinds of events to be observed to quite a small number. They attempt to simplify the situation enough to make relationships among different kinds of events more apparent.

For example, a directed observation might be designed to try to discover the relationship between the amount of knowledge a woman has about birth control and the likelihood that she'll have an abortion. Another directed observation could attempt to discover the relationship between belonging to various churches and the likelihood of obsessional thinking. By discovering such relationships, directed observations make it easier to make predictions about future events.

Some directed observations allow much clearer predictions and conclusions to be derived than others, but most allow clearer conclusions than case studies or surveys. This book is primarily concerned with directed observations, but many of the principles apply to the interpretation and design of case studies and surveys as well.

You might have been wondering why I've avoided using the word "experiment." The reason is that there's considerable disagreement about what constitutes a "scientific experiment." Writers of textbooks for the behavioral sciences often adopt very strict definitions, as we'll discuss further in Chapter 2. They suggest that it's necessary for the experimenter to directly influence one kind of event and to see how that affects other kinds of events.

By that definition, virtually none of the research performed by astronomers, earth scientists, or anthropologists qualifies as experimental, because these researchers rarely exert direct influence over the events they study. Similarly, about half of the work of behavioral scientists is not experimental, since no events are directly influenced. However, in professional publications in the behavioral sciences, the strict definition is rarely applied. All directed observations are referred to as experiments, with qualifications made about the conclusions that can be drawn from different kinds. This is similar to usage in the physical sciences, where virtually any directed observation is called an experiment.

At the risk of offending my colleagues in the behavioral sciences, I'll adopt the more lenient approach, using the word "experiment" rather loosely to describe any directed observation. Starting in Chapter 2, I'll discuss experiments in which events are manipulated or nonmanipulated, but I'll use the word "experiment" to describe either case.

**Experiments,** then, are formal observations designed to make it easier to understand the relationships among a small number of kinds of events. They make it easier to make predictions. By this definition, most of the original observations published in the behavioral sciences are experiments. Only case studies and surveys are excluded, since they aren't designed to show the relationship between events but only to record as many events as possible.

## Reading Experimental Reports

After reading an experimental report you can make predictions, formulate or modify theories, and discover practical applications. But to do any of these things you must understand the limits of the observations and know how to interpret them reasonably. Neither task is easy.

One reason that experimental reports aren't easy to read is because the style is usually very dry and concise. Important facts aren't emphasized as they are in textbooks or popular writing, so you must be constantly alert. Trying to read or listen to experimental reports carelessly with your mind on other things is usually just a waste of time. So the first thing to remember about experimental reports is common to all difficult reading or listening: You must be active, constantly questioning and thinking about the information. This book can help you learn what questions to ask, but asking them takes effort which you'll have to provide.

A second reason scientific reports are difficult is that a lot of jargon is used. Actually, there are two kinds of jargon, and this book can help with only one. One kind of jargon is specific to a particular area of study. If a report concerns how eyes work, for example, it uses different jargon than if it concerns why some people are more aggressive than others. If you try to read or hear a report without knowledge of the language used in that area, you probably won't understand much. This book can't help with that kind of jargon; that's what you learn from books and classes covering different research areas. But there's also a more general kind of jargon common to many areas. Words and phrases such as control group, counterbalancing, three-way interaction, factorial design, confounding, and significant effect are bandied about with great freedom and no explanation because the reader is expected to be familiar with these terms and their implications. This book discusses much of that.

A third reason people have difficulty with experimental reports is that even if they understand all the words in the report, they still don't know how to interpret the observations. They have no way to recognize the limits of the study or to know how different kinds of shortcomings affect the conclusions that can be drawn from the study. The primary goal of each Section A in every chapter of this book is to teach interpretational skills.

## Overall Plan of Reports of Experiments

When experiments are reported, the writers make a formal attempt to distinguish between events they actually observed and their interpretations of these observations. It's common for many experimental reports to be divided into four main sections, although they're not all labeled: the introduction, method, results, and discussion sections. The survey by Melamed had an introduction and a method section and combined the results and discussion sections.

In an ideal experimental report the *introduction* tells the purpose of the experiment. It describes the area of research, giving information about what's already known. Then it tells what the reported experiment should add to what's already known, limiting the investigation to a few well-specified questions. Sometimes these are stated as formal predictions, or experimental hypotheses. In reading the introduction section you should understand all these different aspects. The introduction section is discussed in Chapter 2.

The *method section* tells exactly how the experiment was performed. It's the reader's substitute for actually doing the experiment. You must decide whether the methods are adequate for answering the questions raised in the introduction. The method section is discussed in Chapters 3 and 4.

The *results section* tells what the authors observed. It presents the new findings, substituting for direct observation of the events by the reader. You must be able to interpret the results and decide whether they seem trustworthy. The results section is discussed in Chapter 5.

The *discussion section* tells how the authors interpreted their findings. What were the answers to the questions raised in the introduction? If there were hypotheses based on a theory, the authors point out how the results either support or modify the theory. You must decide whether you agree with the authors and then interpret their findings in accord with your own interests.

In addition, most experimental reports have a summary at the beginning or end. When it's at the beginning of the article, the summary is called an *abstract;* when it's at the end, it's called a summary. Reading abstracts or other brief presentations is an art in itself. It's discussed in Chapter 7, along with information on how to read the article title for as much information as possible. Also in Chapter 7 is an overview of the four main sections and some hints on reading articles that don't have section divisions.

As you can see from this brief description, only the method and results sections contain descriptions of actual observations. The other sections are interpretations.

## Sources of Reports of Experiments

Scientific reports are sometimes presented orally at university colloquiums and at professional meetings. These oral reports are structured the same as printed reports. Thus, by learning to read reports, you learn a good deal about how to listen to them. Written reports are available from many sources. The U.S. government has experimental reports on many scientific areas. Many private and public research agencies distribute reports to the general public. Universities maintain files of original observations made by students for masters' theses and doctoral dissertations.

However, the most common sources of scientific reports are professional journals. They're magazines published periodically. You can order them directly from the publishers, but most people look at them in university library collections. Most libraries bind the issues into large books, but they were magazines originally. You should find a place you can look at journals from the field in which you're interested.

If possible, try to find a library that allows you to browse rather than to have to request each volume individually.

Not all professional scientific journals contain original observations. Some specialize in printing review articles designed to summarize many observations and to interpret them as a coherent group. Some journals deal primarily with issues of experimental design, statistical techniques, new kinds of equipment, or book reviews. Some journals present primarily theoretical discussions about how groups of observations might be interpreted or essays about various areas and approaches that might be used to solve problems. Some journals specialize in issues of personal interest to professionals, such as political events likely to affect the field, job openings, elections in professional societies, and discussion of ethical and moral issues relevant to the profession. In this book we are interested only in descriptions of original observations, with a brief mention of the review and theoretical articles. A list of journals containing descriptions of observations in the behavioral sciences is given in Appendix A.

### How to Find a Specific Journal Article

As you saw in the case study and survey you read, in psychology the journal articles are referred to in the text by the authors' names and the year of publication. Thus the case study you read would be referred to as Cohen and Smith (1976) or (Cohen & Smith, 1976). Other fields use different styles such as numbering the articles or employing footnotes. After seeing the notation, you can turn to the list of references at the end of the article or to the footnote and locate all the information necessary to find the article. In this text all the references mentioned in the articles have been combined alphabetically by author's name at the end of the book. If you'd like to know more about the subject mentioned in one of the articles, you can look at the related experiments.

In psychology, this is the most common format for giving the information needed to find an article:

Authors' names. Title of article. *Journal Name,* year of publication, *volume number,* page numbers of article.

For example, the complete reference to the survey you read looks like this:

Melamed, L. Therapeutic abortion in a midwestern city. *Psychological Reports,* 1975, *37,* 1143–1146.

To locate that article, you find the journal by looking it up in the card catalogue or periodicals list in your library. Then find Volume 37, published in 1975, and turn to page 1143. There it is! Formats for some psychology journals differ slightly from this, and other areas use different formats entirely, but all references include the same information, and you find the article the same way.

### How to Find Experimental Reports on Particular Topics

Good sources of experimental reports about particular topics are advanced textbooks in the area of interest. They can be located by subject in the card catalogue in a good academic library, by browsing in university bookstores, or by asking people likely to know. Many advanced textbooks are actually collections of articles written by several authors reviewing different subject areas. Other textbooks are written by one author and attempt to organize and clarify large numbers of experiments. Introductory and lower level texts are sometimes good sources of references to original articles but more commonly are only summaries of results and interpretations.

Other good sources of references are articles published in professional journals specifically designed to review the literature. In psychology two important journals are the *Annual Review of Psychology* and *Psychological Bulletin*. They contain articles that organize and summarize large numbers of studies performed about specified topics. Fields other than psychology have similar journals.

Another technique for finding experimental reports is simply to browse through journals likely to discuss the research area in which you're interested. Find a recent issue, look down the list of titles, and see if anything looks useful. If it does, you can use not only that article but also its reference list as a source of information.

If you want to locate many articles related to your topic, most scientific areas have one or more sources specifically for that purpose. In psychology, the most commonly available source is *Psychological Abstracts,* a journal containing summaries of articles appearing in other journals. To use it, read the abstracts about your subject; if they're interesting, go to the original journal and read the whole article. Locating all the relevant abstracts requires some knowledge about how *Psychological Abstracts* is organized.

In the library you'll probably find that six issues are bound together in one volume. Within each volume the abstracts are numbered, starting with 1 in the first issue of each volume and going well over 10,000 in the last issue. In each monthly issue the abstracts are organized into broad categories such as "Psychometrics," "Experimental Psychology (Human)," and "Developmental Psychology." Thus, if you have a general interest in an area, you can just turn to that group of abstracts by using the table of contents in each issue to find the area.

If you have a more specific interest, or one that cuts across several broad categories, you can use the subject indexes to locate the relevant abstracts. At the back of each monthly issue is a *Brief Subject Index,* which refers to all the abstracts in that issue by their numbers. You could look up, for example, "marijuana" and find perhaps three abstracts of articles dealing with that topic. If you looked up "drugs," you'd find a list of cross-references to specific drugs. If you look up a topic and find no references at all, try to think of other related words or topics—coming up with the right word is essential.

At the end of each six months a new volume appears called a *Volume Index to Psychological Abstracts*. It contains all the references from that six-month period. Less frequently, *Cumulative Indexes* appear that cover even longer intervals.

Sometimes it's also helpful to use the author indexes that appear in the monthly issues and in combined issues to follow the work of one or more authors whose interests are similar to yours.

A useful technique is to start with a very recent issue of *Psychological Abstracts* and to use it to locate a few articles dealing with the topic in which you're interested. Then the reference lists in these articles can be used to locate other relevant papers. Obviously, that technique is less thorough, but it is also less time-consuming than trying to locate *all* the related articles.

In addition to using *Psychological Abstracts,* a person wanting to perform a thorough literature search might want to contact one of the increasing number of computer services designed for this purpose. *PASAR* and *PsycInfo* were created to facilitate searches through *Psychological Abstracts*. Another service called *Scientific Citations* locates articles using specified words in their titles, articles referring to *other* specified articles, or articles by specified authors or organizations.

### Choose Practice Articles

Now it's time to locate one or two sample articles to practice on while you're reading this book. Choose them carefully, because you'll be using them over and over again. You might want to photocopy them to avoid returning to the library. If you already have some articles you must read, use them. If not, go to the library and locate the journals in your area of interest. Appendix A has the journal names arranged according to the Library of Congress call numbers. By looking at that list you can tell what area of the library is likely to contain journals that might appeal to you. If you have the time and opportunity, just browse through a number of them to see what's available. Look at some review articles and some journals that don't particularly interest you just to see what they are like. Then find some journals with titles that appeal to you. Look through them to see if they contain reports of original experiments. Don't choose journals that present mainly review articles, essays, theoretical position papers, or other nonobservational articles.

After you have selected some journals with experiments, look down the list of titles and turn to any articles that seem interesting. After you find a promising looking article, look more closely at it. Is it an experiment? Do the authors appear to have relatively limited goals? Don't choose a case study or a survey for this sample article. Certainly don't choose a review article that only reports other people's observations or an essay that reports someone's opinion about a topic. For your practice articles choose some with clearly labeled sections. The introduction is never labeled, but the method, results, and discussion sections should be.

Many articles present more than one experiment. That's fine—you can just pick one of them and read mainly about it. Finally, look again and see whether you can understand it, more or less. If not, get rid of it for now. Some articles contain so much specialized jargon that they aren't good practice articles.

Naturally, if you choose articles in an area of some interest to you, you'll enjoy reading them more.

After you've selected one or two articles, answer these questions about them:

1. Who are the authors?
2. What is the title?
3. What's the journal title?
4. What are the year of publication and the volume number?
5. On what pages is the article?

**Section B
Learning from
Your Own
Observations**

Being able to read and interpret other people's scientific reports is very useful but, obviously, many of the observations you might want may not yet have been made in a systematic way. You may want (or be compelled) to perform an experiment for yourself. If so, there are quite a number of things you should know.

**Necessary Skills**

You should know how to read other people's reports so you can find out what's already known about your area of interest. Thus you need all the material in the first sections of all the chapters of this book.

You should have a goal or purpose for your research. That's discussed in the remainder of this chapter.

You must be able to state clearly just what you're trying to find out and to break it down into small enough units so that you can make the appropriate observations to answer your questions. That process is discussed in Section B of Chapter 2.

You must be able to set up an experiment to test what you have in mind. The process of experimental design is discussed in the second sections of Chapters 3 and 4. You also should have some idea of how to actually perform your experiment so that the findings are relatively easy to interpret. That's also discussed in Section B of Chapter 4.

You must know some procedures for summarizing and presenting your observations to others. There are several techniques, some statistical and others not. Some of the most common are discussed in Section B of Chapter 5.

You must be able to interpret your findings. The skills that you acquire in interpreting other people's studies are obviously relevant here, but skills specific to your own experiment are discussed in Section B of Chapter 6.

Finally, to be part of the scientific community, you should know how to report your findings to others so that they can profit from them. Some formats for reporting your experiment to others are discussed in Section B of Chapter 7.

**Sources of Ideas for Experiments**

After performing one or two experiments most people find it easy to think of more. But deciding where to start can be a problem. Different people get started in different ways.

Many people are most highly motivated by experiments that directly test some theory or hypothesis. For example, Cohen and Smith's case study might lead someone to develop a hunch, or hypothesis, like this: Bad life experiences cause people to return to religions they previously rejected. A person might want to try to design an experiment to test this theory. The experiment should test whether that relationship really occurs. If the experiment shows that the relationship does occur, the theory is supported. If the relationship doesn't occur, the theory must be wrong.

Perhaps you know of a formal theory that you'd like to test. Alternatively, you may have formed a theory of your own, perhaps from your own life experiences or from reading or from discussions you've heard. One theory might be "People make fewer typing errors when they're not distracted."

★ Are there any formal theories you know about that you'd like to test?
★ Do you have any informal hunches of your own that you'd like to test?

Some people find it easier to plan experiments to solve practical problems such as "Which of these study techniques leads to the best grades?" or "Which arrangement of dials on the car dashboard leads to the most accurate readings?" Some people prefer to try to answer such questions as "Are people who give lots of advice likely to be happy people?" or "Are turtles or fish better problem-solvers?" If no practical problems or interesting questions come quickly to mind, think about jobs you've had, about materials you've learned in classes, about your hobbies, or about your family.

★ Is there some problem you'd like to solve or a question you'd like to answer?

### Section B  Learning from Your Own Observations

Many first-time experimenters feel that they'd like to learn more about some general area but don't have any specific questions. For example, someone might be interested in learning more about the personalities of successful athletes. Someone else might want to learn about the process by which people recognize an acquaintance in a crowd of people. You might want to learn about self-destructive behavior or have an interest in porpoises or chimpanzees. If so, you could design an experiment to learn something about a small portion of one of these areas. In the next chapter we'll discuss the process of limiting yourself to a manageable question. If you do have an interest in a general area, think about it and try to decide what aspects seem most interesting to you.

★ Is there some general area that interests you?

If there's no theory you want to test, no problem you want to solve, no question you want to answer, and no area of sufficient interest to you that you want to learn more about it, the only reason for you to design and perform an experiment is that you're required to do so for some class. This happens to lots of people. However, they often find that they rather like the whole process once they get into it; with luck, that may happen to you.

If you don't have any area you want to investigate and you have to design an experiment about something, go back to the library. Browse through the journal articles until you find a study you can more or less understand and which is in an area you could possibly work in. Don't choose a study that requires working with great white whales, for example, unless you happen to have access to great white whales. Your experiment can be built on the one you've selected. You may decide to replicate (that is, repeat) the whole experiment or just part of it. You may plan to make some changes in the experiment to see how the changes affect the results. Just reading the experiment might give you a theory to test, a problem to solve, a question to answer, or an interest in a general area. In any case, you should develop some plan for an experiment.

The more specific you can be in your plans for an experiment, the easier it will be to design that experiment. However, at this time you may not be able to be very specific. That's okay.

★ As specifically as you can, state the purpose of your experiment. It might help to start with one of these headings. My purpose is
  To test this theoretical idea:
  To solve this practical problem:
  To answer this question:
  To learn more about this topic:
  To replicate this study:

**Chapter Two
What's the
Research
About?**

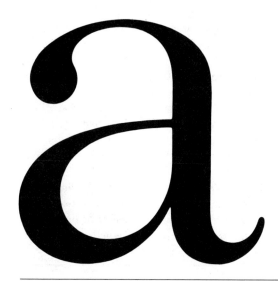

**Section A
Reading
Introduction
Sections**

The introduction section of an experimental report has several purposes. It should orient the reader to the area investigated in that study by telling about past work and thinking. It should explain why the experiment being reported was performed. And it should indicate how the new data will fit into what's already known.

The introduction to a published paper follows. It was written by Herbert J. Cross and Gary L. Davis and is entitled "College students' adjustment and frequency of marijuana use." It was published in the *Journal of Counseling Psychology* in 1972, Volume 19, on pages 65 to 67. The rest of the sections of this article appear in later chapters of this book, but if you'd like to see the whole article at once you can look it up in the library. As you read this introduction section, try to answer these questions:

    **1.** To what area of knowledge does this study relate? Be as specific as possible.
    **2.** What was already known about that area?
    **3.** Specifically what information was the present study supposed to add?

**The Introduction Section of the
Cross and Davis Marijuana Study**

Marijuana use is currently an exciting and controversial topic, especially among educators and those they educate. A research literature has just begun to appear which, in time, might offset the propaganda and opinions from the popular press, underground press, and the government.

Investigators need to address themselves to many questions, of course, but a high priority must be given to studies of drug users' personality characteristics, especially in the college population.

Blum (1969) found drug users to be slightly more dissatisfied and pessimistic than abstainers. Hogan, Mankin, Conway, and Fox (1970) found college student marijuana users to be self-confident, socially poised, and wide ranging in interest, whereas nonusers were described as pleasant, responsible, considerate, dutiful, and rather conventional. Another group, those who said they would not use drugs, were termed principle nonusers and were deferential to external authority, narrow in interests, and overcontrolled. Robbins, Robbins, Frosch, and Stern (1970) surveyed 286 college students and found that 24% had used illicit drugs. The users described themselves as anxious, cynical, bored, rebellious, and restless, whereas the nonusers saw themselves as ambitious, contented, decisive, and secure.

These data are adding to a base of knowledge but do not yet allow satisfying conclusions about the personal adjustment of student marijuana users. The present study addresses itself directly to the questions of whether or not college student drug users are more maladjusted than nonusers and whether or not adjustment is associated with frequency of use.

---

To what area of knowledge does this study relate? The general area, of course, is marijuana usage. However, Cross and Davis had a specific interest within that area. They weren't interested in how people feel when they use marijuana or whether its use leads to harder drugs. A specific answer would be something like this: The area of interest was how the personality characteristics of marijuana-using college students compare with the personality characteristics of non-marijuana-using college students.

What was already known about that area? Cross and Davis mention three papers that compared marijuana users with nonusers and looked at personality differences between the groups. After reading these papers Cross and Davis believed their conclusions and reported them. Naturally you must have some faith in their judgment and in that of the original authors, but it's good to remember that the only way to be sure of these conclusions is to read the other papers and evaluate them for yourself. With that qualification in mind, it's appropriate to try to summarize all the findings mentioned by Cross and Davis. A summary might go something like this: The studies give a general picture of marijuana users as poised, dissatisfied, anxious, and rebellious while nonusers are responsible, conventional, ambitious, and secure. There were also principle nonusers who were narrow and overcontrolled.

What information was this study supposed to add? Cross and Davis said,

"The present study addresses itself directly to the question of whether or not college student drug users are more maladjusted than nonusers." That's the question this study is supposed to answer. We could restate it as a formal prediction, or *experimental hypothesis:* College student drug users are more maladjusted than nonusers. The experiment can be thought of as a test of that hypothesis.

Cross and Davis added that they wanted to determine whether "adjustment is associated with frequency of use." That's a more technical statement of a slightly more general experimental hypothesis. Exactly what does it mean?

### Variables

In the jargon used by experimenters, "adjustment" and "frequency of use" are called **variables.** In any experiment, the events or behavior with which the experiment is concerned are variables; the topic of an experiment is defined by the variables in it. Quite often the title of an article names the most important variables in the study, as was true in the Cross and Davis title, "College students' adjustment and frequency of marijuana use." Some examples of variables in other experiments might be person's height or amount of study or type of weather or time of day.

One characteristic of a variable is that it can be measured in some way. You can measure how much of that variable a person or event has, or you can determine at what **level** of that variable a person or event is located. For example, frequency of marijuana use could be measured in several ways, such as watching people and keeping track or asking them how often they use it and recording their answers. You might be able to think of other ways.

Not all experimental variables can be measured in the sense of assigning meaningful number values. For example, eye color could be used as a variable, perhaps in an experiment to find out who gets better grades, people with blue, brown, green, or gray eyes. Each eye color is the value, or level, of the variable eye color for each person. Obviously, assigning numbers corresponding to blue or brown would be possible, but it wouldn't mean much. If blue = 1 and brown = 2, it wouldn't make much sense to say that blue + brown = 3. Variables like this are sometimes said to be variables that can be measured only on a nominal scale (a scale with names only) and might be called **nominal variables.** Some other examples of nominal variables are sex, type of car driven, or color of room.

A second characteristic of a variable is that it takes different values for different individuals or occasions; that is, it varies. For example, once you've decided how to measure frequency of marijuana use, you can find that some people use it more often than others or that people use it more on some occasions than on others.

Adjustment is also a variable in the Cross and Davis experiment. As you can see, it meets both criteria. It might be measured by using a standardized test or by having experienced judges making ratings of people's adjustment, or by recording what people say about their problems. And if it is measured, some people are more well-adjusted than others, or people are better adjusted at different times.

**Problems**

To be sure that you understand what a variable is, try to answer these questions:

**Q.** List two or three possible experimental variables. Don't try to describe a whole experiment; just name some variables.
**A.** You should be able to judge for yourself whether your answers are correct. If you're not sure, ask "Can this variable be measured in some way?" and "Does this variable vary from individual to individual or from time to time?"

**Q.** Were any of your variables nominal variables? If not, try to write down one or two nominal variables.
**A.** To tell whether a variable is a nominal variable, think about assigning numbers to the different levels of the variable. If the numbers are sensible and meaningful, such as heights or amount of intelligence or amount of liking, then the variable isn't a nominal variable. But if the numbers are completely arbitrary, like calling all males "Level 1" and all females "Level 2," then the variable is a nominal variable.

**Association between Variables**

Next we'll go on to the phrase "associated with," which is also experimental jargon. All experiments in all sciences are designed to discover **associations between variables.** Physicists demonstrate the nature of the association between force and velocity. Chemists search for the association between certain operations and changes in chemicals. Botanists measure the relation between amount of sunlight and amount of growth in many kinds of plants. Ethologists point to the relationships between species and many kinds of behavior such as mating rituals. Sometimes it's possible to express these relationships in the form of mathematical equations, and sometimes not. However, the results of an experiment always can be thought of as a description of the association between variables.

When one variable is associated with another, you can use that information to make predictions from one variable to the other. If adjustment is associated with frequency of marijuana use, then information about a person's frequency of marijuana use tells something about that person's likely level of adjustment, and vice versa. There are other phrases that mean much the same thing as "associated with." Instead of saying that frequency of use is associated with adjustment, an experimenter might say that adjustment "varies with" frequency of use or "is correlated with" frequency of use or "can be predicted from" frequency of use. There are small differences between these phrases, but for most purposes you can think of them as equivalent.

Suppose that after completing their experiment Cross and Davis found that people who use marijuana frequently are less well-adjusted than people who use marijuana rarely or not at all. If that were true, and if you know that Joe Blow uses marijuana very frequently, you'd probably be correct to predict that he'd be less well-adjusted than MaryJane Clean, who never uses marijuana at all.

In fact, as will be discussed fully in Chapter 5, Cross and Davis's results were somewhat equivocal, but if they found anything, it was in that direction. College students who said they used marijuana most frequently received the most maladjusted scores on a standardized test. Is that what you would have expected?

**Problems**

These problems should help you to learn to make predictions based on descriptions of associations between variables.

**Q.** Measured scholastic aptitude varies with school grades such that people who get high scores on scholastic aptitude tests tend to get good school grades. If the students in Rutherford B. Hayes High School tend to get lower scores on that test than the students in Grover Cleveland High School, which students would you predict would get better college grades?
**A.** The students from Grover Cleveland should get higher grades in college, on the average.

**Q.** School grades also are correlated with number of absences from school, such that people who are often absent tend to get lower grades. If Goody Twoshoes is often absent, would you predict that she has higher or lower grades than Jimmy Switchblade, who rarely misses?
**A.** Goody should have lower grades, based on the overall pattern.

**Q.** Suppose an industrial consultant finds that typing ability is associated with absenteeism in such a way that that the typists who make the fewest errors are absent most frequently. If you know that Ms. Hunt is absent more than Ms. Peck, who would you predict makes the most errors?
**A.** Ms. Hunt is absent more frequently. Therefore she probably makes fewer errors.

**Statistical Expressions Related to the Concept of Correlation**

To try to simplify the description of some of these relationships, experimenters and statisticians have adopted some verbal conventions. When high scores on one variable go with high scores on another, as with scholastic aptitude test scores and school grades, they are said to have a **positive association.** You might say that scholastic aptitude test scores are positively associated with school grades or related positively or positively correlated. On the other hand, when high scores on one variable go with low scores on another, as with number of absences and school grades, they are said to have a **negative association.** You could say that number of absences and school grades are negatively correlated or negatively associated; you could also use a different word and say that absences and school grades are **inversely related.** Some sets of scores may be related in more complicated ways. For example, people with very low levels of anxiety and people with very high levels both do relatively poorly on tests, but people with medium levels do well. This might be described as a **nonlinear relationship** or a **nonmonotonic relationship,** because it doesn't go in a straight line.

### Problems

These problems should help you learn to distinguish between positive and negative associations.

**Q.** Suppose a lion-tamer keeps track for 20 years and finds that size of lion can be used to predict aggressiveness: The biggest lions are least likely to attack. Are lion size and aggressiveness related positively or negatively?
**A.** High scores in animal size go with low scores in aggressiveness. Therefore they're negatively associated.

**Q.** Cross and Davis found that people who use marijuana very frequently were less well-adjusted than others. Is that a positive or negative relationship?
**A.** That's a negative relationship.

**Q.** If large values of X go with large values of Y, are variables X and Y associated positively or negatively?
**A.** Positively.

When the relationships are **linear,** that is, generally go in a straight line, statisticians sometimes actually calculate a formula to express that line algebraically. Such a line is called a **regression line,** or a **line of best fit.** It can be used to make predictions from one variable to the other. However, the prediction is never perfect. There's always some error, because real relationships just don't fall exactly along straight lines. To express that amount of error, a different kind of statistic called a **correlation coefficient** has been developed. Correlation coefficients express how much difference there is between the general relationship, as expressed by the regression line, and the actual relationship between the two sets of scores.

Most correlation coefficients can take any value between $-1.00$ and $+1.00$.

A correlation coefficient near .00 means that there's essentially no correlation between two sets of scores. The correlation of adult body weight and intelligence is near .00. A correlation coefficient near +1.00 means that high scores on one variable nearly always go with high scores on the other. A correlation coefficient near −1.00 means that high scores nearly always go with low scores. The nearer to +1.00 or −1.00, the stronger the relationship; that is, the closer the relationship is to really being a straight line. Correlations of measured intelligence with school grades vary between +.40 and +.60. The correlation of length of upper arm with length of upper leg is near +1.00. Correlations between strength of religious values and strength of empirical/scientific values vary between −.30 and −.50. Correlations between weight of car and miles per gallon are near −1.00.

In all these examples both variables being correlated have a large number of levels, but the concept of correlation also applies when there are relatively few levels. For example, suppose these were the data from a hypothetical experiment varying occupation and type of TV show most preferred:

|  | TV Show Most Preferred | |
|---|---|---|
|  | Violent | Nonviolent |
| Pro football player | 22 | 73 |
| Pro baseball player | 61 | 15 |

The number in each cell represents the number of people of each type interviewed. In looking at the table it seems apparent that the two variables are correlated: Football goes with nonviolence, and baseball goes with violence. The appropriate correlation coefficient would show the correlation to be +.56, a reasonably strong positive relationship, but not perfect.

We'll discuss correlation coefficients and the concept of correlation further in Chapter 5.

The next introduction section you'll read uses correlation coefficients to describe the strength of association between variables. It was written by Arthur M. Sullivan and Graham R. Skanes and is entitled "Validity of student evaluation of teaching and the characteristics of successful instructors." It appeared in the *Journal of Educational Psychology* in 1974, Volume 66, on pages 584 to 590. As you read it, try to answer these questions:

1. To what area of knowledge does this study relate?
2. What was already known about that area?
3. What sort of information should this experiment add?
4. Can you name two variables in this experiment?
5. What kind of correlation do you expect Sullivan and Skanes to find between the two variables?

## The Introduction Section of the Sullivan and Skanes Teacher Evaluation Study

Although considerable data have been collected concerning the reliability of student ratings of instructors, relatively little information is available concerning the validity of student evaluations when the criterion is the amount which students learn from a course. Most studies of validity have used correlations with peer ratings or supervisor ratings as the criterion.

When final grade is used as the measure of achievement, approximately half of the studies reported show positive correlations between grades and student evaluations while the remaining half show no correlation or negative correlations between the two (Costin, Greenough, & Menges, 1971).

When a more objective measure of final achievement has been used as the criterion of achievement, for example, score on an achievement test, positive but modest correlations have usually been reported (Elliott, 1950; Morsh, Burgess, & Smith, 1956).

Using a university mathematics course taught by teaching assistants, Rodin and Rodin (1972) found, however, a correlation of −.75 between a rating of overall performance and final grade, as determined by the number of criterion problems passed, and with the effect of prior achievement removed. They concluded that students are not able to judge teaching effectiveness if the latter is measured by how much they have learned. This conclusion has been challenged by Gessner (1973) who found, in the case of sophomore medical students taking a one-semester science course, correlations of .74 and .62, respectively, between student rating of content and organization and of presentation and class performance on national normative examinations. However, no significant correlation was found between student ratings and performance on institutional examinations. The Rodin and Rodin study has also been challenged by Frey (1973), who found a positive correlation between instructor's rating and students' examination performance of .958 and .951 for two courses in mathematics.

In addition to the contradictory nature of these reports, it is difficult to generalize from the studies which have been reported because in most cases only one course was investigated and in many cases the instructors were teaching assistants rather than full-time experienced instructors. Also, the measures of achievement used were often not those commonly used or readily understood.

At Memorial University of Newfoundland, Canada, we have been able to take advantage of circumstances which enabled us to carry out a study involving a large number of instructors, many of whom were full-time experienced faculty members, and an objective and very common measure of achievement.

This study was designed to find out if, under these circumstances, students were able to identify those instructors from whom they learned most.

To what area of knowledge does this study relate? It deals with student evaluation of teachers, an area of interest to students and teachers alike. More specifically, it deals with the relationship between student evaluations and how much the students learned in the class.

What was already known about that area? From the information presented by Sullivan and Skanes, the area is quite confusing. Some studies show positive relationships, while others show negative relationships. The studies differ in many ways from each other, so it's not clear why the different results occur.

What sort of information should this experiment add? Presumably it adds another study showing the relationship between student evaluations and how much the students learned in the course. The study is unique in that it uses a large number of students and a large number of experienced instructors. Sullivan and Skanes also stress that they will use an objective measure of student achievement, that is, how much the students learned.

Sullivan and Skanes made no formal prediction about the direction of the relationship. Their purpose was simply to learn whether there *is* a relationship, whether positive, negative, or nonmonotonic. A formal statement of their experimental hypothesis is simply that there is a relationship between teacher rating and student achievement.

What association do I expect between the variables? I'd like to believe that students rate teachers highly if the teachers are able to get the students to learn a lot of material, but I'm not sure. Maybe students really prefer easygoing teachers who don't ask them to work very hard. If the first occurs, it would lead to a positive correlation between student ratings and measures of student achievement. If the latter occurs, there would be a negative correlation.

Now consider the introduction section of another paper. This one was written by Warren G. Wallace, John J. Horan, Stanley B. Baker, and George R. Hudson. The title is "Incremental effects of modeling and performance feedback in teaching decision-making counseling," and it appeared in the *Journal of Counseling Psychology* in 1975, Volume 22, on pages 570 to 572. As you read it, try to answer these questions:

1. To what area of knowledge does this study relate?
2. What was already known about that area?
3. What sort of information should this experiment add?
4. Can you name two variables in this experiment?
5. What association did Wallace et al.[1] expect between the variables?

---

[1] The abbreviation *et al.* stands for *et alia,* which is Latin for *and the others.* It's used in order to save space when more than two authors write an article.

## The Introduction Section of the Wallace et al. Teaching Method Study

One of the most pressing problems the contemporary counselor faces is helping students learn decision-making skills (Greenwald, 1973; Herr and Cramer, 1972; Krumbolta and Thoresesen, 1969). Thus, instructional units on the rationale and practice of decision-making counseling are often embedded in counselor education programs. However, little or no research exists on evaluating the various methods counselor educators may employ to foster such skills in counselor trainees.

The purpose of the present investigation was to compare three cumulative methods of teaching decision-making counseling to counselor trainees. Method 1 represents the traditional approach; the students were given a lecture and a written handout on decision-making counseling. Since modeling has made extensive inroads into counselor education and practice, the students exposed to Method 2 observed a videotaped model of a counselor engaged in decision-making counseling, in addition to receiving the lecture and written materials of Method 1. Method 3 incorporated some of the features of microcounseling (Ivey, 1971): along with the lecture, handout, and model of Method 2, the students were given the opportunity to practice decision-making counseling and receive feedback on their performance from a counseling supervisor.

Since neither modeling nor performance feedback can exist in isolation, this study was concerned with the cumulative rather than the separate effects of the training efforts. Modeling and performance feedback do require additional classroom time. Unless these procedures produce significant gains, they should not be included in counselor education curriculums.

It was hypothesized that the three teaching methods would be differentially effective in producing student acquisition of decision-making counseling skills. Specifically, Method 2 was expected to be superior to Method 1, and Method 3 was expected to be superior to both Methods 1 and 2.

---

To what area of knowledge does this study relate? Wallace et al. were interested in how to train counselors to be effective in helping students learn decision-making skills. The general area of knowledge is something like teaching methods; the specific area is methods related to these skills.

What was already known about that area? According to Wallace et al., nothing at all was known about the relative effectiveness of the various training procedures in this specific case. However, it seems unlikely that there was no related research with teaching other kinds of material. If they had checked the literature before doing this experiment, they might have learned a great deal.

What sort of information should this experiment add? Wallace et al. were interested in a very specific question: How effective are three different teaching methods in training counselors to help students with decision-making processes? Their experiment was designed to answer that question so that they could decide whether the amount the counselor trainees learned justified the added expense of the more complex teaching methods.

What were the two variables in the experiment? One was the teaching method used. If you thought of that variable as a nominal variable despite the numbers, that's probably good. It had three levels consisting of the three teaching methods. The other variable was "decision-making counseling skills," that is, how much the counselor trainees actually learned.

What association did Wallace et al. expect? They expected that the students taught with Method 1 would learn the least, those taught with Method 2 would learn the next most, and those taught with Method 3 would learn the most. That's their experimental hypothesis based on their expectations.

**Association versus Causality**

If you think about this experiment, you might see that Wallace et al. would be able to assign the three teaching methods to different counselor trainees. They were able to manipulate that variable directly. That's probably the most important difference between this study and the Cross and Davis marijuana study, in which the authors just measured what happened naturally. If Wallace et al. manipulate the teaching methods in the appropriate way, they'll be able to find out whether the change in teaching method *causes* a change in the amount learned, in addition to being able to determine whether amount learned is associated with teaching method. Why is that distinction important?

You should be able to see a clear difference between these two statements: (1) People who drive a Rolls Royce drink more Scotch than people who drive a Chevy; and (2) driving a Rolls Royce causes people to drink more Scotch than driving a Chevy. The first is a description of an association; the second involves a conclusion about causality. The first statement could be true and the second one false, so the two are quite different.

When you know that an association exists between variables you can make predictions based on that information, and sometimes you can make practical use of these predictions. If you know of an association between driving a Rolls Royce and drinking Scotch, and you happen to invite ten Rolls drivers over for cocktails, you could prepare by having a good deal of Scotch in the house. But when you know that a **causal relationship** exists, you can not only make predictions, but you can change the events themselves. If you knew of a causal relationship between Rolls Royces and Scotch, and you wanted to increase Scotch sales in your town, you could do so by having everyone drive a Rolls for awhile.

### Problem

This problem will help you learn how to apply causal relationships as compared to associations.

**Q.** Let us suppose we have two grouchy people. Charley Crank knows that a certain time of day is associated with grouchy feelings; that is, he always feels grouchy at a certain time of day. Sarah Snarl, on the other hand, knows that being hungry causes her to feel grouchy. What kind of action can Charley and Sarah take to minimize the bad effects of their grouchiness?

**A.** Charley is aware of only an association, so the only action he can take is to try to accommodate his life to his predictions. He could avoid seeing people or doing important things at his bad time of day. Sarah, on the other hand, knows of a causal relationship, so she can change the situation. If she wants to avoid feeling grouchy, she can eat something.

### Drawing Causal Conclusions

When an association exists between two variables, it's sometimes true that a causal relation exists between them. How can you tell?

The issue of drawing conclusions about causality has aroused much philosophical debate. It is now traditional in scientific experiments to conclude that one variable has caused a change in another when there appears to be no other reasonable explanation for an association between them except (1) that causal relationship or (2) pure chance or accident.

You can draw conclusions about causality when experimenters arrange things so that initially there's only one variable, like teaching method, that differentiates the levels of the experiment. If there's a later difference in another variable, such as amount learned, the only reasonable explanation is that the first variable caused a change in the second. If there are many initial differences, there will be many reasonable explanations for an association; the cause and effect relationships will be obscured.

The important difference between the Cross and Davis marijuana study and the Wallace et al. teaching method experiment is the number of ways the subjects at different levels of the variables differed from each other. Wallace et al. would be able to assign the counselor trainees to different teaching methods in such a way that there aren't likely to be any differences between the people assigned to different groups. There should be no reason to believe, for example, that the trainees who received Method 1 were any less intelligent or empathic than the trainees who received Methods 2 or 3. A difference like that could occur by pure accident, but there's no reason to expect such a difference. If noticeable differences appear in the amount learned by the people in the different groups, the only reasonable explanation is that the differences were caused by the teaching methods. There's nothing else to explain them other than pure chance.

On the other hand, in the Cross and Davis marijuana study, we could expect many differences between the students who used marijuana frequently and those who used it rarely or not at all. They probably differed in many ways, such as wearing different clothes, coming from different backgrounds, or having a different college major. These differences can be expected because Cross and Davis didn't manipulate the variable directly, making sure that the people at each level of marijuana use were alike in other ways. They just found people who happened to use different amounts of marijuana. The negative association that Cross and Davis found between frequency of use and adjustment *might* be explained by saying that marijuana use *does* cause maladjustment. But there are other equally reasonable explanations. Perhaps maladjustment causes heavy marijuana use. Perhaps some unmeasured variable like type of upbringing or amount of self-confidence causes both maladjustment and heavy marijuana use. None of these explanations is clearly more correct than any other.

All experiments determine whether associations exist between variables, but only some experiments allow you to conclude that an association is causal. It's important for you to know the difference, so this issue will be raised frequently throughout the rest of this book.

### Problems

These problems will help you learn to recognize some experiments in which causal conclusions can and can't be drawn.

**Q.** Let's go back to the imaginary lion-tamer, who found that big lions tend to be less aggressive than small ones. Would he be correct to conclude that being small *causes* lions to be more aggressive based on the evidence?

**A.** Did the lion-tamer assign lions to be different sizes? Did he eliminate other differences between the groups? He didn't, so he can't draw causal conclusions. Being small *may* cause lions to be more aggressive, but this study doesn't allow you to draw that conclusion.

**Q.** Suppose a biologist chose two groups of 10 orangutans that were as similar as possible to each other. She trained one group to smoke cigarettes made of lettuce leaves and the other group to smoke marijuana cigarettes. Both groups smoked 10 cigarettes a day for a year. Then she gave them all a test of orangutan adjustment (whatever that might be) and found that the orangutans smoking marijuana were more maladjusted than those smoking lettuce leaves. Is she correct to conclude that smoking marijuana *causes* maladjustment in orangutans?

**A.** She assigned the orangutans to groups, trying to eliminate other differences between the groups. Therefore, the causal conclusion is warranted. What else could cause the difference in adjustment between the two groups? The only other possibility is accidental differences between the groups.

### Manipulated and Nonmanipulated Variables

In some experiments the experimenter **manipulates a variable** directly, actually deciding when a given level of that variable will occur. In other experiments, no variable is manipulated. The experimenter simply accepts the levels that occur naturally or actively searches out some naturally occurring levels. Manipulating variables allows causal conclusions to be drawn, but it's not always possible to manipulate variables.

One kind of variable that is usually very difficult to manipulate has sometimes been called a **subject variable.** Subject variables are integral parts of the personality or habitual behavior of the individuals serving as experimental subjects. Some subject variables are age, sex, intelligence, type of car owned, and income. In the Cross and Davis study, both frequency of marijuana use and adjustment were subject variables. When **nonmanipulated variables** are used in an experiment, it's never possible to draw really clear causal conclusions, because there are always many differences other than just the variable of interest.

In addition to subject variables, there are many other kinds of variables that are difficult to manipulate. Some are too big, such as the weather or earthquakes or star movement. Some are too dangerous, such as some drugs or exposing people to great terror. Some are too expensive, such as changing the kind of car people drive or the color of their homes or their income. And some are just impossible to consider, such as time of day or year of event. Regardless of the reason, when variables are nonmanipulated there are many differences present that make it difficult or impossible to draw any clear-cut conclusions. Nonmanipulated variables can be used to show associations, but not causal relationships.

### Independent and Dependent Variables

As you read through different experiments, you'll often see the phrases **independent variable** and **dependent variable.** In an experiment in which one variable is directly manipulated or assigned by the experimenter, the variable that was manipulated is called the independent variable and the one measured later is called the dependent variable. In the Wallace et al. teaching method experiment, teaching method was the independent variable and amount learned was the dependent variable. These phrases make some sense when you recognize that the amount learned *depends on* the teaching method used, but the experimenters *independently* manipulated the teaching method.

# Section A  Reading Introduction Sections

**Problems**

These problems will give you practice in recognizing independent and dependent variables.

**Q.** Back to the orangutans and the biologist. She trained one group of orangutans to smoke lettuce cigarettes and the other group to smoke marijuana cigarettes. Then she measured all the orangutans' adjustment. What was the independent variable? What was the dependent variable?

**A.** The variable that the biologist manipulated directly was material smoked or lettuce versus marijuana, so that was the independent variable. Adjustment was the dependent variable.

**Q.** Suppose an advertiser had two apparently equal groups of Cadillac-owners, and he blindfolded both groups and drove them around in a car for an hour. Then they rated the smoothness and comfort of the ride. One group rode in Cadillacs, and the other group rode in Lincolns. What was the independent variable? What was the dependent variable?

**A.** The independent variable was the brand of car. The dependent variable was the rating of smoothness and comfort.

## True Experiments, Pseudo-Experiments, Correlational Studies, and Observational Studies

As mentioned in Chapter 1, some teachers and writers in the behavioral sciences like to use a very strict definition of what constitutes an "experiment." They reserve the words "experiment," "independent variable," and "dependent variable" for studies that manipulate the independent variable. This makes sense when you realize that this is the only time the variables really are *independently* varied. Studies in which no variable is manipulated are called **pseudo-experiments, correlational studies,** or **observational studies.** The phrase *independent variable* may be replaced by **pseudo-independent variable** or **predictor variable,** while *dependent variable* is replaced by **pseudo-dependent variable** or **criterion variable.** Since these substitute phrases aren't common in published articles, this book doesn't use them. It uses the word "experiment" to include all directed observations and uses "independent" and "dependent variable" whenever it seems helpful.

When a variable is manipulated, use of the terms independent and dependent variable is quite clear. However, when variables are nonmanipulated, the usage is less clear. If a study simply measures two or more nonmanipulated variables, as in the Cross and Davis marijuana study, it's not necessary to refer to any variable as either independent or dependent. There's just no special reason to distinguish one variable from the other by giving it a special name.

In other experiments there are reasons to distinguish between the variables even when neither variable was manipulated. Sometimes individuals are *selected* as experimental subjects because they have a particular value of some variable to be investigated. They aren't *assigned* to a level but are chosen because they already belong at that level. For example, an experimenter might select 50 students with high math grades and 50 with low math grades and then give them all a vocabulary test. When an experiment using this kind of selection is published, the selected variable is often called an independent variable and the one measured later is called a dependent variable even though neither was manipulated. In this example math grades are the independent variable and scores on the vocabulary test are the dependent variable. It's very important to notice that even though these words are used, it doesn't mean that we can draw causal conclusions about the effect of the independent variable on the dependent variable.

*To repeat:* There are variables the experimenter manipulates directly, variables on which the experimenter bases selections, and variables that are just measured.

When one variable is manipulated and another measured, the manipulated variable is called the independent variable and the measured one is called the dependent variable. Your teacher might want to call these **true experiments.** Of the studies we've discussed, it seems likely that only the Wallace et al. teaching methods study would qualify. If such experiments don't have other problems, it may be correct to conclude that changes in the independent variable caused any changes observed in the dependent variable. There may be other problems that prevent this conclusion; we'll discuss them in Chapter 4.

In studies in which one variable is selected and the other is measured later, it's common to call the selected variable the independent variable and the one measured later the dependent variable. Your teacher might want to describe these as *pseudo-experiments* or possibly as *correlational studies* and avoid using the terms independent and dependent variable. Later in this chapter you'll read a study in which children of different ages and social classes are selected. These variables could reasonably be called *pseudo-independent* because of that selection. It's rarely possible to draw causal conclusions in this kind of study.

In studies in which both variables are measured, the terms independent and dependent variable aren't usually very helpful, although they're sometimes used anyway. Your teacher might want to describe these as *correlational studies* or possibly *observational studies*. The Cross and Davis marijuana study could reasonably be described as a correlational study in which marijuana usage is the *predictor variable* and adjustment is the *criterion variable*. It's rarely possible to draw causal conclusions in this kind of study, regardless of what words are used.

**Problems**

These problems will give you additional practice in recognizing independent and dependent variables and in deciding when to draw causal conclusions.

**Q.** A researcher approached all the women who entered a grocery store and asked them two questions: "What is your occupation?" and "If you must choose between a brand-name product and a store-brand product, which do you usually choose?" He found the women who said they were homemakers tended to say that they choose brand names more than the women who gave other occupations. He concluded, "Women who are homemakers watch more daytime TV than women with other occupations, so they see more brand-name advertising, which influences their choices. Therefore, being a homemaker causes women to choose brand names." First, name the variables in this experiment. Then decide whether his conclusion is correct and tell why.

**A.** The two variables were occupation and product choice. Both variables were just measured, so there's little reason to call one of them independent or dependent. Since both variables were nonmanipulated, the researcher wasn't correct to draw his conclusions. There are many differences that might explain the results. His completely unsubstantiated comments about watching TV were just a distraction to make the conclusion appear more reasonable.

**Q.** A professor at a large university located 50 engineering majors and 50 psychology majors, all seniors. She gave them a test of music appreciation skills and found that the engineering students got higher scores than the psychology students. She concluded, "Engineering courses sharpen your analytical skills and therefore increase your ability to deal with any activity requiring analytical skills. Since music appreciation requires analytical skills, it's not surprising that taking engineering courses increases music appreciation skills. Therefore, being an engineering major causes an increase in music appreciation skills." Name the variables, and evaluate her conclusion.

**A.** Again, the experiment dealt with two nonmanipulated variables, choice of major and music appreciation ability. Since the professor selected subjects on the basis of their majors, that would probably be called an independent variable and music appreciation would be the dependent variable. Since both variables were nonmanipulated, her conclusion is incorrect. The professor can't conclude that one variable causes the other, because there are many other possible explanations.

All the experiments we've discussed so far have had only two variables. Most experiments in the behavioral sciences are more complicated than that, with several independent and dependent variables.

The introduction section of another article follows. This experiment has two independent variables and one dependent variable. It was written by William F. McDaniel and Leslie C. Vestal and is entitled "Issue relevance and source credibility as a determinant of retention." It appeared in the *Bulletin of the Psychonomic Society* in 1976, Volume 5, on pages 481 to 482. As you read it, try to answer these questions.

1. To what area of knowledge is this study related?
2. What was already known about that area?

**3.** Name the two independent variables and the one dependent variable. (*Hint:* Try the title for help.)

## The Introduction Section of the McDaniel and Vestal Source Credibility Study

The postulate that an individual attends more readily to stimuli congruent with his own beliefs, values, and concepts has encouraged study of the interaction of such variables with opinion, learning, memory, perception, etc. Much of this research has concerned the selective learning and retention of controversial materials. Early research by Levine and Murphy (1943), based on the unconfirmed hypotheses of Edwards (1941), demonstrated superior retention of mildly pro-communist subjects for a pro-Soviet communication in comparison to anti-communist subjects and the superior retention of anti-communist subjects for anti-Soviet learning materials. Thus, the effects of subjects' attitudinal variables on memory were demonstrated, and this finding has tended to be supported by some of the subsequent research (Alper & Korchin, 1952; Jones & Aneshanel, 1956; Jones & Kohler, 1958; Taft, 1954; Zimmerman & Bauer, 1956).

Brigham and Cook (1969), in an experiment paralleling the Levine and Murphy (1943) and the Jones and Kohler (1958) paradigms, failed to confirm the results of the attitude-selective learning hypothesis as well as the postulate that individuals retain plausible statements consonant with the subjects' attitudes better than implausible statements dissonant with the subjects' attitudes. This finding, in conjunction with other similar results (Smith & Jamieson, 1972; Waly & Cook, 1966) conflicting with the Levine-Murphy hypothesis, has seriously questioned the generality of attitudinal mediation in the retention of controversial materials and has suggested that other variables may have been confounded with the earlier findings. Dutta and Kanungo (1967) have shown that the intensity of unpleasant and pleasant affect[2] generated by tasks influences later task retention, while Smith and Jamieson (1972) have found ego involvement to be a major determinant of retention rather than pre-existing attitude.

The present experiment was designed to investigate the effects of stimulus variables contained within the communication upon retention. The roles of relevance of the issue to the subject and credibility of the source reporting the communication in the mediation of retention were examined. It was hypothesized that communications varying with respect to these variables would lead to discernible group differences in retention of issue content.

To what area of knowledge is this study related? The general area is retention, or recall. More specifically it has to do with the effects of preexisting beliefs and feeling on the recall of material.

---

[2] In this context, *affect* equals emotion.

What was already known about that area? McDaniel and Vestal give quite a lot of past information. They quote a number of studies both in support of and in conflict with what they call the Levine-Murphy hypothesis, which seems to be that people tend to recall things they already believe. This experiment was meant to be a continuation of the investigation into the truth of that hypothesis.

The dependent variable in the study is retention. The two independent variables are relevance of the issue to the subject and credibility of the source reporting the communication.

Suppose that McDaniel and Vestal were interested only in the association between issue relevance and recall. To find the nature of that association they could obtain two groups of subjects. One group would be given material of high relevance to recall, and the other group would be given material of low relevance. If there were a difference in recall between the two groups, they would know that it was probably caused by the relevance of the material being recalled.

Now suppose that they were interested only in the association of source credibility with retention. Again, they could obtain two groups and present information to one group from a highly credible source and present information to the other from a less credible source. If there were a difference in recall of the information, they could conclude that it was caused by the source credibility.

But McDaniel and Vestal wanted to answer both questions. To do that as efficiently as possible, they used an arrangement of variables called a **factorial experimental design.**

### Factorial Designs

In a factorial design all the levels of each independent variable are combined with all the levels of the other independent variables. McDaniel and Vestal actually had four groups of people, as shown here:

| Source Credibility | Issue Relevance High | Low |
|---|---|---|
| High | 1 | 2 |
| Low | 3 | 4 |

After assigning people to the four groups, they performed the experiment and measured the recall of each group. The use of the factorial design allows the effects of both independent variables to be investigated simultaneously.

A difference between the people with high and low issue relevance would show that issue relevance caused a change in retention. A difference between the people who heard the information from sources with high and low credibility would show that source credibility affected retention. The fact that two variables are combined doesn't affect the conclusions that can be drawn about each of the variables considered separately. If one were manipulated and the other were not, it would be appropriate to draw causal conclusions about the manipulated variable but not about the nonmanipulated one.

**Problems**

These problems will give you practice in recognizing independent and dependent variables in factorial designs and in deciding when it may be appropriate to draw causal conclusions about their relationships.

**Q.** Suppose an experimenter obtained 30 baby white rats and 30 baby brown rats. He divided each group into three other groups of 10 rats each, keeping the groups as equal as possible. When all the rats were 6 months old, he started them on special diets. One group of each color was fed a diet of white, unenriched bread. Another group of each color was fed a diet of white, vitamin-enriched bread. The third group of each color was fed a diet of regular rat chow. All the groups continued on their diets until they died. The experimenter recorded how long each rat lived.

Name the two independent variables and the one dependent variable. For each of the independent variables, tell whether the experimenter would be able to conclude that changes in that variable *caused* changes in the dependent variable.

**A.** The independent variables were color of rat and type of diet. The dependent variable was length of life. Color of rat was a nonmanipulated, selected variable, so if there were any differences between the white and brown rats, the experimenter wouldn't be able to conclude that color caused those differences. Type of diet was a manipulated variable, and there's no reason to believe that the rats assigned to one diet differed from the rats assigned to a different diet. Therefore, if there were any differences between the rats in the three diet groups, the experimenter would be able to conclude that the diet was probably responsible.

**Q.** Suppose a teacher divided her class of 40 students into four apparently equal groups. Two of the groups were assigned to use one textbook and the other two a different textbook. From each of the textbook groups, one group was assigned 10 homework problems every night while the other group was told to study on their own. After keeping to this plan for 6 weeks, the teacher gave an exam and compared the performance of the four groups.

Name the independent and dependent variables. For each independent variable, tell whether the teacher would be able to conclude that changes in that variable caused changes in the dependent variable.

**A.** The independent variables were textbook and homework problems. The dependent variable was test performance. Both independent variables were manipu-

Section A  Reading Introduction Sections                                                                    49

lated variables, so causal conclusions could probably be drawn in both cases, if there were no other problems with the experiment.

**Q.** Suppose a physical education teacher chose 50 girls who said they had a high interest in sports, 50 girls who said they had a low interest, 50 boys with a high interest, and 50 boys with a low interest. Then he gave them all a test of physical fitness.
Name the variables, and tell whether causal conclusions are likely to be appropriate for each independent variable.
**A.** The independent variables were sex and interest in sports. The dependent variable was physical fitness. Both his independent variables were subject variables, nonmanipulated variables on the basis of which he selected his subjects, so causal conclusions wouldn't be correct in either case.

Experiments often have as many as three or more independent variables. The last introduction section that you'll read describes a study with three independent variables combined in a factorial design. That is, all levels of each independent variable were combined with all levels of the other two. The article was written by Samuel R. Schutz and Evan R. Keislar and is called "Young children's immediate memory of word classes in relation to social class." It appeared in the *Journal of Verbal Learning and Verbal Behavior,* 1972, Volume 11, on pages 13 to 17. As you read it, try to answer these questions:

  1. What area of research is dealt with?
  2. What was already known about that area?
  3. What are the independent and dependent variables in the present study?
  4. For each independent variable, is it likely to be manipulated directly or just selected or measured?
  5. What relationships do the authors expect to occur between the independent and dependent variables?

**Excerpts from the
Introduction Section of the
Schutz and Keislar Word-Class Study**

Of all the recent literature relating to social class differences in language, perhaps no single theoretical formulation has stirred more interest than that of Basil Bernstein (1960, 1961). Although many researchers have suggested that there is a functional difference between the lower and middle class cultures in "standard" language usage (e.g., Deutsch, Katz, and Jensen, 1968), Bernstein proposes an explanation for the apparent disparity. Essentially he suggests that the lower class culture follows a "restricted" linguistic code which hinders the development of vocabulary and keeps thought at a low level of conceptualization. Bernstein proposes that if lower class children were helped to acquire the "elaborative" linguistic code of the middle class, the measured intellectual and academic differences between the groups would be reduced.

To encourage research related to the theory, Bernstein has enumerated various hypotheses capable of investigation. He has suggested that one important distinction is the relative inferiority of the lower class in the use of function words (e.g., conjunctions, prepositions, negation, etc.). . . .

In the present investigation, attention was focused solely on the child's ability to echo words belonging to a certain word class; that is, the words were presented alone, without sentential constraint.[3]

**Problem**

The major purpose of this investigation was to compare the ability of lower and middle class children to recall immediately function words, in relation to differences in recall of two other word classes, nouns and verbs. It was expected that there would be differences between the two socio-economic groups as well as differences between age groups in ability to recall the words of each class. The central concern of the inquiry, however, was focused upon the interaction of socio-economic status and word class.

For the project two major hypotheses were tested dealing with the interaction of socio-economic status with word class and grade:

**Hypothesis 1.** Deficiency in word recall of lower class children in the kindergarten-primary grades is greater for function words than for nouns and verbs in relation to the performance of middle class children.

**Hypothesis 2.** At the kindergarten-primary level, the difference between the social classes for recall of function words, relative to recall of nouns and verbs, is larger for children in the higher grades than for children in the lower grades.

---

What area of research is dealt with? This study focuses on the relationship between social class and the ability to deal with various kinds of words. Schutz and Keislar were especially interested in the ability to deal with function words. In the parts of the introduction section reproduced here, little information is given concerning the real relationships between these variables. The discussion of Bernstein deals with theory rather than with demonstrated relationships. This paper is an attempt to test one aspect of Bernstein's theorizing. In the parts of the section that weren't reproduced some studies were mentioned, but Schutz and Keislar pointed out that there were

---

[3]*Without sentential constraint* means not in sentences.

## Section A  Reading Introduction Sections

some problems with these studies that made their results difficult to interpret. You're not yet equipped to deal with this discussion.

What were the variables in this study? One independent variable was word class. It had three levels, function words, nouns, and verbs. A second independent variable was social class, apparently with two levels, middle and lower. The third independent variable was discussed less, but it was age. It's not mentioned in the introduction section, but it also had three levels, kindergarten-age children, first-grade-age children, and second-grade-age children. Thus there were a total of 2 × 3 × 3 treatments, that is, 18 different treatments, arranged like this:

| Word Class | Middle-Class Children | | | Lower-Class Children | | |
|---|---|---|---|---|---|---|
| | Kindergarten | Grade 1 | Grade 2 | Kindergarten | Grade 1 | Grade 2 |
| Function words | 1 | 2 | 3 | 4 | 5 | 6 |
| Nouns | 7 | 8 | 9 | 10 | 11 | 12 |
| Verbs | 13 | 14 | 15 | 16 | 17 | 18 |

The dependent variable was the ability to recall the words. It was evaluated for each of the 18 treatments.

For each independent variable, is it likely to be manipulated? It seems very likely that Schutz and Keislar will manipulate word class directly. However, they couldn't manipulate either age or social class, so these variables must be either selected or just measured. Only the conclusions about the effects of word class on recall could possibly be causal.

What relationships do the authors expect between the variables? The main points of the relationships of interest are described in their formal hypotheses. As you can see, these relationships are relatively complex, dealing with a technical relationship called an **interaction**. We'll discuss that concept in Chapter 5. In addition to the relationships mentioned in the hypotheses, the authors expected several other associations. One was that the middle-class children would recall more than the lower-class children. Another was that older children would recall more than younger children. As you see, the questions asked can be quite complicated in studies with three independent variables.

Now it's time to turn to your own practice articles and to read the introduction sections. Try to answer all the questions that follow. Your articles are probably more difficult than the sample ones in this book—most articles are. Try not to get discouraged.

1. To what area of knowledge is this study related?
2. What was already known about that area?
3. What kind of information is the present experiment supposed to add?
4. Name all the variables in the experiment. There may be several independent and dependent variables and perhaps some you can't define. Remember to check the title for possible help.

**5.** Did your experiment have clear-cut independent variables that were either manipulated or selected? If so, name them. Then tell whether the experimenters manipulated them. If they did, you may be able to draw causal conclusions about the effects of these variables on the dependent variables. However, you can't tell for sure until you finish Chapter 4. If they didn't manipulate the variables directly, you probably can't draw causal conclusions.

**6.** Did your experiment have clear-cut dependent variables? If so, name them.

**7.** If your experiment had variables that weren't either independent or dependent, was it because the experimenters were just measuring several variables to see how they were associated, such as in the Cross and Davis marijuana study? If so, name those variables here.

**8.** Do you have any variables left over that you didn't note in your answers to questions 5, 6 or 7? Are they really variables in which the experimenters were interested and which they varied in the experiment on purpose, or are they just extraneous variables not really measured in this experiment? If this experiment didn't measure them or look at their relationships with other variables, then cross them out of your answer to question 4. They're not variables in this experiment. If the experiment *did* measure these variables, then figure out what they are: independent, dependent, or just measured.

**9.** For all the variables listed in questions 5, 6, and 7, tell what kind of association the authors expected to occur among them, if you can. If no clear prediction was made, then the authors probably just predicted that *some* association would occur. Your answer to this question is a statement of the experimental hypothesis in your experiments.

If you understood everything you read and could answer all these questions and know that you're right, that's good. If you're not sure your answers are right, find someone you can check with and ask. If there's no one to ask—well, this will get clearer with practice.

If you feel like you understood what this chapter was saying and you answered the practice questions without trouble but couldn't understand the introduction section of your article, you probably should get rid of that paper. You may have picked up a paper that is not an experiment at all, or the article might be too specialized for you to understand.

If you could understand the introduction section of your article but couldn't answer the questions, there are two possible explanations. One is that the answers really aren't in the introduction section at all. Look at the rest of the paper, and try to find them. If you can't find them there, you may need a new article. The other possibility is that you don't quite understand what the questions are asking. If you think this might be true, go back and reread the appropriate parts of the chapter. If that doesn't help, then try to find someone to talk to.

**Extra Library Assignment.** Try to find examples of both manipulated and nonmanipulated independent variables in articles in the library.

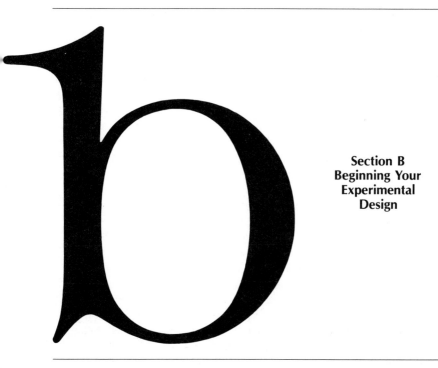

**Section B
Beginning Your
Experimental
Design**

In this section you'll make the first decisions toward limiting your experiment to a manageable question. You'll select your variables, designating some as independent and some as dependent, and decide whether to manipulate, select, or just measure your independent variables. In making these decisions, you'll do some library research to find out what's already known. After you've selected your variables, you'll state as clearly as you can the questions your experiment should answer and then formally state your experimental hypothesis.

**Select Your Variables**

Now that you know something about different kinds of variables, you should have a clearer idea of how to express what your proposed experiment is about. What variables will you have in your experiment? You needn't know how to measure them; we'll discuss that in the next chapter. But you should know the *names* of your variables. Otherwise you won't know what you're doing.

The variables you choose determine the topic of your experiment, so you'll want to make this choice carefully. The smallest number of variables you can have is two; your experiment will investigate the relationship between them. At the other extreme, few experiments in the behavioral sciences have more than three independent variables and three dependent variables. More than that leads to experiments that are too complicated for even the researchers to understand. You probably should limit yourself to no more than two of each type at this time, if that many. As you continue designing the experiment, you may find that you want to add extra variables.

The more well-defined your problem, the easier to choose your variables. If you've decided to replicate someone else's experiment, your choice is quite simple: Use the same variables. If there are too many, choose the ones you most want to investigate. Your experiment will tell you whether you find the same relationship among the variables as the original workers found.

If you have a well-specified practical problem, such as wanting to know which of three teaching methods is most effective, the process of choosing your variables is again relatively simple. You choose the ones you want to know about, such as teaching method and effectiveness.

Similarly, if you started with curiosity about a well-defined question such as "Is adjustment related to marijuana use?" the process of choosing your variables is direct. Your question dictates that choice, so that you might choose marijuana use and adjustment as variables. To make the experiment more complex and general, you might add educational level as an additional variable.

If you started with a vague interest in a research area rather than a specific question, you should think about the area and try to choose some variables arbitrarily from all the possibilities. Rather than asking "What are athletes like?" you might limit yourself to the three variables of, say, interest in athletics, mathematical ability, and interest in show business. Instead of asking "What is sleep good for?" you might choose to investigate the relationship between amount of sleep and irritability. Limiting yourself this way often seems frustrating. You want to know all about a big topic, and all you can do is attack one little corner of it. Scientists deal with that frustration by trying to answer lots of small questions, one or two at a time. They hope that the answers to the small questions will form a pattern so that big topics eventually will be understood.

If you've decided to try to test a theory, either a published one or one of your own, the theory should guide your choice of variables. For example, Schutz and Keislar's study was trying to test a theory that predicted a particular pattern about the relative difficulty of function words as compared to other words for lower- and middle-class children. Therefore they chose as variables word type, social class, and difficulty. You might have formed a theory of your own that you'd like to test such as "People make fewer typing errors when they're in a quiet place." You could choose as variables amount of noise and number of typing errors.

If you're planning an experiment to answer a specific question or to test some aspect of a theory, you should be careful to choose your variables in such a way that they'll really do the job. You'll be better equipped for making that choice after you've completed every Section A in this book. If you haven't, just do the best you can for now.

Suppose you had a theory stating that women are more likely to be affected by emotional arguments, but men are more likely to be affected by rational arguments.

### Section B  Beginning Your Experimental Design

An experiment might be performed in which some men's and women's attitudes toward economic growth are measured and found to be about equal. Then both groups could be exposed to a very emotional, nonrational speech against economic growth, and their attitudes remeasured. This experiment is not a test of your theory, because it has nothing to do with the effect of rational arguments on anyone. You can't draw conclusions about untested comparisons.

To test this theory there must be two independent variables: sex (men and women) and type of argument (emotional and rational). To draw conclusions about a particular comparison requires that you make that comparison.

Choosing variables to test theories and answer questions in useful ways is challenging. The theory should make a specific prediction about the relationship that will occur between variables. Your experiment should test that prediction. If your experiment shows results in accord with the prediction, then the theory is supported. If not, then some aspect of the theory must be inadequate. Thus you should choose your variables to test some specific prediction from the theory.

**Problems**

These problems should help you learn to choose appropriate variables to test different theories.

**Q.** Suppose you had a theory that said "Hungry shoppers buy more food than other shoppers." What would be wrong with the following experiment as a test of that theory?

As people enter a grocery store, ask them whether they're hungry. If they say yes, watch the register as they check out and record their bill.

**A.** That experiment tells nothing about the amount bought by nonhungry shoppers. There's no way to compare the hungry shoppers to others. In fact, this "experiment" has no independent variable at all, since there was no variability in hunger.

**Q.** How might an experiment to test that theory be set up?

**A.** One possible setup would be simply to record the bills both of people who said yes *and* of people who said no. Then the results could be compared. Hunger or answer to question would be the independent variable, and amount of bill would be the dependent variable.

**Q.** Suppose you had a theory that suggested people who call themselves "day people" are likely to eat a higher proportion of their day's food at breakfast than people who call themselves "night people." What would be wrong with the following experiment as a test?

Obtain two groups of people, one who identify themselves as "day people," and the other as "night people." Find out how much each group ate for breakfast, and compare the two groups.

**A.** This experiment will tell you which group ate the most for breakfast but won't tell you about the *proportion* of food eaten at breakfast.

**Q.** How might an experiment to test this theory be set up?
**A.** One possible setup would be to ask everyone what they ate all day, at each meal and for snacks. Then calculate the porportion of the total food reported eaten at breakfast and compare the two groups. In both experiments the independent variable was the same: type of person. To make the experiment adequate, it was necessary to change the dependent variable from *amount* eaten at breakfast to *proportion* eaten at breakfast.

**Q.** Suppose you had a drug you thought might help cure people with organically caused skin problems but not people with psychologically caused skin problems. What would be wrong with the following experiment as a test of the drug's effectiveness with both groups of people?
Find two groups of people, one identified as having organically caused skin problems and the other as having psychologically caused skin problems. Give them all one pill a day for a month. Compare the amount of improvement in each group at the end of the month.
**A.** The problem here is that there's no way of knowing what would have happened in each group without the drug. This experiment will allow you to know which group improved more at the end of a month but wouldn't allow you to conclude anything about the effectiveness of the drug. The drug wasn't a variable that could be evaluated separately from other variables.

**Q.** How might an experiment to test the drug's effectiveness be designed?
**A.** The important change is to add drug versus no drug as an independent variable. One way to do this would be to find four groups of people, two with each type of problem, and give the drug to only one group of each type. As you'll see in the next chapter, there are other ways to accomplish much the same thing, but they all involve comparing the levels drug and no drug.

★ Name your variables.

Now that you've chosen some variables, think about them. First, does it seem possible to you to investigate them? For example, you may want to study psychosis, but if you have no access to psychotic people, that area of research is closed to you. You need new variables. Second, are you interested in knowing the relationship between the variables you've chosen? Are you already virtually certain of the nature of the relationship? If you're not interested or you already know the relationship, doing the experiment can still be good practice, but you won't care very much about the outcome. You may want to change your mind and tackle some variables of more interest to you.

### Library Research

Next, go to the library and look up information about your variables using the material from Chapter 1. Have other researchers investigated the relationship between your variables? What did they find? What else is known about your variables?

If you have trouble finding published work about your variables, it could mean one of several things. The most likely possibility is that you're looking in the wrong place. Think of synonyms for your variable names and try again. Ask someone likely to know about those variables where to find information. Another possibility is that the variables you've chosen aren't of interest to the professional "establishment," which determines what research is of sufficient importance to warrant publication. If the area is of interest to you, then go to it. This is often the best kind of problem for people who aren't caught in a "publish or perish" bind. A third possibility is that you've chosen a difficult problem so that other researchers haven't been successful in their attempts to answer questions in that area. You may want to think twice before continuing, but don't stop automatically—difficult problems will never be solved if no one tries. Yet another possibility is that you've thought of a completely new idea—but don't surmise that until you've exhausted the other possibilities.

If you look in several places and ask around and still can't find anything, then stop and get back to your experiment. Don't get bogged down in the library.

A more likely event is that you'll be inundated with more related information than you would have dreamed possible. Try not to be intimidated. Keep in mind that this means a lot of people have found the area rewarding. Organize the papers into similar subareas, and choose a subarea that looks interesting to you. If you are interested in maze learning, you might find literature for rats, chimpanzees, children, normal adults, and mentally retarded adults. If you're only interested in normal adults, then you could choose that subarea. Eliminate the papers in other subareas so that you get the relevant material down to a manageable amount. It's not necessary for you to include all the information with any bearing on your variables. Include only the most closely related information.

You may find that the problem you began with has already been analyzed into many subproblems. If so, you should plan to choose new variables to tackle one of the subproblems.

Another thing that could happen is that you might discover someone has already answered your question. Then you may decide that you'd like to switch problems or to extend the existing findings. Perhaps the original researchers used college students as their experimental subjects. Would you obtain the same results with non-college adults? Or the researchers might have used an expensive machine in their experiment. Would you obtain the same results if you used a cheaper technique? If you do decide to extend the findings, you should continue your literature search to find out if other researchers have made similar extensions on this or related problems.

After you find five or so articles fairly closely related to the problem you're working on, try to summarize each one in two sentences. The first sentence should describe what the experimenters did; the second should describe what they found. You'll learn more about how to do this in later chapters, but do the best you can for now.

When you have your summaries, try to organize them in some way. Do they form a pattern? Do several seem to show the same sort of thing? Do one or two seem to show something different? If so, can you figure out what might account for the difference?

For example, suppose you were originally interested in studying the variables of anxiety and problem-solving ability. In the literature you might find three articles showing that the most anxious people are better problem-solvers and two articles showing the opposite. What might account for the difference? Looking carefully at all five articles, you might find that the three showing the high-anxious people to be superior all used very easy problems, while the two showing the opposite used very difficult problems. That would suggest that problem difficulty might account for the different patterns of results.

If you do find a pattern like that, you're in a position to build a small theory. It might go something like this: With easy problems, high-anxious people are better problem-solvers; with difficult problems, high-anxious people are worse problem-solvers. After building your theory, the next thing to do is to see whether other articles in the library support it. Can you find another article with difficult problems that shows that high-anxious people are inferior? Perhaps you can find an article that compares high- and low-anxious people on both difficult and easy problems. That experiment would be a direct test of your theory.

If you find no such test, you might want to change your experiment to make it into a test of your theory. To do that in this case you'd have to add another variable. Your experiment would need two independent variables, anxiety level and problem difficulty; problem-solving ability would be the dependent variable.

Regardless of what you find in the library, stop looking after awhile. Do see if there's interesting information about your variables. You might change your thinking about your experiment and want to add variables or to change your experiment altogether. But get to your experiment soon.

★ After all this searching, do you want to change your variables or add other ones? If so, write down the names of your new variables.

## Decide Whether Your Variables Are Nominal Variables

Having read about your variables, you should be able to decide whether they're nominal variables or whether you'll be able to assign meaningful numbers to different levels. To decide whether your variables are nominal variables, remember that the question to ask is whether you can assign meaningful numbers. If you're work-

Section B  Beginning Your Experimental Design    59

ing with variables such as color of eyes, sex, ethnic background, type of weather, type of teaching method, or removal of a part of the brain, you probably can't assign numbers in a meaningful way. On the other hand, variables such as height, age, problem-solving ability, amount of anxiety, amount of traffic, length of time since eating, attractiveness, or likability can be defined in ways that allow meaningful assignment of numbers. If you have a choice, you should use numbers for statistical reasons rather than having a nominal variable.

 ★ For each of your variables, decide whether it's a nominal variable.

**Designate Independent
and Dependent Variables**

If you haven't done it already, you should decide which of your variables will be treated as independent variables. In many cases the choice is obvious. If you want to know which kind of background music makes people want more food, it's clear that type of music should be the independent variable. However, in some cases the choice may be more arbitrary. For example, an experiment investigating the relationship between social class and style of dress could use either variable as an independent variable. Which one is chosen may not affect the experiment a great deal, but it does change the emphasis somewhat. The results of the experiment enable you to predict the dependent variable by knowing the independent variable. If social class were the independent variable, the experiment would be expressed so that you could predict a person's style of dress by knowing that person's social class. If the independent variable were style of dress, the experiment would emphasize predicting a person's social class from that person's clothing. Personally, I'd find the second emphasis more interesting.

You should think about your variables and determine which relationship would be more interesting to you. If that doesn't enable you to decide, there's one lesser consideration: If one of your variables is a nominal variable, the usual pattern is to designate it as an independent variable. For example, in an experiment investigating the relationship between academic major and mathematical ability, the usual pattern would be to treat academic major as the independent variable. The custom is dictated mainly by statistical considerations and shouldn't influence your choice strongly.

If you still can't decide which of your variables should be independent variables, then just flip a coin. The decision needn't influence your experiment a great deal, but it will facilitate communication through the rest of this book.

 ★ Name your independent variable(s).
 ★ Name your dependent variable(s).

## Manipulation versus Nonmanipulation

Your next decision concerns whether you're going to manipulate the independent variables in your experiment. If you manipulate a variable and don't encounter other problems, you'll be able to draw cause-and-effect conclusions about the effects of that variable on the dependent variables. If you don't manipulate, it's unlikely that you'll be able to draw such conclusions. As we discussed in the first part of this chapter, it's more powerful to be able to draw causal conclusions; they enable you to control events more directly. But there are several good reasons for not manipulating independent variables.

### Practical Considerations

For one thing, it's impossible to manipulate many variables directly. Most subject variables, such as the sex or race or physical or mental attributes of your experimental subjects, are simply beyond your control. Environmental variables such as the weather or time of day also can't be manipulated. Other variables theoretically could be manipulated but are beyond your practical control. You probably can't really manipulate a person's income or grades or color of house, even though they theoretically could be manipulated by someone.

★ Is it possible for you to manipulate your variables directly?

If you answered "no," you must content yourself with the kinds of conclusions that can be drawn from just measuring variables. You'll be able to make predictions but not to draw clear-cut cause-and-effect conclusions. If you answered "yes," the next issue is whether manipulation of your independent variables will give you the information in which you're really interested.

### Theoretical Considerations

If your real question were "Who has a better vocabulary, people who habitually dress well or people who dress carelessly?" then you don't want to manipulate the kind of clothes the people wear. Manipulation would give the answer to a different question: "If a person is made to wear a certain kind of clothes temporarily, does that cause the person to have a change in vocabulary?" The process of knowing what question is being answered by an experiment is sometimes a bit tricky, so answer a few practice questions before answering for your own experiment.

## Problems

These problems will help you to recognize what question is being answered by an experiment.

Section B   Beginning Your Experimental Design                                    61

**Q.** A dentist has just heard that the sugar in plums is different from other fruit sugars. He wants to know whether eating plums causes less tooth decay than eating other fruits. Assuming he's able to manipulate plum-eating, should he do so or should he locate people who already eat a lot of plums and compare them to people who eat other fruits?

**A.** Since the dentist wants to know whether plum-eating causes less tooth decay, he should indeed manipulate plum-eating. If he doesn't, he won't be able to answer his question, even if he finds much less tooth decay in the people who eat a lot of plums. There might be other differences in their diet or even in their physiology that account for the different amounts of decay.

**Q.** An insurance company wants to know who is more likely to have accidents, people with self-contained recreational vehicles or people with pickup trucks with camper shells. Assuming that the company could manipulate the type of vehicle driven by people, should it do so, or should it compare people who normally drive the two vehicle types?

**A.** The insurance company isn't interested in knowing which kind of vehicle causes more accidents. It wants to know which kind is most likely to be in an accident, no matter what the cause. Perhaps poorer drivers choose one kind of vehicle more than another. Therefore, the insurance company shouldn't manipulate the variable.

★ If you manipulate each of your independent variables, will you get the kind of information you want?

If you answered "no," you should measure rather than manipulate. If you're able to manipulate your variables, and if it's theoretically desirable to do so, you must decide whether it's ethically permissible.

*Ethical Considerations*

Scientists have become increasingly sensitive to ethical issues raised by their research. At the risk of oversimplification, this book will offer guidelines. To have more complete understanding of some of the issues of most concern to psychologists, you should read at least two booklets available from the American Psychological Association, which can also be found in many libraries: *Ethical principles in the conduct of research with human participants* and the *Rules regarding animals*. A summary of the principles for human participants is in Appendix B. Other professions have publications on the same topics.

We'll discuss several ethical issues as they arise during the design and performance of an experiment. At this time the question concerns whether to manipulate independent variables in your experiment.

When you manipulate a variable you're attempting to cause a change in some other variable. If you're successful, you cause a change in some individuals or in the world. If these changes are long-lasting, they're more serious than if they're short-lived.

Suppose a teacher wanted to investigate the effects of test anxiety on test performance. Anxiety could be manipulated by giving the students different instructions about how much the test would affect their grades. To keep the grades fair, however, the teacher plans to count the test the same for all the students. Should the teacher do the manipulation? There are two ethical issues raised. One is that the experiment will cause some of the students at least temporary discomfort, although it's unlikely that the effects would be lasting. The other is that the teacher plans to deceive the students by the instructions. Both are questionable.

No reputable experimenter would perform an experiment likely to cause even temporary discomfort if the question weren't important or could be answered in some other way. But many experimenters are willing to cause their subjects some discomfort *if* the research is important, *if* there's no other way to answer the question, and *if* they believe that the damage done to the subjects won't be very bad. Other researchers prefer not to perform any experiment that makes the participants uncomfortable, even though that means they won't be able to answer some questions at all.

Deceit is used fairly often in psychological research, because in some cases telling people the real purpose of an experiment changes their behavior. They might try too hard to cooperate or might try to foul things up. Either leads to data irrelevant to the question the experimenter wants to answer. However, deceit should be avoided except when it's absolutely necessary. On moral grounds, it is, after all, a lie. Even if we ignore that, we should recognize that whenever deceit is used it causes the individuals who are deceived to mistrust the experimenters when they're finally told the real purpose of the experiment. If they're not told, the results are even worse: They're left with false beliefs. If subjects are deceived often, they become suspicious of all experimenters and produce irrelevant data whether or not deception is actually used.

You should keep in mind that subjects who agree to participate in an experiment often tend to be extremely cooperative, willing to put up with a great deal of discomfort even to the point of damaging themselves or others. It's your responsibility to protect participants from physical and mental discomfort, harm, and danger. A research procedure must not be used if it's likely to cause serious or lasting harm to human participants, and efforts should be made to minimize the discomfort of animals. You should also keep in mind that deceit should be avoided if it's at all possible.

★ Is manipulation of your independent variables likely to cause any long-lasting changes in your experimental subjects?

★ Will manipulating your variables hurt your subjects in *any* way at all?

★ Would manipulating your variables involve deceiving your subjects in some way?

If you answered "no" to all three questions, think again, to be sure. If you're right, there are probably no serious ethical obstacles to manipulating your variables. If you answered "yes" to any of the questions, then you should determine whether there's another way to go about answering your questions. Perhaps you could explain the purpose of the experiment honestly and ask the subjects to role-play, that is, to pretend they feel certain ways. Role-playing can give surprisingly accurate results in many cases, though of course it's not the same as the real thing in all respects. The teacher we discussed before could use that approach or could avoid deceiving the students by actually weighting tests differently at different times during the course and comparing the students' performances. This experiment would be harder to interpret, however, because there might be other differences between the tests as well or in how well the students studied for them. You may have to compromise sometimes in order to make your experiment ethically acceptable.

If you suspect that manipulating your variables might cause long-lasting changes in the experimental subjects and can't think of a way to change the experiment, you should consider whether the changes are for the better or the worse. If they're for the worse, then you shouldn't manipulate them with human subjects and should try to avoid manipulating these variables even with animals. If you think they're for the better, think very carefully about whether *all* your experimental subjects are likely to agree with you. Remember, most people don't like to be changed by others, even for the better. Unless you're sure that all your subjects will be pleased, or at least feel neutral about the changes, you probably shouldn't manipulate that variable.

If you suspect that your experiment will cause your subjects some temporary discomfort and you can't think of a way to change the experiment, then you just have to decide whether *your* research is worth *their* discomfort. Is the research worth the harm it will do? If not, you shouldn't manipulate the variable.

If you can't think of a way to avoid using deceit in your experiment, you must weigh the value of your research against the mistrust you'll cause in your subjects and decide whether your research justifies it.

A good rule for experimenters to follow when making decisions about ethics is "When in doubt, don't." Moreover, search carefully for doubts.

★ Taking into account all practical, theoretical, and ethical considerations, which (if any) of your independent variables will you manipulate?

**Selection versus Measurement**

For any independent variable you've decided not to manipulate you must decide whether to select events or subjects at several levels or simply to measure all the events that happen to occur.

It's easier to perform the experiment if you simply measure. However, it's likely that you'll observe a large number of events at some levels and few or none at other levels. That makes it difficult to interpret the results and means that the statistics you can use are less powerful. If you decide to select events, you'll have to work harder to make sure that you have a reasonable number of events at each level.

★ For which (if any) of your independent variables will you select events at different levels?

★ Which (if any) independent variables will you simply measure?

**State the Purpose
of Your Experiment**

If you've chosen to manipulate an independent variable, then your experiment is an attempt to answer this question: Does a change in *the independent variable* cause a change in *the dependent variables?* (Fill in the names of your variables.) If you've chosen not to manipulate an independent variable, then your experiment is an attempt to answer this question: Is a change in *the independent variable* associated in a regular, predictable way with a change in *the dependent variables?*

★ For *each* independent variable, write down the question your experiment should answer.

Look at the questions you've just written. Are they questions in which you're interested? Is that what you started out to test? If not, perhaps you should reconsider your choice of variables. However, your experiment may be testing more complex relationships among your variables than these simple questions, particularly if you designed your experiment to test a theory which predicts a complex relationship. We'll discuss more complicated relationships in detail in Chapters 5 and 6. For now, just do the best you can to figure out whether your experiment is designed to answer questions in which you're interested and to state these questions clearly.

★ Is there a more complicated relationship you'd like your experiment to test? If so, try to state it.

**State Your Experimental Hypotheses**

Your experiment *is* a test of some experimental hypotheses, whether or not you planned it that way. If you have no theoretical reason to believe that your results will come out one way or another, then your experimental hypotheses are just declarative forms of the questions you just phrased:

A change in the independent variable causes a change in the dependent variable.

The independent variable is associated with the dependent variable.

If you do have some reason to predict a particular pattern of results, then your experimental hypotheses should reflect that prediction:

An increase in the independent variable causes an increase in the dependent variable.

An increase in the independent variable is associated with a decrease in the dependent variable.

A more complicated hypothesis might look like this:

A change in independent variable A causes a change in the dependent variable when independent variable B is present, but not when B is absent.

An increase in variable A is associated with a decrease in variable B at low levels of C, but with an increase in B at high levels of C.

You'll learn more about these complex relationships in Chapters 5 and 6.

★ Try to state your experimental hypothesis. If you have more than one independent variable, you may have several hypotheses or one complex one.

## Some Preliminary Statistical Considerations

After you've performed your experiment, you'll probably want to analyze your results statistically. When you're more familiar with statistics, it's often a good idea to plan your statistical procedures as you design the rest of the experiment. For now, I'll limit the discussion to just two points.

One point is that it's important to have clearly stated experimental hypotheses. These hypotheses determine the relationships you'll be looking at in your results. New experimenters are often tempted to collect mountains of data and to look for interesting relationships afterward. As we'll discuss in Chapter 5, this means that those relationships have been discovered **post hoc** (after the fact) rather than actively sought. Such relationships are less trustworthy and therefore are less valuable scientifically. If you were able to state your experimental hypotheses clearly, you should have no difficulty with this statistical issue.

The second point concerns the nature of your experimental hypotheses. They may say there's *no* relationship expected between two variables or that you predict *no* effect of one variable on another or make some other negative statement. Such a statement of "no difference" is sometimes called a **null hypothesis.** If your experimental hypotheses are null hypotheses, you'll encounter serious statistical and interpretational difficulties. As we'll discuss in Chapter 5, null hypotheses can never be given clear-cut statistical support; you can never show the null hypothesis to be true. Also, there are interpretational problems: A badly run experiment will tend to give results that show no difference, that is, that agree with the null hypothesis. This means that if your results turn out exactly as you predict, with no difference, an unfriendly critic can say this occurred because your experiment wasn't sensitive enough to detect any difference that was present.

In Chapter 6, Section B, we'll discuss some partial attempts to counteract this problem, but there's no real defense. A null hypothesis is a perfectly valid scientific hypothesis. Unfortunately, experiments can't be designed that test the null hypothesis properly.

If your experimental hypotheses predict no relationship, no effect, or no difference, what should you do? It's perfectly all right to perform such an experiment. You can interpret the results usefully for your own purposes. The drawback arises when you try to tell other people what you've found. It's very hard to convince them that your results are trustworthy. If that matters to you (as it does to many scientists who want to contribute maximally to their field), then you should change your experiment to one in which you expect to find a relationship, effect, or difference.

This is a review of the steps you've taken in this chapter toward designing your experiment:

**1.** Choose a small number of variables to investigate.
**2.** Find out what's already known about your variables through library research.
**3.** Decide whether any of your variables are nominal variables.
**4.** Designate your variables as either independent or dependent variables.
**5.** Decide whether to manipulate your independent variables based on practical, theoretical, and ethical considerations.
**6.** For any nonmanipulated variables, decide whether to select or just to measure.
**7.** State as clearly as possible the questions your experiment should answer.
**8.** State your experimental hypotheses.

**Chapter Three
How's the
Research Done?**

**Section A
Reading Method
Sections**

The purpose of the method section in an experimental report is to inform the reader about the experiment. It's the reader's substitute for performing the experiment and controlling the procedures directly. A good method section gives enough information so that the reader could perform the experiment using the same procedures.

In this chapter you'll learn how to understand method sections and interpret the information in the context of other subjects, experimental situations, and real-world situations. The method section also reveals any problems there may be in an experiment that make it difficult to interpret the results. These issues are discussed in Chapter 4.

There are four major elements in a description of an experiment: (1) the nature of the experimental subjects, (2) the nature of the experimental situation, (3) the precise meaning of the variables in the experiment, and (4) a description of the experimental design. Scientists assume that if all those conditions are repeated, the results of the experiment will be the same. If any changes are made in the kind of subjects or procedures, the results *might* be the same if the changes are small or irrelevant, but it's less certain. Thus the method section tells those limited circumstances under which the results observed in the experiment can be expected to recur. Once you know those limits you may decide that the results could reasonably be expected to recur under somewhat different circumstances; that is, you may decide to stretch the conclusions of the study to include situations not encompassed by the experiment. In this section we'll discuss how to read method sections in order to discover the limits and some of the issues involved in going beyond the limits.

Any method section should include information about the experimental subjects, the general situation, and the meaning of the variables in the experiment. The method section from the Cross and Davis marijuana study follows. As you read it try to answer these questions:

1. Who were the experimental subjects?
2. What was the general experimental situation?
3. *Exactly* what was meant by frequency of marijuana use?
4. *Exactly* what was meant by adjustment?

### The Method Section of the Cross and Davis Marijuana Study

One hundred seventy-eight students, 91 males and 87 females, participated in this survey. One hundred fourteen were volunteers from general psychology courses, and 64 were from advanced courses. They were administered the Rotter Incomplete Sentences Blank (ISB; Rotter & Rafferty, 1950) and an extensive questionnaire about drug use as part of a larger battery. Testing was anonymous, and subjects were assured that the data would only be used by the investigators. The drug questionnaire asked about marijuana, LSD, amphetamines, barbiturates, narcotics, and alcohol. Since marijuana is clearly the most popular drug, and all other drugs, except alcohol, were used by only a small minority of subjects, the marijuana data are discussed here. One hundred, or 56% of the subjects, had tried marijuana at least once. Five categories of marijuana users, roughly paralleling the classifications of Nowliss (1969), were developed. A description of each category, with the percentage of subjects, follows:

Adamant nonusers (22%) have never considered using a drug and state that drug use should be prohibited.

Nonusers (21%) have seriously considered using a drug but have not actually done so. They state that drug use might be allowed with proper controls.

Tasters (24%) use marijuana less than once a month.

Recreational users (15%) use marijuana from one to four times per month.

Regular users (17%) use marijuana more than once a week.

The ISB was scored by three judges. Judge A scored all the inventories, and Judges B and C scored half each. Reliability[1] between A and B was .91 and

---

[1] A measure of *reliability* is a special kind of correlation coefficient and tells the extent to which two sets of scores agree with each other. In this case, if the judges agreed perfectly, the reliability would be +1.00.

between A and C was .93. Any inventory on which there was more than a 7-point discrepancy between the two judges was discussed and the differences resolved. The final ISB score was the mean[2] of the two judges' scores.

Who were the subjects? They were 178 college students in psychology who volunteered to be in the experiment. Of them 91 were male and 87 were females; 114 were introductory students and 64 were advanced. All those facts are important because they describe the subjects from whom the data were taken.

What was the experimental situation? We're not told clearly, but it was probably a general laboratory situation with the students filling out the questionnaires in classrooms, in groups. If the situation were different, Cross and Davis would have described it more carefully.

Exactly what was meant by frequency of marijuana use? A standard definition might read something like this: how often marijuana is used during some time interval. However, in experiments, a different kind of definition is used, an **operational definition.**

### Operational Definitions

An operational definition tells what operations were performed to define a particular term. Operational definitions aren't general, like dictionary definitions, but are limited to particular experiments. To operationally define "frequency of marijuana use," the operations used by Cross and Davis were to give the subjects a questionnaire about drug use as part of a larger battery of tests. The students were assured of anonymity. On the basis of the subjects' answers on the questionnaire, their frequency of use was determined.

Note that we can't be sure the answers reflect real drug use. The questionnaire could be faulty in some way so that the questions were misunderstood. The students could have made mistakes in marking their answers or even could have lied. Nevertheless, the answers to the questionnaire are exactly what Cross and Davis mean by frequency of marijuana use in this experiment, no more and no less. The operational definition of frequency of use is this: answers given to a particular drug questionnaire, administered as part of a larger test battery, and under conditions assuring anonymity. In fact, a complete operational definition would include a detailed description of the questionnaire itself, the kinds of questions on it, the ways the students were

---

[2]*Mean* refers to a statistic you probably know as the average. Specifically, it's the number you get if you add up all the scores in a set and divide the sum by the number of scores. In this case, they added up the scores from the two judges who scored each person's ISB and divided that sum by 2.

given to respond, and similar information. However, there are practical limits on how much information can be included in any report.

Using their operational definition of frequency of use, Cross and Davis classified their subjects into five groups: adamant nonusers, nonusers, tasters, recreational users, and regular users. Each group, in turn, was operationally defined. A regular user was defined as a person who reported that he or she uses marijuana more than once a week. Any conclusions drawn about regular users are based on that definition.

Operational definitions prevent misunderstandings about the exact meaning of some phrase. You know exactly what Cross and Davis mean when they refer to frequency of use or to a "regular user." However, you must decide for yourself whether you think a particular operational definition is adequate or would be equivalent to a different operational definition of the same term. You may feel reasonably content with Cross and Davis's definitions, but you should be aware that some definitions are very questionable. For example, suppose Cross and Davis had operationally defined frequency of use as the frequency estimated by the *teachers* of the experimental subjects. Suddenly the experiment would look completely different, although the rest of it might be unchanged. When an operational definition is inadequate for your purposes you must recognize that the experiment isn't measuring what you want, even if the names of the variables made it look like just what you were after.

Exactly what was meant by adjustment, that is, what was the operational definition of adjustment? It was the mean of two judges' scores of the person's responses on the Rotter Incomplete Sentences Blank, a standardized test of adjustment. To decide whether that test is an adequate definition of adjustment you'd have to learn more about it by looking up the reference given for that test. The reference would tell you what the test was like so that you could decide how good it was. It's clear that the test is scored in a way that leaves room for disagreement between different scorers, so the adjustment score was defined as the mean of two judges' scores. The reliability figures of .91 and .93 suggest that Judges B and C agreed pretty well with Judge A, so apparently there weren't *big* differences between different scorers. On the other hand, we're told that if there were differences as big as 7 points, the judges discussed the difference so they could make it smaller. We don't know how often that happened. Moreover, it's possible that the two judges agreed with each other and were *both* wrong about that individual's adjustment. Without looking up more information about the ISB, it's hard to decide whether it's a good operational definition of adjustment. However, we at least know what the definition is and could duplicate this aspect of the experiment pretty well based on the information given.

Notice that according to the operational definition, each subject's adjustment is equal to his or her score on the test. If Cross and Davis had chosen to, they could have grouped the scores into levels of adjustment just as they did for marijuana use. For example, they could have said that anyone with a score lower than 135 on the ISB was exceptionally well-adjusted, anyone with a score between 135 and 145 was average, and anyone with a score above 145 was maladjusted. That (or any other) method of grouping the scores might lead to different results.

**Problem**

In any experiment the variables are necessarily defined operationally. For practice recognizing operational definitions, try the following problem.

Q. Suppose an experiment were designed to investigate the effects of stress on stomach ulcer formation. For 3 hours every day for 2 months 20 dogs were placed in an apparatus in which they must watch a light display and press one of three levers, depending on which lights were lit. If they failed, they received an electric shock. On the average they had to respond once every 10 sec. (seconds) while they were in the apparatus. Another 20 dogs were placed in the same apparatus every day for 2 months, but the machine was turned off. At the end of 2 months, all the dogs were killed and their stomach linings investigated for lesions. Any lesion larger than 1 mm (millimeter) in diameter was called an ulcer.

Find the operational definitions of stress, no stress, and stomach ulcer formation.

A. For stress the definition includes everything from "For 3 hours every day" to "while they were in the apparatus." All of that is what these experimenters meant when they said stress. Notice that this doesn't look anything like a dictionary definition. For no stress the definition is the sentence about the other dogs. In this case no stress includes being confined in a small space for 3 hours every day for 2 months. Clearly a different operational definition of no stress might lead to different results. The definition for stomach ulcer formation is the last sentence. If the subjects were people, the operational definition would have to be different.

Good operational definitions are reasonable definitions of the variables being defined. In an experiment designed to measure a product's attractiveness to men and women it's necessary to define product attractiveness. An experiment using a poor operational definition might be to keep track of the number of men and women purchasing the product. That experiment might not measure the attractiveness of the product but just the number of men and women who happen to be in the vicinity. A better operational definition might keep track of men and women both purchasing and not purchasing the product and define attractiveness as the ratio of buyers to nonbuyers. Still better might be a definition keeping track of buyers, serious lookers, casual lookers, and nonlookers and defining attractiveness as some combination of those categories.

A second property of good operational definitions is that they can be described clearly enough so that other experimenters can duplicate the operations if they want to. They don't depend on subjective judgments by the experimenter. For example, a poor operational definition of "amount of anxiety shown" might be "rating by the experimenter on a scale from 1 to 10." Some better operational definitions might be these: "number of times the subject crosses or uncrosses his or her legs," "amount of perspiration recorded by a machine," or "average rating of five graduate students watching the subject."

Experimenters lacking precise operational definitions can unconsciously influence their judgments to support their original ideas, as we'll discuss further in Chapter 4. If they try to avoid this, they may unconsciously overcompensate and slant their judgments the opposite way. An imprecisely defined variable could be duplicated only by an experimenter unconsciously slanting judgments the same way. Thus operational definitions should be precise and complete.

### Extending Conclusions beyond the Experimental Method

The conclusions to be drawn from any experiment are, in a strict sense, limited to the type of experimental subjects, the experimental situation, and the operational definitions used in the experiment. To go beyond these strict limits requires intelligent guessing. To apply the experimental findings to the real world requires a guess that they will apply not only to the subjects actually used in the experiment but to other kinds of subjects as well. It requires a guess that if changes are made in the experimental situation, or in how we define the variables, the relationships between the variables would be much the same. Of course, these guesses may be incorrect. The only way to be certain is to perform more experiments to determine the limits under which the relationships do in fact occur.

*Generalizing from Samples to Larger Groups*

In the Cross and Davis marijuana study the subjects were 178 college students in psychology classes who volunteered to be in the experiment. When you think about it, it's obvious that Cross and Davis weren't interested in only these 178 students. They were really interested in knowing about people in general, and they'd like to *generalize* their findings to a much larger group. In their experiment they may have found that high marijuana use was associated with maladjustment with their 178 subjects. They'd like to conclude that's true not only for those 178 subjects, but also for others. The question is, what others? You must answer this for yourself. The following comments might help.

In a technical, statistical sense, the only situation in which generalization to a larger group is permissible is a situation in which a **random sample** has been selected from a larger **population.** When a random sample is used, it's statistically correct that whatever is true of the sample probably is also true of the whole population. That is, it's correct to generalize from the sample to the population. For example, if an experiment were performed using a random sample of 100 children selected from all the schoolchildren in Boise, Idaho, that showed one teaching method to be better than another, it would be statistically correct to expect that teaching method to be superior for *all* the schoolchildren in Boise.

It's important to know that in scientific publications the word **random** is a technical word and doesn't just mean haphazard or accidental as it often does in popular writing. By the technical meaning, if a random sample is selected from a population, each member of that population has an equal chance of appearing in the sample. Obtaining a random sample requires considerable effort; it doesn't happen accidentally. In Section B of this chapter there's discussion about how to obtain a random sample. For reading purposes it's sufficient to know that random is a technical word and that if the word doesn't appear with respect to sample, the sample is almost certainly *not* random.

The 178 students in Cross and Davis's study weren't a random sample from *any* population. They're certainly not a random sample from the whole population of human beings. For example, a small child or a Soviet business person had no chance of appearing in this sample. Nor do these students represent a random sample of American college students—a student not taking psychology or a student at a different school, for example, had no chance of being in the sample. These 178 students aren't even a random sample of psychology students at that school; nonvolunteers weren't represented. Because this isn't a random sample from any larger population, the results aren't strictly generalizable to *any* people other than the ones actually in the experiment.

But most readers would agree that this group of 178 students probably are representative of some larger group. To how large a group are *you* willing to generalize the results? Do you believe the same results would be obtained for a volunteer group of psychology students at a different college? Do you think the results would be the same if nonvolunteers were somehow included? students in physics classes? college-age factory workers? older people?

My own decision is that I'd be very comfortable generalizing these results to social science student volunteers at most American colleges. I'd be less comfortable including students not taking social science classes and even less comfortable trying to generalize to noncollege people. I'd be unwilling to generalize to people much older than college age or to non-Americans of any sort. But that's just my opinion.

**Problems**

To decide how far to generalize the results from an experiment you must take into account both the subjects used in the experiment and the variables under investigation. These problems should clarify some of the issues.

**Q.** Suppose the subjects in an experiment were 100 fourth-graders in one school. The school is in a suburban district including both upper and lower middle-class families. Half of them were taught long division using a technique involving rote memory of a series of steps; the other 50 used a technique involving understanding of how long division works. After 4 months all the children were given the same test to measure how well they could perform long division problems. To what larger group would you be willing to generalize the results of this experiment?

**A.** The 100 fourth-graders aren't a random sample from any larger population, so, in a statistical sense, no generalization is warranted. Nonetheless, I'd be willing to believe that whichever technique was superior for these 100 fourth-graders probably would also be superior for fourth-graders in suburban schools throughout the United States. I'd be hesitant about generalizing to urban or rural schoolchildren or to severely disadvantaged children. I'd be unwilling to generalize to children from cultures other than western. (Of course, I could be wrong in my guesses.)

**Q.** Suppose the subjects in an experiment were 60 volunteer male convicts in a federal prison. One third were given tablets containing 50 mg (milligrams) of a new allergy drug, 20 were given tablets containing 25 mg, and the other 20 were given tablets containing no active ingredient at all. Their sleep was monitored during the next 24 hours to find out the effects of the drug on sleep. To what larger group would you be willing to generalize the results of this experiment?

**A.** Again, no statistical generalization is possible. Moreover, these subjects are clearly not representative of the population at large. However, because of the nature of the variables, I'd be willing to believe that any effects of this drug on the sleep of these convicts probably would also occur for most adult males. I'd be a bit cautious in generalizing to females. Of course, if the convicts were taking other drugs illicitly, my generalization could be quite wrong.

**Q.** Suppose the subjects in an experiment were 30 severely retarded adults in one institution. Half were subjected to a 6-month treatment in which they were given small tokens every time they performed some self-care task such as dressing or feeding themselves. The tokens could be traded for cigarettes or candy after enough were collected. The other 15 subjects were given no special treatment. After 6 months the self-care behavior of both groups was examined and compared. How far would you generalize?

**A.** I'd be willing to generalize these results to severely retarded adults in other institutions. The same results also might be true for severely retarded children, but I'd be cautious about that. I'd also be cautious about generalizing to severely retarded adults living at home rather than in institutions.

**Q.** Suppose the subjects in an experiment were 200 adults randomly selected from all the registered voters in New Mexico. All of them completed a questionnaire asking (1) their attitudes toward environmental issues and (2) their attitudes toward organized religion. The point of the study was to investigate the relationship between the two variables. How far would you generalize the results?

**A.** In this case the experimental subjects *were* randomly selected from a larger population, so it would be statistically appropriate to generalize these results to all the registered voters in New Mexico. However, I'd be willing to go beyond that population and perhaps include all the registered voters in the Southwest or perhaps even a larger group.

*Generalizing beyond
the Experimental Situation*

The Cross and Davis experiment was probably performed in classrooms. Strictly speaking, the conclusions drawn from the experiment apply only to that situation. Do you think the results would have been the same if the students had been stopped as they walked across campus and asked to complete the questionnaire or if they had taken the questionnaires home and completed them there? I suspect that these changes would make little difference in this experiment.

Generalization across situations isn't always very broad. Suppose an investigator wanted to study people's reactions when they're yelled at by a large, unpleasant-looking person. If the experiment were performed in a laboratory setting, I'd be unwilling to believe that the results could be generalized to include a bar setting. For variables for which generalization across situations isn't broad, experiments are often performed in the situations of interest. To learn about behavior in bars, some experiments must be performed there. Experiments performed out in the "real world" rather than in the laboratory are often called **field studies.** It's important to remember that even results from field studies may not generalize very widely. Results found in a bar full of happy people might not generalize to the same bar when it's nearly empty.

With many behavioral variables it's not safe to generalize from a situation in which the subjects know they're being observed to a situation in which they don't. Drivers at a stop sign change their behavior if they see someone sitting with a clipboard. When people or animals in an experiment change their behavior simply because they're being observed, the behavior observed is sometimes called a **reactive measure,** indicating that it's a reaction to being measured. When reactive measures are likely to reduce the generalizability of the experiment, experimenters sometimes try to collect data without letting the subjects know they're participating in an experiment. Observations taken in this way are sometimes called **nonreactive** or **unobtrusive measures.** Collecting data without the subject's knowledge or consent raises some ethical questions that are discussed in Section B of this chapter. For reading purposes it's sufficient to recognize that problems of generalization sometimes do arise from reactive measures.

### Problems

As with generalizing from one set of subjects to others, generalizing from one situation to others requires consideration of both the situation and the variables. These problems should give you some practice in generalizing from the experimental situation to others.

**Q.** Suppose an efficiency expert wanted to know how much time was used each day by a group of workers for various activities. She followed each worker around for 3 days, keeping careful track of all activities. To what situations do you think it would be reasonable to generalize her results?

**A.** I'd expect her findings to hold true only for situations in which workers are followed around by efficiency experts.

**Q.** Suppose a fourth-grade teacher wanted to know how time of day was related to test performance. For half the year he carefully planned his tests for different times of day and kept track of class performance. To what situations do you think it would be reasonable to generalize his results?

**A.** I'd expect his results to generalize to most other classroom situations at other schools.

**Q.** Suppose a social psychologist wanted to know the degree to which having one person smile at another causes the recipient to smile back. She arranged a lab experiment in which there was a memory task supposedly being performed cooperatively by two subjects. One was a real subject, but the other was an accomplice, who smiled at some subjects and not at others. To what situations do you think it would be reasonable to generalize her results?

**A.** I'd expect her results to generalize over a wide variety of situations in which two strangers are forced to interact with each other.

*Generalizing beyond*
*Operational Definitions*

Again, speaking strictly, the conclusions drawn from an experiment apply only to the operations actually performed. Cross and Davis wanted to investigate the relationship between frequency of marijuana use and adjustment. Their experiment, however, was limited to the variables as they were operationally defined. They were limited to a definition of frequency of use that included a particular questionnaire administered under certain conditions, one of which was assurance of anonymity. And they were limited to a definition of adjustment that included a particular standardized test, administered a particular way, and scored by two judges under certain conditions.

Do you think the results would be the same if the students weren't assured of anonymity? if only one judge scored the adjustment test rather than two? if adjustment were defined as "peer ratings of adjustment"? The only way to answer these questions with certainty is to perform the appropriate experiments and see whether the results in fact do change. In the absence of data, the only recourse is to try to guess intelligently. Obviously, the more you know about an area, the more intelligent your guesses are likely to be.

I can offer little advice here other than to suggest that you try to recognize when you generalize beyond the operational defintions so that you can adjust the amount of faith you place in your guesses.

Section A  Reading Method Sections                                                                79

**Problems**

These problems should give you some practice in generalizing beyond operational definitions.

**Q.** Suppose an experiment were performed to investigate the relationship between amount of caffeine in coffee and amount of coffee-drinking. In that experiment amount of caffeine was operationally defined by adding varying amounts of pure caffeine to decaffeinated coffee in an office coffee machine. Amount of drinking was operationally defined as the number of cups of coffee taken from the machine each day. Would you believe that the same results would occur if amount of caffeine were defined as the amount that occurred naturally in different brands of coffee as measured by a chemical test designed to detect caffeine?

**Q.** Suppose an experiment were performed to investigate the relationship between intelligence and problem-solving ability and that it defined intelligence as the score on a standardized test and problem-solving ability as the number of problems in formal logic the person was able to solve in a 30-min. (minute) period. Would you believe that the same results would occur if problem-solving ability were defined as the number of patented inventions produced by each person?

**A.** I can't give you a straight answer in either case, although it certainly seems safer to assume the equality of the two operational definitions of caffeine amount than of problem-solving ability. The point is that you should recognize the limitations of the particular definitions used in an experiment and not go beyond them without due consideration.

Now it's time to read a method section from an article you haven't seen before. This paper was written by Anthony D. Lutkus and is called "The effect of 'imaging' on mirror image drawing." It appeared in the *Bulletin of the Psychonomic Society*, 1975, Volume 5, on pages 389 to 390. In mirror image drawing an apparatus is used that forces the subject to reach beneath a barrier and try to trace with a pencil some figure that can't be seen directly. What *can* be seen is a mirror reflecting the drawing and subject's hand. Of course, the mirror reverses the image, so the task is far from trivial. In this experiment there were two independent variables combined in a factorial design and two dependent variables. As you read, try to answer these questions:

**1.** Who were the subjects? To what larger group would you be willing to generalize?
**2.** What was the experimental situation? To what other situations do you think the results might generalize?
**3.** Name both independent variables, and name all their levels. Give the operational definition of each level.
**4.** Name and operationally define the two dependent variables.

## Most of the Method Section of the Lutkus Mirror Image Drawing Study

*Subjects*

The subjects were 64 underclassmen from Princeton University. They were divided equally by sex. All were right-handed.

*Apparatus*

A standard mirror-drawing apparatus (Woodworth & Schlosberg, 1954; Starch, 1910) shielded the subject's hand from his line of sight and allowed him to see the outline of the figure he was tracing only in mirror image.

*Procedure*

The subjects were randomly divided into four groups in a 2 by 2 design. Half received standard instructions to trace the mirrored figure as quickly and accurately as possible. The other group was instructed to "image" the figure and to let their image guide their tracing as much as possible. Two figures, a diamond with 2-in. sides and a zig-zag made up of line segments and angles equal to the diamond, were used. A given subject traced only one of the figures for 10 trials in either the Imagery or Standard condition. Both time and errors were recorded for each trial. Errors were defined as touching or tracing outside the $\frac{1}{8}$-in. wide path made by the double outline of the figure to be traced.

---

Who were the subjects in this experiment? They were 32 male and 32 female underclassmen, all right-handed. They weren't a random sample from any larger population but were apparently those that the experimenter was able to get. To what larger group could the results be generalized? I'd be willing to generalize the results to all right-handed college undergraduates in the United States. I'd be a bit concerned about extending the results to noncollege young people or to children or older people. I'd be more concerned about extending the results to left-handed people. You may feel differently.

What was the experimental situation? The experiment was performed in a laboratory, and the individuals knew that they were participating in an experiment. To what other situations could the results be generalized? I've found that people behave much the same during a mirror image tracing task regardless of the situation. If they were at home or on the street, it wouldn't make much difference.

How were the independent variables operationally defined? One of the independent variables was supposed to be what the subject was thinking about while trying to trace the figure. This was operationally defined by using two kinds of instructions. "Standard" instructions were to trace the figure as quickly and accurately as possible; "image" instructions were to image the figure and use the image as a guide. Presumably, most subjects would follow the instructions they were given and therefore would think of different things while tracing the figures, although that's not certain.

The other independent variable was the type of figure to be traced. Both figures were drawn with a double outline with ⅛ in. between the outlines. Both figures had four 2-in. line segments and the same angles. The "diamond" was a picture of a diamond, and the "zig-zag" was some other shape; we're not told precisely what it was.

How were the dependent variables operationally defined? Time and errors were both measured across 10 trials, so the measures are probably some average across those 10 trials. We're told nothing more about time, but it seems safe to assume that it just means the amount of time from starting a figure to finishing it. Errors were more carefully defined as either touching or crossing the outline. Other possible operational definitions of errors might have counted retracing (that is, going backward) as an error or not have counted touching the line.

Both time and errors are common operational definitions of "task difficulty," although Lutkus never used that term. The assumption is made that more difficult tasks require more time or cause more errors. If the image instructions reduce the time required, the assumption can be made that they made the task less difficult. Lutkus could have used a somewhat different approach to defining task difficulty. He could have set a standard of performance, or *criterion,* which each subject must attain. One possible criterion would be to complete the figure in 30 sec. or less with 2 or fewer errors. Then he could have counted the number of trials or measured the length of time each subject required to reach that criterion. Then task difficulty would have been operationally defined as "trials to criterion" or "time to criterion." Both are used frequently.

### Some Elements of Experimental Design

We've now dealt with reading method sections to discover the nature of the experimental subjects and the operational definitions of the variables. The other factor is the experimental design. In order to discuss experimental design, more distinctions are needed to distinguish various designs. One important distinction concerns between-subject variables and within-subject variables.

*Between-Subject and Within-Subject Variables*

In both the Cross and Davis marijuana study and the Lutkus mirror image drawing experiment the independent variables have been defined in such a way that there were different subjects at each level of the variable. In the Cross and Davis study there were some subjects classified as adamant nonusers, others as tasters, and so forth. Each subject appeared at only one level of the independent variable. In the Lutkus mirror image drawing experiment each subject had one set of instructions or the other; no subject received both sets of instructions. Similarly, each subject had only one figure. Independent variables defined in this way are called **between-subject variables,** because the experimenter looks at differences *between* subjects at different levels of the independent variable.

Sometimes all the levels of an independent variable are used with each subject in an experiment. For example, Lutkus could have measured each subject twice: once with the diamond and once with the zig-zag. Variables defined in that way are called **within-subject variables,** because the experimenter compares differences that appear *within* the same subject's performance. Notice that within-subject does *not* mean anything about inherent qualities of the subject, such as intelligence or adjustment. It's just a technical term describing how the independent variable was defined. Between-subject variables use *different* subjects at each level of the independent variable; within-subject variables use the *same* subjects at all the levels of the independent variable.

### Problems

These problems should help you learn to recognize between-subject and within-subject variables.

**Q.** Suppose a clothing manufacturer wanted to know which style of plastic raincoat was preferred, one with snaps or one with a zipper. She found 30 people and gave them one raincoat of each kind to wear for 6 months. At the end of that 6 months she had all the people rate their liking for each raincoat on a 5-point scale. Was the independent variable, style of raincoat, varied between subject or within subject?
**A.** Each person rated *both* styles; therefore it was a within-subject variable. If it had been a between-subject variable, there would have been two groups of people, one group rating each style of raincoat.

**Q.** Suppose a community mental health agency was trying to determine which kinds of people were likely to miss their appointments. When calls for appointments came in from people who had never been seen before, the workers receiving the call categorized it on several variables: sex of caller, apparent degree of agitation (high, medium, or low), and whether the caller mentioned suicide. Then they kept track of whether each caller kept their appointment. For each independent variable, sex, agitation, and mention of suicide, tell whether it was varied between or within subjects.
**A.** Clearly each person could belong to only one sex, so this must be a between-subject variable. Each call was categorized at only one level of agitation, so

## Section A  Reading Method Sections

that was also a between-subject variable. Similarly, each caller either mentioned suicide or didn't (not both), so all variables were between-subject variables.

**Q.** Suppose a merchant wanted to prepare for the Christmas season and decided to investigate people's preferences scientifically. He showed 16 pictures of Christmas trees to 30 people and asked them to rate how much they liked each one. Eight of the pictures showed trees with garlands of tinsel, and the other eight showed trees with silver icicles. Four of each of these types were decorated with multicolored lights, and the other four had all white lights. The rest of the ornaments differed slightly, but in no particular pattern. Name the independent variables, and for each one tell whether it varied between or within subject.
**A.** The two independent variables might be called garland versus icicle and color versus white lights. Both were within-subject variables.

**Q.** Suppose a nursery school teacher wanted to know which of two techniques helped her children memorize poems better. One technique consisted of reading the poem aloud and having the child repeat each line. The other technique was to read the whole poem aloud, show a picture for each line, and have the child try to remember each line in turn. She used two equivalent poems and had all the children learn both poems, using first one technique and then the other. For half the children the reading technique was first; for the other half the picture technique was first. Name the independent variables, and tell whether they varied between subject or within subject.
**A.** The two independent variables of interest mght be called teaching technique and order of technique. The first was varied within subject and the second between subject. If you were alert, you also might have recognized that poem was another within-subject variable.

**Q.** Suppose a personality theorist found 100 people and gave them all a questionnaire to measure their levels of aggressiveness, degree of religious interest, and degree of authoritarianism. Her goal is to look at the relationships (correlations) among the variables. Are these variables varied between or within subject?
**A.** This question may be confusing because there are three variables, and they can all have many levels. That is, there might be as many as 100 different levels of aggressiveness with only one subject at each level. It may not be very helpful to categorize this as either between or within subject, but the most reasonable choice between the two is to think of all the variables as between subject. There are different subjects at each level of aggressiveness; no subject is at more than one level. The same is true of degree of religious interest and degree of authoritarianism. This is really the same design as the mental health agency problem or as the Cross and Davis marijuana study. The only difference is that the subjects weren't divided into relatively few groups or levels.

*Cross-sectional and
Longitudinal Variables*

Some nonmanipulated variables also can be investigated as either between-subject or within-subject variables. The most common example is "age." It can be investigated either by looking at several groups of individuals of different ages (between subject) or by following one group of individuals as they get older (within subject). Some jargon has developed here so that for nonmanipulated variables for which there's a choice, between-subject variables are called **cross-sectional** and within-subject variables are called **longitudinal.** These terms aren't used for nonmanipulated variables for which there's no choice (such as sex or intelligence), and they're rarely used with manipulated variables. However, they do appear with enough frequency so that you should know what they mean.

**Problems**

This problem illustrates the difference between cross-sectional and longitudinal variables.

**Q.** Suppose a grocery store-owner performed an experiment in which she obtained the cooperation of a number of her regular customers. Some of the customers usually shopped with lists, while others didn't. Each time one of the customers entered the store, the store-owner had them mark off a rating scale from 1 to 5 on how hungry they felt. She kept track of how much they bought on each trip. Name the independent variables, and for each one tell whether it was varied between subject (cross-sectionally) or within subject (longitudinally).

**A.** There were two independent variables, presence of list and hunger. Presence of list had two levels and was varied so that the store-owner looked at a cross section of her customers. Hunger had five levels and was varied longitudinally, from one occasion to another. Both variables were clearly nonmanipulated.

The next method section describes an experiment in which one of the independent variables was a within-subject variable; the other is a between-subject variable. They were combined factorially. This experiment was performed by Elizabeth F. Loftus and Guido Zanni and is entitled "Eyewitness testimony: The influence of the wording of a question." It appeared in the *Bulletin of the Psychnomic Society,* 1975, Volume 5, on pages 86 to 88. As you read the experiment, try to answer these questions:

**1.** Who were the subjects? To what larger group would you generalize?
**2.** What was the situation? To what other situations would you generalize?
**3.** Name one independent variable and its levels and give operational defini-

tions for all the levels. Then tell whether it was a between-subject variable or a within-subject variable.

**4.** Do the same things for the other independent variable.

**5.** Name and operationally define the dependent variable.

## The Method Section of the Loftus and Zanni Question-Wording Study

One hundred graduate students participated in this experiment, in groups of various sizes. All subjects were told that they were participating in an experiment on memory and that they would be shown a short film followed by a questionnaire. The content of the film was not mentioned.

The film itself depicted a multiple car accident. Specifically, a car makes a right turn to enter the main stream of traffic; the turn causes the cars in the oncoming traffic to stop suddenly, causing a five car bumper to bumper collision. The total time of the film is less than 1 min, and the accident itself occurs within a 4-sec period.

At the end of the film, the subjects received a questionnaire asking them to first "give an account of the accident you have just seen." When they had completed their accounts, a series of specific questions was asked. Six critical questions were embedded in a list totaling 22 questions. Half the subjects received critical questions in the form, "Did you see *a* . . . ?" and the other half of the subjects received them in the form, "Did you see *the* . . . ?" Three of the critical questions pertained to items present in the film and three to items not present. Subjects were urged to report only what they saw and did so by checking "yes," "no," or "I don't know." Each subject received a different permutation[3] of the questions.

---

Who were the subjects? They were 100 graduate students. We're told nothing else. Without more data, I'm uncertain as to what larger group to generalize. I believe that most other graduate students would probably behave like this 100, but beyond that I feel too ignorant to guess intelligently. Are graduate students more observant than other people? or perhaps less observant? more willing to commit themselves when they're uncertain? I really have no idea, so I don't know how widely to extend the findings of this study.

---

[3] A *permutation* is an order. All possible permutations of the items P, Q, and R are PQR, PRQ, RPQ, RQP, QPR, and QRP. Since there were 22 questions on the questionnaire, there are a great many possible permutations—about $1.124 \times 10^{21}$ or 1124 followed by 18 zeros. That's certainly enough for each of 100 subjects to have a different order.

What was the situation? Again, the experiment was performed in a laboratory situation, probably in a classroom. Loftus and Zanni are clearly interested in generalizing their results to other situations in which eyewitness testimony is relevant, as in witnesses of real auto accidents. Do you believe that such a generalization is warranted? I have serious doubts that witnesses of real accidents would respond just as these experimental subjects do. These subjects know that the people asking the questions know the correct answers and will find out if they're wrong. Real eyewitnesses know that the people asking the questions *don't* know the correct answers; if they knew, they wouldn't be asking. I'm not sure how that difference would affect the answers given, but I feel confident that there would be some change. I'd be comfortable only in generalizing these results to other situations in which the eyewitnesses knew that they were about to be tested on what they saw by people who knew the correct answers.

What were the variables in the experiment? One independent variable might be called something like "the leadingness of the questions." It was operationally defined by the phrasing of six questions: one level used the indefinite article *a*, and the other level used the definite article *the*. Presumably questions phrased with *a* are less leading than questions phrased with *the*. Leadingness was a between-subject variable since half the subjects received *a* questions and half received *the* questions; no subject received both kinds.

The other independent variable had to do with whether the items in the questions were present in the film the people had seen. To give it a name we might call it "present versus not present" or just "presence." This variable was operationally defined in terms of the relationship of the six questions to the film viewed by all the subjects. Three of the questions asked about items that were present in the film, and three asked about objects that weren't. This was a within-subject variable, since each subject was asked about both kinds of item, present and not present.

The dependent variable was whether the subjects reported that they had seen the objects asked about in the critical questions. This was operationally defined as whether they checked "yes," "no," or "I don't know" for each object on a printed list of questions given to them after the film and after they'd already written a description of the film contents.

*Before-After Variables*

It may have occurred to you that the experiments we've been discussing bear little resemblance to experiments in the physical sciences. In a typical chemistry experiment, the chemist takes some collection of chemicals, goes through some operations he or she expects to change the chemicals in some way, and observes what happens. How does that fit into the scheme of experimental design we've been discussing?

Consider for a moment the various techniques that a chemist might use to demonstrate that heat causes a certain powder to crystallize. In this case the subject is a small pile of that powder or, if you prefer, the subjects are the grains or even molecules of powder. He could use an approach in which heat is similar to a within-subject variable with two levels: heat and no heat. He already knows that under the level no heat, if everything else remains constant, the powder just sits; no crystallization occurs. So his experiment consists of adding heat and keeping everything else constant. When he observes that crystallization occurs, he's justified in concluding that the heat caused the crystallization, because he *knows* that the change wouldn't have occurred without the heat.

If he weren't certain, he could use an approach in which heat would be treated as a between-subject variable. He'd need two piles of powder, one assigned to "heat" and the other to "no heat." If only the heated one crystallized, he'd be justified in concluding that the independent variable caused the change in the dependent variable. However, in most circumstances that level of control would be unnecessary because of the chemist's knowledge of the behavior of chemicals in the absence of any special treatment. Normally a chemist would use the first kind of design, which I'll call a **before-after design,** because the level of the dependent variable *before* the independent variable is compared to the level *after.*

The important aspect of a before-after design is that the experimenter must be certain of what would happen to the dependent variable if the independent variable *didn't* occur. If the experimenter is uncertain, then a before-after design is inappropriate.

There are a number of areas in the behavioral sciences in which enough is known to use before-after designs. For example, we know that a rat will press a metal bar a few times each hour the rat is awake. It will continue to press the bar at about the same rate indefinitely if nothing changes. If we now add a factor, say, every time the bar is pressed some food is delivered, it's possible to draw certain conclusions. If the rat suddenly increases his rate of bar-pressing, we can assume that the change in that dependent variable must have been caused by the independent variable, food delivery. We can draw this conclusion because we assume that the rat otherwise would not have increased his bar-pressing rate. If we weren't quite sure of that assumption we'd have to perform an experiment that guarded against that possibility. We could have some rats who receive food and compare them to others not receiving food.

Suppose a first-grade teacher measured the reading skill of his pupils both before and after a 2-month period during which he gave them reading lessons. If he observed an increase in reading skill, he couldn't be terribly confident that it wouldn't have occurred without his intervention. The reason for his lack of confidence is his lack of knowledge about how the children would have performed in the absence of his teaching. They might have learned to read anyway for other reasons. Thus **before-after variables** have some interpretational difficulties not encountered with between- and within-subject variables, even when they're manipulated variables.

Sometimes nonmanipulated variables are written about as if they were before-after designs. For example, an experimenter could measure the level of depression in a group of people both before and after an earthquake. If a change occurred, it wouldn't be appropriate to conclude that it was caused by the earthquake, because we can't be sure what would have happened to the people's depression without the earthquake. We'll discuss this further in Chapter 4, which deals with causal conclusions.

The next method section is from an experiment that used a before-after design. It was written by Linda M. Yeatts and John C. Brantley and is entitled "Improving a cerebral palsied child's typing with operant techniques." It appeared in *Perceptual and Motor Skills* in 1976, Volume 42, on pages 197 to 198. The before-after variable was the modification or treatment given. The term **baseline** refers to the intervals when the treatment was *not* occurring. There was also an additional independent variable, the technique for presenting the words. It's probably best thought of as a within-subject variable. As you read, answer these questions:

1. To what larger group of subjects would you generalize?
2. Find the operational definitions of the treatment given, that is, the before-after variable.
3. Find the operational definitions of the two presentation techniques.
4. Find the operational definition of the dependent variable.

**The Method Section of the Yeatts and Brantley Typing Study**

The subject was a 7-yr.-old hospital inpatient, under treatment for spasticity of the legs and arms. Intellectual ability was within the average range (WISC IQ = 92), and fine motor performance was characterized by poor printing, reversal of letters, sequencing errors, and difficulty in left and right localization.

A baseline, modification, baseline design was used. Target behaviors were defined as typing on an electric typewriter from a model and from dictation. Words were taken randomly from the Slosson Oral Reading Test and printed as a model. Words for dictation were taken from appropriate difficulty levels of Wide Range Achievement Test. Observations were made for 10 min. each day during baseline. During modification, 30 min. were spent on task during the morning twice a week. Instructions were that one raisin would be given for every word typed without letter or spacing errors and for each correct carriage return. Raisins were presented immediately but not consumed until the end of the session. From 10 to 20 raisins were earned during these sessions. In a return to baseline condition, reinforcement was withdrawn, and observations were continued for 3 wk. (the middle week was excepted due to vacation).

Correct words were those without letter or spacing errors. Errors were substitutions, omissions, insertions, no space, extra space, failure to return carriage, and words or letters not on the same line. Word rates (Rate = Number Correct/Time) were computed as measures of performance for 5-min. intervals on each task twice each week.

Who were the subjects? A complete answer here includes the whole first paragraph. In short, it was one 7-year-old girl with cerebral palsy. To whom could one generalize? That's a very difficult question. With just one subject, I feel reluctant to generalize at all. However, the results might generalize to other similarly affected children.

For "treatment" the operational definition included everything from "During modification . . ." to "were earned during these sessions."

The two presentation techniques are left rather unclear; we're told only that there were two levels, model (presumably written) and dictation, and we're told where words were obtained for each level. Oddly, the words for written presentation were taken from an oral reading test, and it may be that the words for dictation were taken from a printed reading test!

The dependent variable was typing rate, defined by the procedures outlined in the last paragraph of the article.

Notice that in experiments in which before-after variables are used, the primary emphasis is on describing the operations defining the independent and dependent variables. That's analogous to the detailed descriptions of procedures seen in experiments in the physical sciences.

### Description of Experimental Designs

In an experiment the variables are combined with each other in particular ways. A summary of that pattern is a description of the **experimental design.** The important factors in the summary are the number and nature of the independent and dependent variables; the number of levels of each independent variable; whether they varied as between-subject, within-subject, or before-after variables; and how they were combined.

In the Cross and Davis marijuana study the experimental design was quite simple. There was one between-subject independent variable, frequency of marijuana use. It had five levels. There was also one dependent variable, adjustment.

The Loftus and Zanni question wording experiment had two independent variables, leadingness and presence. Leadingness was a between-subject variable, and presence was a within-subject variable. Each had two levels, combined factorially. It might be described as a 2 by 2 mixed design, meaning that there were two variables, each having two levels, one between subject and one within subject.

The Yeatts and Brantley design had two independent variables, treatment and mode of presentation of the words. The number of levels of treatment isn't clear, so that could just be described by saying that treatment was combined factorially with mode of presentation.

It wasn't helpful to point it out before, but the Lutkus mirror image drawing experiment actually had a third independent variable in addition to instructions and figure. Each subject had 10 trials on tracing the figure. This could be thought of as a within-subject variable with 10 levels. It was combined factorially with instructions and figure, each of which were between-subject variables with two levels. Therefore, it could be described as a 2 by 2 by 10 mixed design.

These descriptive terms are useful for many experiments in the behavioral sciences, but not all. Read the following portions of the method section of the Sullivan and Skanes teaching evaluation study. Notice who the subjects were, and notice how and where the experiment took place. You also might want to find the operational definitions of "student achievement" and "teaching evaluation." However, you should make a special effort to understand the experimental design. How were the subjects distributed over the levels of the independent variable, course achievement? How were the subjects grouped?

**Excerpts from the Method Section of the Sullivan and Skanes Teacher Evaluation Study**

*Setting and Materials*

The Junior Division of Memorial University is an administrative unit within the university which is responsible for the planning, presentation, and evaluation of first-year university and special preparatory (foundation) courses. . . .

Successful teaching experience is weighed heavily in making appointments to the division and, therefore, most faculty members have considerable experience in teaching in the high school system. . . . A few faculty members have no previous teaching experience. . . .

Students are usually taught in classes or sections with between 30 and 40 members. . . .

Many first-year subjects have examination committees. The committees usually include some Junior Division faculty members and some Senior Division faculty members who are involved in the teaching of the second year courses. Early in each semester, the examination committee makes up and circulates a model of the final examination, but the final examination itself is not circulated and its specific content is not known to any except the committee members until the examination is written. The committee also sets out guidelines to be used on marking each answer. All answers are board-marked with a small group of faculty members marking one question on all papers. The marks on the final examination are then tabulated simply by adding the raw scores on each question. The mark on the final examination counts 50% of the final grade. . . .

During the first semester of the 1971–1972 academic year, a student evaluation form was prepared. The form contained an evaluation section which included the following questions:

    **1.** As an overall rating, would you say the instructor is
                    5  4  3  2  1
               Excellent          Poor

**2.** Specifically, how would you rate his (her):
   (a) interest in students
   
   5  4  3  2  1
   
   (b) ability to present material in a clear and easily understood manner
   
   5  4  3  2  1
   
   (c) ability to get students really interested in the subject
   
   5  4  3  2  1

The other section contained specific statements which described the performance of the instructor and were intended to provide feedback to him.

Therefore, it was possible to obtain from each student a relatively objective measure of achievement, that is, the final examination mark; and since the rating had been done anonymously, we were able to find the mean rating for each section on the items which measured overall competence, the three items which measured specific characteristics, and the mean proportion of students who answered yes to each of the descriptive items.

*Procedure*

In September of 1971, approximately 2,300 first-year students registered and selected their courses. In each course, students were assigned randomly to one of the available sections. During the tenth week of the semester, the course evaluation form was completed anonymously. At the end of the 13-week semester, all students wrote a final examination and received a grade for the course.

The following subjects used a final, common examination which was board-marked and which counted 50% toward the final grade: Biology 100F, Chemistry 100F, Mathematics 100F (special preparatory courses for those students whose Grade 11 mark in the subject was less than 70% or who had not taken the subject in high school); Biology 1010, Chemistry 1000, Mathematics 1010, Physics 1050, Psychology 1000 (regular first-year courses); and Mathematics 1050, Science 115A (special first-year courses for those intending to become teachers of primary and elementary grades).

The final examination mark of each student was tabulated. For each section, the mean mark on this measure and the mean rating on Question 1 (overall competence) of the evaluation were calculated. Correlations were then obtained between mean instructor rating and mean final examination marks.

Since students had been assigned to them on a random basis, it was reasonable to assume that sections were similar in initial achievement, and, in fact, remarkably little variability among sections was found in average high school marks in those subjects for which these marks were readily available.

---

The confusing aspect of this experimental design concerns how the subjects were grouped. Sullivan and Skanes could have obtained each student's score on both variables and calculated a correlation coefficient showing the relationship between them. In this case it might have been reasonable to think of course achievement as a between-subject variable with a very large number of levels: as many levels as scores in the final exam. However, they performed the experiment somewhat differently. They calculated the mean final exam grade of *all* the students in each of the 130 sections, and they calculated the mean rating given to the teacher of each section. Then they correlated these two sets of scores.

There's no short-cut to describing an experimental design like this. You just have to describe the whole thing. Name the two variables and describe how the scores were grouped by sections and then correlated.

Now it's time to turn to the method section of your sample experiment and try to answer these questions.

1. Who were the subjects?
2. To what larger group are you willing to generalize?
3. In what kind of situation was the experiment conducted?
4. In what other situations do you think the same results might occur?
5. Name all the independent variables and name their levels. Then determine the operational definition either of the whole variable or of each of the levels.
6. Tell whether each independent variable was varied between subject, within subject, before-after, or in some other way.
7. Name and operationally define the dependent variables.
8. Summarize the experimental design.

**Extra Library Assignment.** Find examples of articles with between-subject variables, within-subject variables, and before-after variables in the library.

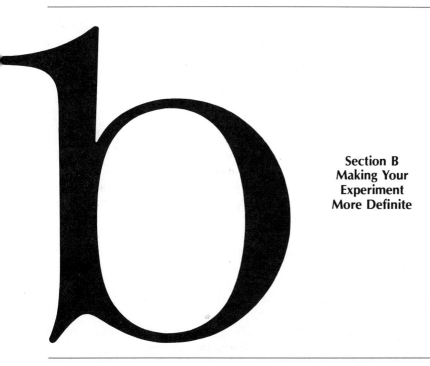

Section B
Making Your
Experiment
More Definite

In this section you'll make a number of decisions that will limit your experiment to very explicit questions. You'll decide on your experimental situations and the nature of your experimental subjects. You'll decide whether to vary your independent variables as between-subject, within-subject, or before-after variables. Finally, you'll operationally define your variables. All these decisions involve ethical, practical, and theoretical considerations.

**Decide Your Experimental Situation**

There are two related issues in deciding your experimental situation. One regards whether you'll conduct your experiment in the real world (the field) or in a laboratory. A lab needn't be very elaborate; in many cases a quiet room in your home can serve nicely. What distinguishes a lab from the rest of the world is that you set it up especially for conducting your experiment, making sure there won't be any interruptions and that all your equipment is ready.

The other issue regards whether you'll tell your subjects that they're participating in an experiment. Normally, of course, subjects in a lab know they're in an experiment, but sometimes experimenters arrange to have one of the variables occur in the waiting room or after the session. At the time of the actual experiment the subjects might not know they're participating, even though they're in a lab. In the field it's often possible to collect experimental data without the subjects ever knowing they've been in an experiment. For example, the relationship between "age of driver" and "speed of driving" could be investigated just by having someone sit beside the road and estimate both variables as drivers pass.

*Practical Considerations*

Some experiments can be conducted only in the field. You could hardly look at the effects of "time of day" on "seating patterns in buses" in the lab. Other studies can be conducted only in a lab, either because they require specialized equipment or because they require a lot of control over the environment. If an experiment requires people to hear extremely soft sounds, you could perform only in a place that's quiet; a lab would be an obvious choice.

★ Can you think of any reason why you *must* conduct your experiment either in the field or in a lab?

If you answered "yes," then that decision is made. If not, you have more freedom to decide.

In some field situations, it's impossible to tell your subjects that they're participating in an experiment, as in the age/driving speed example.

★ Is it *possible* for you to tell your subjects that they're participating in an experiment?

If it is possible, you can decide on the basis of theoretical and ethical considerations whether to inform your subjects. If it's not possible, then you must consider the ethical issues and decide whether it's appropriate to perform your experiment at all.

*Theoretical Considerations*

The reason for conducting experiments in the field or without the knowledge of the subjects is that those variables being observed might be expected to change in the artificial situation of a formal experiment. Results gathered in the lab often fail to generalize to real-world situations because the behavior of subjects who know they're being observed may not generalize to more private situations. The reason for conducting experiments in the lab is that you have more control over what happens so that unexpected events don't make it difficult to interpret your results. The reasons for informing subjects of their participation are largely ethical. We'll discuss them shortly.

★ Is it likely that your results would be different in the lab than in the field?

If you answered "yes," you must decide whether you'd be more interested in the results from the lab or from the field. To which kind of situation are you most interested in generalizing your findings? If you answered "no," you can choose between the two situations. Given this choice, the lab has a theoretical edge because of the additional control you have there.

★ Is it likely that your results will be different if your subjects know they're being observed than if they don't?

Again, if you answered "yes," you must decide which kind of results would be more interesting. If you believe that uninformed subjects would give more interesting results, you must consider the ethical implications of your choice. If you answered "no," informing the subjects is definitely to be preferred because of the ethical considerations.

*Ethical Considerations*

It's clearly ethically preferable to inform research subjects of their participation and to give them a chance to decide whether to take part. The experimenter should tell prospective subjects all features of the research that might reasonably be expected to influence willingness to participate. Then the people should decide freely whether to participate. This is called **informed consent.**

Clearly, subjects not given a chance to decline to participate must receive extra protection. You should recognize that many lay persons and scientists regard even the most innocent collection of data from unaware subjects as an invasion of privacy and feel that such research is unethical. Other people feel that such research is permissible if it's important and if proper precautions are taken to guard the rights, welfare, and dignity of the subjects. You should think carefully about this issue.

Having given the matter thought, you may decide not to inform your subjects of their participation. If so, there are some guidelines I suggest you observe. One is that you not record data from people who don't expect to be observed by other people. When people are in a restaurant or shopping or sitting in a lecture, they expect to be observed by others. It's a relatively mild invasion of privacy if you record some of their behavior systematically rather than observe it unsystematically as people usually do. There are other times when people don't expect to be watched, such as when they believe they're alone in a room or at home in their backyard. Recording their behavior under these circumstances is a much worse invasion of privacy and is rarely justified by the value of the research.

Another guideline is to keep your data anonymous. That's a good idea in any case but is especially important when subjects are unaware of their participation. You should make sure that individuals' identities can't be discerned from your records. That normally eliminates the use of cameras or tape recorders and limits you to paper and pencil. Of course you should *never* reveal information that could be traced back to any individual, no matter how harmless that information appears to you, unless the individual agrees in advance.

★ If you'd like to conduct your experiment without the knowledge of the subjects, do you feel its value justifies any invasion of privacy that may occur?

If you answered "yes," you should reconsider your answer from the point of view of the subjects. If your experiment might cause inconvenience, annoyance, or harassment of innocent people, it should be very valuable indeed. If it will invade their privacy, it should be worth considerably more than satisfying idle curiosity—even of a scientific nature.

If you answered "no," you should change your study. You might be able to figure out a way to inform the subjects before they participate and allow them to refuse. If that won't work, you might try to learn the same information in the artificial setting of a lab, perhaps with fully informed, role-playing subjects. If that won't work either, it may be necessary to abandon the whole idea. However, there is one possible approach that might work: don't observe individuals at all, but instead observe objects or other evidences of behavior. For example, an experiment could investigate the relationship between ostentatious display and aggressiveness by operationally defining both variables in terms of automobile characteristics. Ostentatious display could be defined in amounts of chrome or knickknacks, for example. Aggressiveness could be defined as dents, scratches, and obvious repairs. The whole study could be performed in a parking lot. Graffiti, trash, evidences of vandalism, and the like have all proved useful as unobtrusive or nonreactive measures of various human emotions and activities.

★ Taking all the practical, theoretical, and ethical issues into consideration, will you conduct your experiment in a lab or in the field?

Having made that decision, you must decide more precisely where to conduct your experiment. It must be a place where you can get subjects and which will allow you to observe the variables you've chosen.

★ Where will you conduct your experiment?
★ Taking all the practical, theoretical, and ethical issues into consideration, will you obtain informed consent from your subjects before they participate?

**Decide What Kind
of Subjects to Use**

The subjects you choose must show various levels of the variables you're investigating. To study the effect of instructions on ability to memorize lists of words,

# Section B   Making Your Experiment More Definite

you certainly wouldn't use animals, and you probably wouldn't use children. To manipulate removal of the lower portion of the brain you'd be limited to animal subjects. To look at subject variables such as type of car driven or age, you must locate individuals showing different levels of the variable.

★ What kind of subjects do you need for your experiment?

*Practical Considerations*

For many experimenters, practical considerations completely determine the choice of subjects. In psychology the most common subjects are introductory psychology students and white rats. You might be able to use a group of friends, co-workers, passers-by in a public area, people studying in libraries, pets, or animals in zoos. Depending on your experiment, one of these relatively easily available groups might be sufficient to your purposes.

★ What kind of subjects can you get who might be useful in your experiment?

*Ethical Considerations*

Assuming that you're obtaining informed consent from your subjects, it should be kept in mind that human subjects shouldn't be coerced into participation. This is especially important to remember that when the experimenter is in a position of direct or indirect power over the potential subjects, as when a teacher requests consent of his or her own students or when the boss requests participation of an employee. If you're in a position of power over your potential subjects, you should be certain that they won't be punished in any way if they refuse to participate, and you should make sure they understand that.

★ What kind of subjects can you get without coercing them?

*Theoretical Considerations*

In selecting subjects it often happens that theoretical considerations directly conflict with practical and ethical considerations. Theoretically, subjects should be selected to permit the widest possible generalization. For example, results based on a random sample of all American adults are more generalizable than results based on a group of introductory psychology student volunteers from Podunk U. Nonetheless, the only ethically available subjects may be these volunteers. Similarly, you might be interested in generalizing your observations on the mating behavior of elephants to elephants in the wild, but the only subjects available to you may be those in the zoo. If that's so, then you just can't perform the study you really want to perform.

To increase the size of the group to whom it seems reasonable to generalize your findings, you should increase the diversity of your sample, insofar as that's possible. Commonly, that's all that's done about increasing the generality of results. However, it's sometimes very important to be able to generalize the results of a study to a specific larger population. That's the problem faced by pollsters, who must determine public opinion by measuring the opinions of only a small sample of the public. As you learned in Section A of this chapter, they should select a sample randomly from the population. It's unlikely that you will do that for your experiment, but you should know how it's done.

*Selecting a Random Sample from a Population*

To select a random sample from a population, you must first specify the population and then use a selection technique that gives each member of that population an equal chance to be selected.

To use a small example, let's specify as a population a class of 100 introductory psychology students. How can we select a random sample of 20? It wouldn't be random to select the first 20 people who volunteered or who signed up for a particular time. It wouldn't be random to select the 20 people in the last two rows or those first in the alphabet. It wouldn't even be random to select the person in each fifth seat, but that probably would be close enough to "random" for most purposes, if everyone was present that day.

The most straightforward technique would be to write each of the 100 names on a slip of paper, put all the slips into a container, mix them around a bit, and draw out 20 names. If that procedure were followed there would be statistical justification for believing that the results of an experiment performed on these 20 people also would occur if all 100 people were used.

A different technique for obtaining a random sample would be to use a table of random numbers such as in Table C.1 in Appendix C of this book. You must arrange all 100 names in some order. Start anywhere in the table of random numbers, in some nonsystematic way, perhaps at the twenty-second row of the tenth column. Then go in any direction from there—up, down, across, or diagonal—and write whatever digit you find next to each person's name. Since you know you want to select one-fifth of the people, you could select anyone with a 0 or 1 beside their name, since that's one-fifth of the digits. If you wanted to select one-fourth of the people, you'd have to figure out beforehand how to make things come out even. You might decide to skip all the 8s and 9s in the table and assign only the digits 0 through 7. Then you could select all the people with 0s and 1s, because that would be one-fourth of the digits used.

## Section B  Making Your Experiment More Definite

There are other tricks to using a table of random numbers, but this should be enough to give you a general idea.

If a random sample is large enough, it's representative of the population as a whole. However, in many cases a small random sample doesn't look very representative. For example, you might find that there were only 2 males in your sample of 20, even though the whole class has 40 percent males. To prevent that from happening, you can *stratify* the population *before* selecting the sample. To stratify the population, you divide it into subpopulations and select from these. In this example, we could take the 40 males and select 8 randomly and then take the 60 females and select 12 randomly. The population could be stratified by other variables as well. Notice that random selection always *follows* the stratification. You don't keep selecting "random" samples until you get one that "looks random."

Once you've finally selected the random sample you somehow must insure that the people in it all participate in the experiment. If they refuse or are absent, your results will not generalize to people likely to refuse or to be absent.

Obviously, selecting a random sample is a lot of work. The difficulties are multiplied if a population larger than 100 people is specified. Where do you find a list of "all human beings"? Typically, random samples are obtained from populations consisting of voter rolls, telephone books, or passers-by. Results obtained on such samples apply only to these populations. Refusals to cooperate and missing people are a big problem.

This explanation probably makes it clear why random, representative samples are so uncommon in behavioral research, although they're obviously theoretically desirable.

★ Is there anything you can do to increase the size of the population to which it seems reasonable to generalize your results?

If so, then you should do it. If not, keep the limitations of your results clearly in mind.

★ Taking all these practical, ethical, and theoretical factors into account, what kind of subjects will you use in your experiment?

### Estimate the Number of Subjects

There is more discussion of this question at the end of the next chapter. However, you should have these figures in mind while you're making the following decisions.

There are statistical techniques for estimating the minimum number of subjects you need for results which can be trusted, but they're beyond the scope of this book. To put the matter simply, increasing the number of measurements in an experiment increases the trustworthiness of the results; it also increases the amount of work. I'll give you some very rough guides for the *minimum* number of subjects you should use in different situations.

With one between-subject independent variable, using fewer than about 20 human subjects at each level probably leads to unreliable results. Thirty or even more would be better in many experiments. Thus, if you have one between-subject variable with three levels, you need a minimum of about 60 subjects. With animals, where there may be less difference between individuals, fewer subjects may be used.

Within-subject variables require fewer subjects, because each subject is measured more than once. The more times you measure each subject on each level, the fewer subjects you need. If you measure each subject once on each level, 20 to 30 subjects are about what you need. In some areas, most commonly in studies of sensory processes, each subject may be measured hundreds of times on each level. In such cases you can use only two or three subjects, enough to show that the results aren't peculiar to just one individual.

Before-after variables also commonly use relatively few subjects. In situations where the experimenter can control the whole experimental situation very closely, three or four subjects may be sufficient. Where less control can be exerted, numbers up to 20 or more may be used.

### Decide on Between-Subject, Within-Subject, or Before-After Independent Variables

This decision concerns how you'll vary each independent variable. Will you use different subjects at each level, measure each subject during all levels of the independent variable, or limit yourself to just observing the subjects before and after the occurrence of the independent variable? The decisions are based on different considerations for nonmanipulated than for manipulated variables.

*Deciding for
Nonmanipulated Variables*

With some variables you have little choice. If you're dealing with a subject variable such as sex, intelligence, or area lived in, these can be investigated only as between-subject variables. Similarly, if you're dealing with a variable that occurs only rarely, such as effects of earthquakes, you'll be limited to a before-after variable, despite the interpretational problems that arise with such variables.

In some cases, however, you have a choice. Subject variables such as age, year in college, or feelings of depression can be investigated as cross-sectional (between-subject) or longitudinal (within-subject) variables. Similarly, some environmental variables such as time of day, type of weather, or season also give you a choice.

You can either follow the same group of individuals through the various levels or choose different individuals to observe at each level. How can you choose?

Cross-sectional variables often require less time to complete, especially with subject variables. If you wanted to compare 10-year-olds to 50-year-olds in a longitudinal study, you'd have to wait 40 years! With environmental variables the time savings are less important, but a cross-sectional design is still easier because there's no need to keep relocating the subjects.

One additional factor may cause you to decide against a longitudinal design. In any within-subject design the subject must be able to be measured more than once on the dependent variable. In some cases, this isn't possible. For example, if you wanted to determine how the ability to solve a certain kind of problem increases with age, you might not be able to give the problem to the same people more than once. In that case, you should use a cross-sectional design.

However, longitudinal studies require fewer subjects and may be less work on that account. Moreover, they allow different kinds of questions to be answered. For example, by varying "year in school" as a cross-sectional variable you investigate the difference between, say, freshmen and seniors in 1980. By performing a longitudinal study you can perceive how the same individuals change over that 4-year interval. If you're investigating an environmental variable such as type of weather as a cross-sectional variable you may be investigating the difference between the kind of subject you locate in the rain and the kind you locate in fair weather. By performing a longitudinal study you investigate changes within each individual. You may be more interested in one kind of relationship than in another.

★ If you're using a nonmanipulated independent variable, you should now decide: Will you use a between-subject, within-subject, or before-after variation?

*Deciding for Manipulated Variables*

If you're using a manipulated variable, one goal that you have is to show that a causal relationship exists between that variable and the dependent variables. This is discussed in detail in the next chapter. You'll learn there that between-subject variables are usually the easiest to understand and have the fewest problems of interpretation. However, they also require large numbers of subjects and often are inefficient in other ways. Because of that inefficiency, experimenters with limited time and resources often prefer within-subject or before-after variables. How can you decide?

If you know enough about an area, you may know enough to use a before-after design. In order to draw causal conclusions you must know with some certainty how your subjects would behave in the experimental situation if your independent variable weren't added. For most psychologists, the only areas in which that's likely to be true are animal behavior modification or physiological investigations. Even in these cases other controls are usually added.

★ Do you know enough to use a before-after design?

If so, then this is the most efficient design to use. If not, you must decide between between-subject and within-subject manipulation. As with nonmanipulated variables, if each subject can't be measured more than once on the dependent variable, you must use a between-subject variation. For example, an experiment measuring speed of maze learning as a function of hunger would probably not be able to manipulate hunger as a within-subject variable because each subject could be measured only once on speed of maze learning.

★ Can your subjects be measured more than once on your dependent variable?

If not, you must use a between-subject variation. As you'll learn in more detail in the next chapter, use of a manipulated within-subject variable requires one additional thing in order to allow causal conclusions: the levels of the independent variable must be able to be presented in *any* order. That means that you couldn't use a within-subject design to investigate the effects of destroying a portion of the brain, because the order of treatments can't be reversed. Similarly, to investigate the effect of a teaching method on learning to read, a between-subject variation must be used. Once the teaching has occurred, it can't be undone.

★ Can the order of the levels of your independent variable be changed?

If not, you won't be able to draw clear-cut causal conclusions from a within-subject manipulation, so you should use a between-subject design. If you can vary the order of presentation of the levels, you then can choose a between-subject or within-subject manipulation as a matter of convenience.

The following points argue in favor of between-subject variations. (1) The effects of between-subject variables are usually easier to interpret. (2) Each subject is required to participate for only a short period of time. (3) Choice of a within-subject variable will require you to add another independent variable, thereby making it more difficult to complete the experimental design and perform your statistical analyses.

The following points argue in favor of within-subject variables. (1) They require fewer subjects. (2) If the subjects must perform individually, the total experiment will require less experimenter time. (Obviously this last point requires further discussion. It's treated fully in the rest of this chapter and in the next chapter. You might want to read quickly through that material and then return to this point and continue designing your experiment.)

★ Will you vary each of your independent variables between subject, within subject, or before-after?

## Operationally Define Your Variables

As you'll recall from the Section A of this chapter, the operational definitions actually determine the point of the experiment. They should be reasonable definitions of the variables you're trying to investigate, they should be complete enough so that other experimenters could duplicate the operations, and they should be sufficiently precise to prevent the experimenters from unconsciously biasing their judgments.

In addition to these theoretical points there are ethical and practical issues. The operations shouldn't embarrass human subjects or make them angry or hurt them. It should be possible to maintain confidentiality about each individual's performance. With animal subjects, the operations should be as humane as possible.

Practically, you must develop operations you can really use. Tests or machines that you need must be developed or located. It mustn't take too long to measure any one subject. You can make life easier on yourself if you develop operations that allow you to have more than one subject perform at a time, but this is not always possible. Talking aloud or moving around are generally too disruptive to allow group performance. This issue is discussed further in Chapter 4.

Before working on your own operational definitions, try some practice problems.

### Problems

These problems will help you learn how to make up good operational definitions.

Q. An experiment is being designed to discover what changes room temperature causes in learning ability. The experiment is to be conducted in a classroom using student volunteers as subjects. Develop an operational definition for room temperature.

A. Another experimenter should be able to go to a room, apply your definition, and *know* the room temperature. If someone else can't do that, your definition is inadequate. A good definition might be this: "Adjust the thermostat until a thermometer known to be accurate within .25 degree Celsius reads a particular number. That number is defined to be the room temperature." Someone could apply this definition in any room and know the temperature.

If you said something such as "Read a thermometer; whatever it says is the room temperature," that *is* an operational definition. However, it wouldn't allow the experimenter to manipulate the variable directly, so he wouldn't be able to draw causal conclusions.

If you said something such as "How warm it was in the room," that is *not* an operational definition. There's no way to apply it and know the temperature in the room. There were no operations.

Similarly, if you said something such as "the amount of movement in the molecules in the air of the room," that's not an operational definition unless you include some operations for measuring molecule movement.

**Q.** An experiment is being designed to determine the relationship between a child's age (between 3 and 16 years) and the child's beliefs about the feelings of animals. The experiment will use neighborhood children as subjects; it will be conducted in the home of the experimenter. Develop an operational definition of age.

**A.** Perhaps a good kind of operational definition of age is something like this: Obtain a copy of each child's birth certificate, and count the number of years and months between that date and the date on which the experiment is performed. The age assigned is the nearest whole year. However, obtaining the birth certificate could be a problem. Asking parents would probably be as accurate, but then you'd have to get in touch with all the parents as well as the children.

If you said something such as "ask the child his or her age and the response defines the age," that *is* an operational definition. However, it might not be a good definition, because the younger children might not know their ages, and some of the children might lie.

If you said something such as "the number of years since the child was born," that's not an operational definition. There are no operations, no measurement technique to be applied.

Obviously the problems would be different if you are dealing with adults, but no less difficult.

**Q.** An experiment is being designed to determine the relationship between facial attractiveness and job rank. The experiment is to be performed using the experimenter's co-workers at a large factory. Develop an operational definition of facial attractiveness.

**A.** A big problem here is to develop a definition that doesn't allow the feelings of the experimenter to intrude, making it impossible to duplicate the situation. One approach to this problem is to carefully define a list of criteria for each level of attractiveness such as measurements of length of nose or size of eyes. That might be difficult to develop, but it would be easy for others to apply. Another approach would be to arrange a panel of judges who don't know the subjects and have them rate the physical attractiveness of each subject from photographs taken by the experimenter. They might use a 7-point scale, and each subject's physical attractiveness score could be the mean of all five judges' ratings.

You've already made several decisions relevant to operationalizing your variables. You've decided what the variables are, designated which are independent and which are dependent, and decided whether to manipulate, select, or measure your independent variables. For each variable you've decided whether it's a nominal vari-

## Section B  Making Your Experiment More Definite

able or whether numbers can be assigned to its levels in a meaningful way. You've decided where to carry out your investigation and what kind of subjects to use. And you've decided whether to vary each independent variable between subject, within subject, or before-after.

There are two intertwined decisions remaining. One concerns the number of levels of each variable to be distinguished; the other concerns the operational definition of each level. Sometimes it's easier to make those two decisions in one order and sometimes, the other. You'll define each of your variables in turn, using whichever order is appropriate.

For nominal variables and for independent variables that are manipulated or selected you should use the first sequence: decide how many levels of the variable to distinguish, and then operationally define each level. For other variables you should use the second sequence: operationally define a technique for assigning numbers, and then decide how many levels to distinguish.

★ Name all the nominal variables and independent variables you'll manipulate or select. Use the first sequence for these variables.

★ Name all your remaining variables. Use the second sequence for them.

Now go through the appropriate sequence for each variable.

*Sequence 1:*
*Designate Levels and Then*
*Operationally Define Each Level*

First decide how many levels to distinguish. The more you use, the more you'll learn, but the harder you'll have to work. For nominal variables and manipulated or selected independent variables most experiments have from 2 to 5 levels; few have more than about 8.

Suppose an experiment were using "injection of drug RST" as a between-subject independent variable. The experimenter might decide to use only two levels, 0 and 5 mg. Clearly, less would be learned than if five levels were used, say 0, 2, 4, 7, and 10 mg.

Similarly, compare two experiments manipulating the independent variable "new teaching technique" as a before-after variable. One experiment looks at performance on the dependent variable only immediately before and immediately after the technique is applied. The other experiment looks at performance 2 weeks before, immediately before, immediately after, 2 weeks after, and 2 months after the technique is applied. Again, the experiment with more levels is more informative, but it requires more effort.

Now consider an experiment that is investigating the relationship between two nominal variables, ethnic background and type of car driven. How many levels of each variable should be distinguished? Should there be a few broad categories such as European/non-European and family car/sports car/truck/other? Or should there be more, narrower categories? If there are to be narrower categories, how narrow should they be? Should Austrians be separate from Germans? Should Datsuns be separate from Toyotas?

In deciding how many levels to distinguish, you can use as a rough guide the concept that with between-subject variables each level requires about 20 extra subjects. Within-subject and before-after variables don't require extra subjects for additional levels but do require extra time and complexity.

★ How many levels will you distinguish?

Next you'll decide what the levels are. That determines the nature of your experiment and therefore is an extremely important decision, but it's not one that gives many experimenters trouble. It follows naturally from your previous thinking.

★ What will the levels be?

The next step is to develop an operational definition for each level. For manipulated variables you'll develop operations that determine each level of the independent variable when they're performed. Another experimenter should be able to duplicate your variable by performing the same operations. For example, an experiment manipulating fear as an independent variable might define three levels: low, medium, and high. These could be the operations: High fear = hooking a person to a machine with batteries, coils, and other electrical equipment and a sign saying "Danger! High voltage!" Medium fear = hooking a person to the machine but without the sign. Low fear = not hooking the person to the machine but having it nearby. Another experimenter could duplicate the variable by repeating these operations.

For nominal variables or selected independent variables you'll develop operations that allow you to measure what level of the variable is occurring. Another experimenter should be able to duplicate your variable by using the same measurement operations. For example, an experiment using style of clothes as a variable might define three levels this way: Dressy = suit with matching jacket and pants or sportcoat with tie or turtleneck. Casual = jeans with anything or slacks with shirt or sweater. Other = other combinations. In defining the levels of nonmanipulated variables special care must be taken to make them complete enough to place events where the experimenter believes they belong. In the clothes example given, an old tramp wearing a ragged suit with no shoes would be classified as dressy, suggesting that the operational definition isn't as complete as it might be.

★ Operationally define each level of your variable.

*Sequence 2:*
*Develop the Operational Definition,*
*and Then Decide the Levels*

This sequence is appropriate for variables to which meaningful numbers can be assigned, except for independent variables that are manipulated or selected. The first step is to operationally define the variable as a whole. For example, an experiment measuring body weight as an independent variable could define that variable as the amount shown on a Brand Z doctor's scale when the subject is fully clothed. For a second example, the humor of various cartoons could be operationally defined as an independent variable by having 10 college students rate each cartoon on a scale from 1 to 5. Based on the ratings, the experimenter could then rank order the cartoons from funniest to least funny. For a third example, an experiment measuring maze difficulty as a dependent variable could define that variable as the number of errors made by each rat or the time taken to complete the maze or the number of rats completing the maze within 10 min. (minutes).

★ Operationally define the variable as a whole.

The next step is to decide how many levels to distinguish. For these variables, no extra work is required to use a large number of different levels. The weight variable could have as many levels as there were weights without any extra work from the experimenter. If it seemed more useful, however, the experimenter could group the scores into fewer levels by using some additional operational definitions. For example, light weight = 100 lb. (pounds) or less, medium light = 101 to 130 lb., medium = 131 to 160 lb., medium heavy = 161 to 200 lb., and heavy = more than 200 lb.

Similarly, the cartoons could be grouped into fewer levels and the maze difficulties could be categorized. The only effect of grouping the scores is on the kinds of statistics you'll use to present and analyze your results, which needn't concern you now. With this type of variable, just choose whatever number of levels seems most useful. If there's no reason to group the scores, then just have as many levels as there are scores.

★ How many levels will you distinguish?

*Recheck Your Definitions*

Now go back and look carefully at your operational definitions. Make sure your definitions seem reasonable. Be sure they're phrased so they could be followed by someone else. Be sure there's nothing ethically objectionable about your definitions and that you can actually perform the operations. If your independent variable is supposed to be manipulated, make sure the operational definition involves manipulation.

For any variable, make sure your definition is explicit enough to prevent you from unconsciously slanting your interpretations about which level is occurring.

This is a review of the steps you've taken in this chapter toward making your experiment more explicit:

1. Decide on your experimental situation.
2. Decide on the nature of your experimental subjects.
3. Decide whether to vary your independent variables as between-subject, within-subject, or before-after variables.
4. Operationally define your variables.

**Chapter Four
Determining the
Causes of
Relationships**

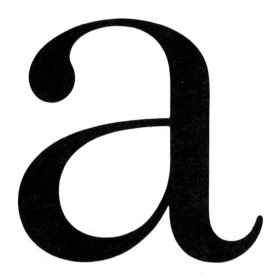

**Section A
Reading Method
Sections for
Confounds**

In Chapter 3 you learned to recognize the subjects used in an experiment, the operational definitions of the variables, and some aspects of experimental design. You learned that any conclusions drawn from an experiment are, in a strict sense, limited to the kind of subjects and operations used in the experiment and that you may extend the conclusions to other situations with care, as long as you recognize that your generalizations may be unwarranted.

In this chapter we'll discuss whether the conclusions drawn from an experiment should imply that a causal relation exists between the variables. To a great degree, this depends on the experimental design. One aspect you learned in Chapter 2 was that causal conclusions can be drawn when variables are manipulated *if* there are no other problems with the experimental design. It's by reading the method section that you discover whether "other problems" exist.

There's one basic rule: to demonstrate a causal relationship between two variables, an experiment must be designed so that the only consistent difference between the levels of the independent variable is the independent variable itself. Then, if a difference occurs in the dependent variable, we say that it either was caused by the independent variable or was completely accidental. Clear-cut causal conclusions can't be drawn from experiments in which the levels of the independent variable differ in more than one way at the beginning of the experiment.

For example, suppose an experimenter selected two equal groups of rats. To one group he gave a special diet *and* also extra attention such as petting and handling.

The other group was given the normal laboratory diet *and* no special attention. If he found a difference between the two groups in some dependent variable, he couldn't be sure whether it was caused by the diet, the attention, or both.

For a second example, suppose an experimenter had one group of rats. First he fed them their normal diet for 4 weeks in February and measured them on some dependent variable. Then he fed them a special diet for 4 weeks in March and measured them on the same dependent variable. If he found a difference in the dependent variable between the first 4 weeks and the second 4 weeks, he couldn't be sure whether it was caused by the diet, by the change in time of year, or by both.

### Confounded Variables and Control Techniques

Another way to state that levels of the independent variable differ in more than one way is to say that the independent variable is **confounded** with another variable. A variable that coincides with or overlaps the levels of the independent variable is said to be confounded with the independent variable. In the two rat/diet examples the independent variable presumably was diet in both cases. In the first example, it was confounded with "attention." In the second, it was confounded with time of year. Whenever the independent variable is confounded with some other variable it's difficult to draw clear-cut causal conclusions.

There are two basic techniques for avoiding confounds. One is to hold other variables constant. The other is to distribute the levels of other variables equally over the levels of the independent variable. Both can be described as techniques that **control** the possibly confounded variables, that is, attempt to eliminate the confound. Notice that the variable isn't eliminated; only the confound is eliminated. In the rest of this chapter you'll see many instances of both control techniques.

### Possible Sources of Confounding Common to All Independent Variables

There are three possible sources of confounding common to all independent variables: extraneous-variable confounds, subject bias, and experimenter bias.

*Problem:*
*Extraneous-Variable Confounds.*
*Solutions: Hold Variables Constant*
*or Distribute Them Equally*

Confounds with variables unrelated to the experiment can occur while the experiment proceeds. For example, in a between-subject design with instructions as an independent variable, all the subjects with one set of instructions could perform in the morning while all the subjects with another set could perform in the afternoon. In a within-subject design using amount of light as an independent variable, the experimenter could encourage the subjects while the light is dim but treat them coldly when the lights are bright. Any such confounded variables make it difficult to interpret the results. There is no generally accepted term for this kind of confound. We'll call these **extraneous-variable confounds.**

Careful experimenters control as many extraneous variables as they can. One technique is to hold the extra variable constant, for example, treat the subjects the same regardless of the amount of light. Another technique is to distribute the levels of the extra variable more or less equally over the levels of the independent variable, for example, some subjects with each kind of instructions perform in the morning and some in the afternoon.

In the method section, experimenters should report any confounds they know about, and they may report some of their attempts to avoid confounds. When experimenters say that all the subjects performed in the same room, heard the same instructions, and so forth, they're telling the reader that they controlled many extraneous variables by holding them constant. Similarly, when experimenters say that the levels of the independent variable were presented equally often in different rooms, by different experimenters, and so forth, they're saying that these possible extraneous-variable confounds were avoided. Obviously the experimenters can't report all the variables they've controlled; they just describe the general steps they've taken. The reader should be able to decide which variables have or haven't been controlled by these steps.

**Problems**

These problems should give you some practice in recognizing extraneous-variable confounds.

**Q.** A clinical psychologist wanted to know which of two competing methods of therapy was superior. She obtained the cooperation of leading practitioners of each method and arranged to have 20 patients with apparently equal difficulties assigned to each method of therapy. After six months of therapy she measured the amount of improvement and found that it was noticeably more in one group of patients than in the other. She concluded that that method of therapy was superior in that it caused more improvement. Was she correct? Why?

**A.** There is a confound present in the experiment. The groups differed not only in therapy but also in therapist. It's not clear whether the difference in improvement was caused by the therapeutic technique or by differences in skill or personality between the therapists. If you were a patient trying to decide which kind of therapy to undergo, this confound would matter little to you. But if you were a therapist trying to decide which kind of therapy you should attempt, it's quite important.

**Q.** A dentistry student wanted to know which color of wall caused people to be more relaxed, yellow or gray. Searching through his university he found on the fourth floor a room identical in every way to one on the first floor except that one was yellow and one was gray. Overjoyed, he proceeded with his experiment. He had 30 students go first into one room and then into the other, and he measured their level of relaxation in each room using a machine designed for that purpose. Half the students were measured first in the yellow room, while the other half were measured first in the gray room. Finding a difference in the level of relaxation, he concluded that it must have been caused by the room color. Do you agree? Why?

**A.** Color wasn't the only difference between the rooms. They also differed in which floor they were on. It seems likely that walking up three flights of stairs would affect the subject's relaxation level.

**Q.** A tire manufacturer wanted to demonstrate that his product was superior to that of his competitor. Accordingly, he found 40 families with the same kind of cars and driving habits and assigned them to two approximately equal groups. Working through a middleman so the people wouldn't know the purpose of the experiment, he gave the families in one group his best tire. He gave the other group his competitor's best tire. After 6 months he had the two groups exchange tires. Then, after another 6 months, he measured the amount of tread wear and found that his tire had worn *more* than his competitor's. Chagrined, he concluded that his competitor's tire caused less wear to occur and filed the results away in a back cabinet. Was his conclusion warranted? Why?

**A.** There's no particular reason to believe that any confounds occurred. What could they have been? The groups were treated equivalently throughout the experiment. Thus his conclusion was warranted. (People just learning about confounds tend to see them everywhere, but some experiments *are* okay.)

*Problem: Subject Bias.*
*Solutions: Placebos*
*and Single-Blind Studies*

Subjects in an experiment may influence their behavior either to support or disagree with what they think are the experimenter's wishes. Thus the subjects' expectations or intentions may be confounded with the independent variable. This is called **subject bias.** For example, subjects given a drug may *say* that their pain has subsided because they think that's what the experimenter wants. Even more complicated, their pain may *really* decrease, not because of the drug, but because of the subjects' expectations. In the case of a nonmanipulated variable such as social class, subject bias also can occur if the subjects know they're being compared to others.

Experiments in which it's easy for the subjects to understand the purpose and to influence their behavior are said to have obvious **demand characteristics.** Subject bias is almost always a problem in such experiments. However, even in experiments with less obvious demand characteristics subjects develop hypotheses of their own and influence their performance, so attempts to avoid subject bias are made even in these cases.

To protect against subject bias, subjects are sometimes kept ignorant of the purpose of the experiment or the level of the independent variable they're experiencing. In drug studies it's common to include a **placebo** treatment, in which there's a substance that looks like the drug but has no active ingredient. The drugs are administered in such a way that the subject can't tell the difference between the drug and placebo by looking at them or tasting them—that is, the subject is kept "blind." These are called **single-blind** studies.

In nondrug studies analogous procedures are sometimes used. For example, an experiment could compare a particular therapy with a "placebo" therapy. The use of placebos tends to distribute the subjects' expectations equally across the levels of the independent variable. With nonmanipulated variables, experimenters sometimes try to minimize subject bias by keeping the subjects blind as to the purpose of the experiment or they hold subject bias constant by telling all the subjects the same experimental hypothesis.

*Problem: Experimenter Bias.*
*Solution: Double-Blind Studies*

A related source of confounding arises when an experimenter's attitude or behavior is confounded with the independent variable. An experimenter quite unconsciously may treat subjects differently depending on the level of the independent variable, thereby presenting one set of demand characteristics at one time and a different set at other times. In cases where **experimenter bias** is likely, experimenters sometimes arrange the situation so they don't know what level of the independent variable is occurring. With manipulated variables, the variable is administered by someone else or in a coded form, so the experimenter can't tell what level is occurring. With nonmanipulated variables such as socio-economic class, the experimenter purposely avoids learning which class the subjects come from until after the experiment is completed. Since both the subjects and the experimenter are ignorant, this is called a **double-blind** procedure.

Even if the experimenter manages to avoid treating subjects differently, care must be taken to avoid systematic errors in recording, copying, or interpreting the experimental data. Sometimes it's important that individuals unaware of the purpose of the experiment perform these functions to avoid experimenter bias.

The method sections of two of the articles whose introductions you read in Chapter 2 are reproduced here. The first is from the Wallace et al. teaching method study. Recall that the purpose of that study was to compare the effectiveness of three different teaching methods for counselor trainees. The second is the McDaniel and Vestal source credibility study, in which the purpose was to discover the effects of varying levels of source credibility and relevance on recall of presented information.

## Problem

**Q.** Read these sections over, and see if you can find any cases of extraneous-variable confounds or any reason to believe that the results might be affected by subject or experimenter bias.

## The Method Section of the Wallace et al. Teaching-Method Study

*Subjects*

The subjects for this study were 54 beginning graduate students at the Pennsylvania State University. There were enrolled in a three-credit, master's level, Counseling Theory and Method course during one of two consecutive 10-week terms (spring and summer, 1974). In each term the students were randomly assigned to one of the three teaching methods. Each teaching method was applied to 8 students during the spring term and 10 students during the summer term.

*Teaching Materials and Methods*

The lecture and handout on decision-making (used in all three methods) were taken from Herr, Horan, and Baker (1973). The process of helping clients arrive at wise decisions is operationally defined in terms of a number of specific sequential counselor behaviors. The 20-minute model tape used in Methods 2 and 3 depicted a counselor engaging in each of these behaviors while helping a client resolve a choice conflict (continuing or withdrawing from school).

In each of the two academic terms, the entire subject pool received the lecture and written material at the same time during a regular class period. The subjects taught by Methods 2 and 3 observed the model tape during another time period. Method 3 was conducted during a third time period in a small group format and was video-taped. Each subject had the opportunity to counsel a role-playing client from within the group through at least one stage of the decision-making process. The group then viewed the video-tape, and each member received performance feedback (positive reinforcement) from the group leader whenever his counseling behavior approximated the standard. (All students were eventually given the opportunity to receive Method 3; however, the time periods were staggered to allow comparisons between each teaching method as if each method alone were being administered.)

*Measurement of Decision-Making Skill*

After receiving one of the three teaching methods, the students participated in two audiotaped interviews with coached clients, each role-playing a different

## Section A  Reading Method Sections for Confounds 117

decision-making concern. The students were told that they would have a maximum of 30 minutes per interview and that they should accomplish as much as possible. Most interviews ranged from 15 to 20 minutes in length. The audiotapes were coded and then rated by two independent judges on an instrument designed especially for this study. The judges were doctoral students blind to the teaching methods received by over 95% of the subjects (voice identification and consequent treatment identification occurred in several cases with one of the judges).

The measuring device was essentially a rating form consisting of 11 operationally defined counselor behaviors adapted from the Herr, Horan, and Baker (1973) model. Briefly, the counselor behaviors were (a) define the problem as one of choice; (b) explain the decision-making paradigm; (c) identify possible alternatives; (d) gather relevant information from the client; (e) present relevant information to the client; (f) request that the client identify advantages and disadvantages for each alternative; (g) present any additional advantages and disadvantages for each alternative to the client; (h) request that the client select the most promising alternative; (i) verbally cue and reinforce the client for gathering additional information about the most promising alternative; (j) help the client implement the alternative; and (k) maintain neutrality concerning what the client ought to do. On each behavior the subjects could receive a score of 0 (complete absence of the behavior), 1 (behavior was present but weak), or 2 (behavior was clearly displayed). Each subject's decision-making counseling performance was simply the sum of his scores on each item for both judges and for both coached clients. The range of possible scores was 0–88 with higher scores indicating better performance. The interrater reliability coefficient on the instrument was .89 (Pearson $r$); internal consistency was found to be .85 (Kuder-Richardson).

---

**A.** In this study there was an opportunity for extraneous-variable confounds to occur, although it's not clear that any in fact happened. Since the students receiving different teaching methods were in different sections of the course, it seems likely that they also received other kinds of differentiating treatment. They may have been taught somewhat different things, they may have met at different times of day, the teacher might have been in a better mood with some classes than with others, and so on. Whenever individuals at different levels of an independent variable are treated as groups, anything that happens to one group and not to another (such as a chance thunderstorm during class) is confounded with the independent variable.

Subject bias also could be a problem. If the subjects recognize the differences between their treatments, those in Method 1 might decide to perform rather badly, because they knew they weren't expected to do very well. Those with Method 3 might feel favored and elated and actually work harder than the others.

Warren et al. did attempt to eliminate one source of experimenter bias by using blind judges to rate the counselor trainees' performance. They were largely successful in that. However, it's still possible that during the class periods the instructor gave more attention to the students in Method 3 or behaved in other ways that affected their behavior differentially.

Despite all these confounds, this is still not an unreasonable study. It could have been better, but Warren et al. were working within the constraints of trying to perform the experiment without disrupting the class too much.

# The Method Section of the McDaniel and Vestal Source Credibility Study

*Subjects and Materials*

Forty-four introductory psychology students of both sexes served as subjects and were assigned randomly to four groups.

Four 300-word written communications, identical except for information pertinent to the variables source credibility and issue relevance were used. Highly credible communications were defined as those ascribed to "Dr. John A. Wood, PhD, JD," while low credibility communications were defined as those ascribed to rumor. High relevance was defined as a referendum against current laws forbidding the sale of alcoholic beverages in the local county (Watuaga, North Carolina). Low relevance was defined as the same referendum, but proposed in a different county in a distant state (Spillwell, Idaho).

Twelve fill-in-the-blank questions were used to assess recall of communication content. These questions were divided into two separate six-item questionnaires which were randomly distributed among subjects as a measure of immediate retention. All 12 items comprised the questionnaire which was used after the communication presentation and the immediate retention test in order to provide the two measures of long-term retention. One measure was the score achieved on the items that the subject received on the immediate retention test, while the second measure consisted of the score obtained on the new items. This was done as a control for possible confounding results from the interaction between the independent variables and question rehearsal.

*Procedure*

Upon entering psychology classes, the experimenters explained that they were interested in studying aspects of social attitudes. The subjects were offered bonus grade points to participate. The communication pamphlets were randomly distributed, and the subjects were asked to read the communication silently. After a 10-min. interval, the communications were collected and the six-item questionnaires distributed. Subjects were asked to complete the questionnaire within 5 min., after which the questionnaires were collected and the subjects informed that the experimenters would return in 1 week for additional research. One week later, the 12-item questionnaire was presented, and, after 10 min., all questionnaires were collected, and the subjects were debriefed.

Scores on both retention tasks were determined by two paid individuals not associated with the study according to answers specified by the experimenters. One point was assigned for each correct response.

---

**A.** Since subjects in all the treatments were run simultaneously, there's no way for some event to have occurred for individuals at one level and not for those at another. It seems unlikely that the subjects could have figured out the design and shown any subject bias. And the experimenters protected themselves against experimenter bias by having "blind" judges score the answer sheets.

### Sources of Confounding with Nonmanipulated Variables

*Problem: Variable-Connected Confounds.*
*Solutions: Hold Constant or Match*

Nonmanipulated variables can be confounded with extraneous-variable confounds, subject bias, or experimenter bias. Careful researchers try to avoid as many such confounds as they can. However, a nonmanipulated variable like frequency of marijuana use is by its nature confounded with a multitude of other variables. The reason why clear-cut causal conclusions can't be drawn about the effects of that variable on other variables is that any of the variables confounded with frequency of use, such as social class or ethnic background or attitude toward authority, might cause any observed differences. These confounds don't have a commonly accepted name; I'll call them **variable-connected confounds.**

If a variable is nonmanipulated, an experimenter can't draw causal conclusions about its effects on other variables because of variable-connected confounds. However, the experimenter could try to control the confounding variables one at a time. One researcher might look at the relationship between marijuana use and adjustment in a group consisting of only white Anglo-Saxons. He would thereby control ethnic background by holding the variable constant. Another researcher might control ethnic background by making sure she had the same proportion of members of each ethnic group in each of her frequency of use groups, say 85 percent Anglo-Saxons, 10 percent blacks, 3 percent orientals, and 2 percent other. That approach is called **matching** on a variable, in this case matching on ethnic background.

If there's a relationship between frequency of marijuana use and adjustment when the confound with ethnic background is eliminated, either reseacher would know that the relationship was *not* caused by ethnic background. But they'd still have a lot of other variables confounded with frequency of marijuana use. For all practical purposes, a single experiment can't be designed to show that marijuana use does or doesn't have a causal relationship with adjustment, although it can show the pattern of relationships between the variables. However, in a series of experiments more and more variables can be controlled. As the list gets longer, the argument that there's a causal relationship becomes more attractive.

The example we've been using, frequency of marijuana use, has been treated as a between-subject variable, with different groups of people being compared. Within-subject variables have different problems, but if they're nonmanipulated, they necessarily have variable-connected confounds. For example, a teacher might want to investigate the relationship between level of anxiety and test performance. She could measure the level of anxiety of her students before each test and relate each student's test performance to his or her anxiety level. If she found a relationship, she wouldn't be able to conclude that the change in test performance was caused by the change in anxiety level, because other variables are confounded with anxiety level, such as time of year, amount of study, or performance on last test. Any of these could be responsible for a change in test performance.

Thus far we've mentioned two subject variables, frequency of marijuana use and anxiety level. Nonmanipulated environmental variables, such as weather type, also have variable-connected confounds. Suppose an urbanologist wanted to investigate the relationship between weather type and walking speed. He might elect a between-subject study in which he observed one set of people on a sunny day and a different set on a rainy day. If there were differences in walking speed, he couldn't know whether they were caused by the weather or by the fact that different kinds of people come out in different kinds of weather. On the other hand, he might elect a within-subject study, observing the same people on both sunny and rainy days. In this case the problem would be that the rainy days come at one season or that rain generally occurred in the morning. These might also be confounds in the between-subject study.

Before-after variables can also be nonmanipulated. For example, a political scientist might measure attitudes toward politicians both before and after a presidential election. A change in attitude wouldn't necessarily mean that it was caused by the election, because many other variables are confounded with that variable, such as other current events, change of seasons, change in the economy, and the sheer passage of time.

A combination of between-subject, within-subject, and before-after designs can clarify the relationships between nonmanipulated variables somewhat. In addition, with all the designs the experimenter can try to control one or more confounded variables, either by holding them constant or by matching. However, there are always some confounded variables remaining. It's never possible to control *all* the variable-connected confounds when a variable is nonmanipulated.

## Sources of Confounding
## with Manipulated Variables

As with nonmanipulated variables, there may be extraneous-variable confounds, subject bias, experimenter bias, and variable-connected confounds. As you'll see, the control techniques for variable-connected confounds are much more powerful with manipulated variables. In addition, there are some sources of confounding specific to between-subject, within-subject, and before-after variables that should be controlled if clear-cut causal conclusions are to be drawn.

*Problem: Variable-Connected Confounds.*
*Solution: Control Treatments*

Suppose a researcher wanted to manipulate the variable "presence of THC (the active ingredient in marijuana) in the bloodstream" by injecting the drug into rats. When you think about it, you'll see that in order to manipulate that variable the researcher also must manipulate a number of other variables. It's necessary to catch the rats, handle them a bit, probably carry them to a new place, stick them with a needle, and carry them back home. If these rats are compared to rats with none of those experiences and are found to differ on a dependent variable, we can't be sure whether the difference was caused by THC injection or by the confounded events.

To avoid variable-connected confounds with manipulated variables, experimenters often add one or more **control treatments** to their independent variable. In the THC example, an appropriate control treatment would be to catch, handle, carry, inject, and return the rats. The only difference between the experimental treatment would be the substance injected. It should be one that doesn't cause a change in the dependent variable, a placebo. Thus another application of placebo treatments is in controlling variable-connected confounds.

Variable-connected confounds can occur regardless of whether the variable is manipulated between subject, within subject, or before-after. With between-subject variables, careful experimenters usually compare the experimental groups to one or more **control groups,** groups receiving control treatments. With within-subject variables, a control treatment is added to the other levels of the independent variable. For example, the behavior of rats injected with THC on some occasions might be compared with the behavior of the same rats injected with a placebo substance on other occasions.

With before-after variables the problem of variable-connected confounds can be very sticky. Clearly, if we take a group of untreated rats, observe the dependent variable, and then go through the whole injection procedure and observe a change in the dependent variable, we can't be sure which of the experiences caused the change. If we conclude that the drug caused the change, we must assume that we know how the rats would behave *if* they were injected with a placebo. If we can't make that assumption, the results can't be interpreted clearly. A careful experimenter would use a different kind of design unless he were willing to make that assumption.

*Between-Subject Problem:*
*Nonequivalent Groups.*
*Solutions: Hold Constant,*
*Match, and Random Assignment*

When an independent variable is manipulated between subjects, one possible source of confounding is peculiar to that situation: the subjects may be assigned to the treatment groups in such a way that the groups are **nonequivalent.**

Recall the mirror image experiment by Lutkus. The subjects were 64 right-handed students, 32 males and 32 females. They were randomly divided into four groups in a 2-by-2 design that varied instructions (standard versus image) and figure (diamond versus zigzag). Is it reasonable to believe that the subjects in the various groups were equivalent?

Suppose that instead of assigning the subjects as he did Lutkus had assigned all the males to the standard instructions and all the females to the image instructions. If he then had found that the subjects with image instructions made fewer errors and took less time than the subjects with standard instructions, he'd certainly *not* be able to conclude that instructions caused the difference, because the instructions were confounded with the sex of the subjects. To avoid that confound, Lutkus probably arranged his experiment so that each of the four experimental groups had matched numbers of males and females.

Lutkus also avoided confounding handedness with either independent variable by limiting his experiment to right-handed people only, holding that variable constant.

But how can we be sure that the subjects assigned to the diamond figure weren't by some accident more artistically inclined than those assigned to the zigzag or perhaps had better vision or were for some reason more practiced at dealing with mirror images? In fact we can't be sure that none of these variables is confounded with the independent variables—Lutkus couldn't control everything. But he did do one thing to try to distribute all those variables more or less equally across the levels of the independent variables: he used **random assignment** to assign the subjects to the four groups. As before, "random" doesn't mean accidental. It's a technical term, meaning that each subject had an equal chance of being assigned to any of the four groups. Some common techniques of random assignment include flipping coins, drawing numbers out of hats, and using tables of random numbers. With random assignment there's no reason to believe that people with better vision were assigned to one group more than to another. It could happen, but there's no reason to think it's likely.

Experimenters can try to make their groups more equivalent by matching some variables (as Lutkus did with sex) or holding variables constant (as Lutkus did with handedness), but there are always many left over. For these, random assignment is

Section A  Reading Method Sections for Confounds            123

an attempt to avoid confounds by distributing all the uncontrolled variables more or less equally across all the levels of the independent variable.

**Problems**

The problems will help you learn how to recognize some instances of confounding with between-subject variables.

**Q.** Suppose the warden in a prison for delinquent girls noticed that the prisoners seemed to spend almost all their free time watching violent programs on TV. She wanted to know whether watching such programs caused the girls to act more violently. To find out, she randomly selected 50 prisoners and told them their recent behavior had been unacceptable, so that they were to be deprived of TV privileges for a month. Another randomly selected 50 prisoners weren't treated in any special way but were allowed to continue watching violent programs. The warden kept track of the number of incidents involving the girls in each group for the next month and found that the group watching the violent TV programs indeed was more frequently involved in violent incidents. She concluded that watching violent programs caused girls in prison to behave more violently. Was this conclusion warranted? Why?
**A.** The variable was manipulated, and the girls were assigned randomly to the different levels, but the conclusion isn't warranted. The warden intended to vary only whether the prisoners saw violent TV programs but, in performing the experiment, added other extraneous variables. The prisoners not watching the violent programs also were told that their behavior had been unacceptable and that they were being punished. That treatment alone could have caused the difference. Moreover, they were deprived not only of violent TV programs but of any programs at all. What were they doing with their time? Perhaps the other activities actually caused a decline in violent behavior.

**Q.** Suppose an army general wanted to know the effects of performing various jobs on willingness to volunteer for dangerous duty. He arranged to have 200 enlisted men assigned randomly to one of five different jobs: clerk, laundry, grounds, machine repair, and building maintenance. After all the men had worked on their jobs for 6 months, they were given a chance to volunteer for unspecified "dangerous duty." He found that all the groups showed a high volunteer rate, but the clerks showed the highest rate of all. He concluded that being assigned to clerk duties caused enlisted men to volunteer at a very high rate for dangerous duty. Was this conclusion warranted? Why?
**A.** The men were assigned randomly to jobs. There were no apparent other variables confounded with the independent variable. Therefore the conclusion does appear to be warranted. The only reason to suspect it is that perhaps the men got together, figured out the experiment, and manipulated their behavior accordingly. However, that doesn't seem very likely.

**Q.** A psychologist working at a college wanted to know the effects of different kinds of memory techniques on memorizing lists of words. He used the students in the introductory psychology class as subjects, since they were required to serve in a certain number of experiments. The students could sign up to come for one of three sessions on one day. The students in the first session were told to try to memorize the list as if they were studying for an exam. They then were given a list of 30 words to study for 30 sec., waited for 30 sec., and then wrote down all the words they could recall. The students in the second session were told to study the words by making a story or picture using the words. Otherwise they went through the same procedure as the first group. The students in the third session were told to memorize the list by repeating it over and over again to themselves. The psychologist found that the students in the picture group recalled more words than the ones in the other two groups and that these two groups didn't differ. He concluded, "Studying the list by trying to make a picture out of the words caused better recall than either of the other two techniques." Was this conclusion correct? Why?

**A.** The students weren't assigned randomly to the three groups; they assigned themselves. There may be differences other than instructions between the groups in terms of preferences for time of day or hunger levels. In general, whenever experiments are performed with subjects at each level being run together in separate groups there may be confounds.

**Q.** A ski instructor wanted to know which of two possible teaching techniques is most successful. One involves a great deal of demonstration by the teacher; the other emphasizes constant skiing by the learner. The instructor also wanted to know who learns more quickly, people with water skiing experience or people without. Each time that instructor was hired to give an individual lesson, she asked about the person's water skiing and then flipped a coin to randomly assign the person to one or the other teaching technique. She developed a detailed operational definition of "amount of improvement during the lesson," which served as the dependent variable. She found overall that people taught by the method emphasizing constant skiing improved more than people taught by the other method. She also found that people with water-skiing experience improved more than people without it. She concluded, "Emphasizing constant skiing caused more improvement in skiing than lots of demonstration. Also, having water-skiing experience caused more improvement in skiing than not having it." Are each of these conclusions warranted? Why?

**A.** The conclusion about the teaching techniques is warranted. The learners were assigned randomly to the different techniques, and there's no reason to believe that any other variables were confounded. The only point of concern may be experimenter bias. However, the conclusion about water skiing is *not* warranted, because many variables are confounded with the levels of that independent variable, such as interest in sports or income bracket. This is a nonmanipulated variable. Notice that in a factorial design, the conclusions drawn about each independent variable can be evaluated separately.

**Q.** A psychology student wanted to find out how much films really help students recall things. He had 50 students come to a classroom, and he gave them a short lecture on experimental design. Then he randomly assigned the students to two groups. One group was excused, and the other stayed and watched a film covering much the same material as the lecture. The students were all supposed to come back 1 week later to be tested but, as often happens, several didn't return. In fact, only 20 of the 25 students in each group were there a week later. He gave these 40 students the test and found that the students who had seen the film got 80 percent of the questions correct, on the average, while the other students got only 75 percent correct. He concluded, "Watching the film caused an increase in the amount recalled 1 week later." Was that conclusion warranted? Why?

**A.** Although the students were assigned to the groups randomly, it isn't clear that they *remained* in the groups randomly. Perhaps watching the film in addition to hearing the lecture caused those students who really disliked the material to stay away from the experiment and if they had come, their scores would have lowered the average for their group. Whenever subjects disappear during an experiment, the original random assignment becomes questionable and the conclusions to be drawn from the experiment become less clear.

*Within-Subject Problem:*
*Practice or Fatigue Effects.*
*Solutions: Order As a*
*Between-Subject*
*Variable, Repeated Measures*

When an independent variable is manipulated within subjects, again, there is a source of confounding that's peculiar to that situation. There isn't a generally agreed-upon name, so I'll refer to that source as **practice or fatigue effects.** These effects are sometimes called **order effects, position effects, time effects,** or **repeated-measures effects.**

When more than one level of an independent variable is given to the same subject, the levels must be presented in some order: one level is first, another second, and so on. As subjects progress through an experiment they become more practiced, fatigued, confused, and older and probably change in other ways. I'll call all these possible changes practice or fatigue effects. Levels of an independent variable occurring early in the experiment might differ from later levels just because of the time differences and not because of the independent variable. If practice or fatigue effects are confounded with an independent variable, interpretation of the effects of that variable is difficult. Therefore, whenever within-subject variables are used, a primary concern is avoiding such confounds.

Recall the question-wording experiment of Loftus and Zanni. In it the subjects were 100 graduate students. One independent variable was the leadingness (a versus *the*) of the questions; that was a between-subject variable, with the subjects presumably assigned to *a* or *the* questions randomly. Since it appears that the only difference between the groups was the leadingness of the questions, it's appropriate to conclude that that variable must be responsible for any observed differences between the groups' answers.

The other variable was the presence of the items being asked about in the film which had just been seen by all the subjects. Some questions asked about objects that were present, and others asked about objects that weren't present. This was a within-subject variable. For the rest of this discussion I'll abbreviate the two levels as P and NP. If there are differences in the answers to the P and NP questions, is it appropriate to conclude that the differences were caused by the independent variable presence? The answer is yes, but the reasons require discussion.

Suppose Loftus and Zanni had performed their experiment by having each subject first answer three P questions and then three NP questions. If they found a difference, they couldn't tell whether it was caused by presence or by the fact that the P questions were first and the NP questions second. Presence would be confounded with practice or fatigue. To avoid that confound, Loftus and Zanni gave each subject a different permutation, or order, of the questions. Some subjects were asked P questions first and some were asked NP questions first and most probably had mixed orders such as P P NP NP NP P. If there were any practice or fatigue effects, they'd be distributed more or less equally across both levels of presence when the scores of all the subjects were combined. If practice or fatigue caused a change in performance it should have affected both kinds of question equally. As usual, the practice or fatigue effects aren't eliminated; they're controlled. Only the confound is eliminated.

The technique used by Loftus and Zanni for controlling practice or fatigue effects can be described by saying that they used "order of treatment" as a between-subject variable. It had 100 levels, with one subject at each level. They weren't interested in the effects of order on performance, but they added it as an independent variable to control the practice or fatigue effects.

When order is used as a between-subject variable it's not necessary to have a different order for each subject. For example, Loftus and Zanni could have used just 10 different orders and assigned 10 subjects to each order. They could even have limited themselves to two orders if the orders had been chosen carefully to distribute practice and fatigue effects equally across the two levels of presence. Suppose that 50 subjects received the 22 questions in an order in which the 6 critical questions appeared in the sequence P NP P NP P NP and the other 50 subjects had them in the order NP P NP P NP P. There'd be no reason to think that practice or fatigue effects would fall more heavily on P questions than on NP questions, because these effects should be balanced about equally over both levels. These two orders would be called **counterbalanced** orders, which means that, across the two orders, each question type appears equally often at each time. When order is used as a between-subject variable, the two common techniques for distributing practice and fatigue effects are either to use a large number of different orders or to use a smaller number of counterbalanced orders.

A different approach to controlling practice and fatigue effects is to distribute the levels of the independent variable equally across different positions *within* each subject's performance. For example, Loftus and Zanni could have made certain that each subject had one P question and one NP question near the beginning, middle, and end of the list of 22 questions. That approach would be more convincing if there were, say, 10 questions of each type distributed across the various positions rather than just 3. In order to represent each level at several positions, it's necessary to present each level more than once to each subject. The more repetitions of a level, the more likely it is that the level is distributed across practice and fatigue. Variables manipulated in this way are often called **repeated-measures variables,** and experiments in which they appear are called **repeated-measures designs.** Since Loftus and Zanni used a repeated-measures approach in addition to varying order as a between-subject variable, their experiment might be called a repeated-measures design.

When repeated measures are used there are several techniques for equating the distribution of levels across practice and fatigue including **counterbalancing, block randomization,** and **complete randomization.** These are discussed in Section B of this chapter. For reading purposes, it's generally sufficient to recognize these words as showing attempts to control practice and fatigue effects in experiments with within-subject variables.

### Problems

These problems will help you learn to recognize some instances of confounding with within-subject variables.

**Q.** A pediatrician had a pet theory that whenever he held his patients' hands their heart rate was higher. To check the correctness of his hunch, he varied whether or not he held the patient's hand in a repeated-measures design. I'll abbreviate hold and not hold with H and N. He selected 8 children that he knew would be in the hospital for 12 days. For one half of the children he used the counterbalanced order HNNH HNNH HNNH; for the other half he used the order NHHN NHHN NHHN. He found that on H days their heart rate actually was lower than on N days and concluded that, somehow, the hand-holding caused a decrease in heart rate. Was he correct? Why?

**A.** There are no obvious confounds in this experiment. However, it's possible there was an experimenter bias effect. Heart rate is a rather suggestible reaction, and the pediatrician may have unconsciously treated the children to influence their behavior against his own hypothesis. That sometimes happens to experimenters anxious to avoid experimenter bias. Aside from that possibility the conclusion appears warranted.

**Q.** The safety officer of an airport wanted to investigate the limits of attention of air traffic controllers. She arranged to test the controllers without their knowledge by selecting one at a time and feeding false radio and radar messages to that controller's equipment. At first the controller had to process information from only one plane, but over a period of an hour the number gradually increased to 10 planes. Since the information was carefully chosen, it was easy to detect errors. She learned that the controllers made very few errors up to four planes but that errors increased with larger numbers until they were dangerously high with 7 or more planes. She concluded that no controller should ever try to deal with 7 or more planes and that 4 or lower caused many fewer errors. Was she correct? Why?

     **A.** Number of planes was confounded with fatigue effects. Perhaps the controllers were becoming tired and this caused the change in performance. If so, the correct action would be to shorten the work periods rather than to limit the number of planes.

     **Q.** Now read the method section below of the Schutz and Keislar word-class study for which you read the introduction in Chapter 2. Recall that the purpose of this experiment was to investigate how well children of different ages from different social classes were able to recall nouns, verbs, and function words such as *the* and *but*. Obviously, the first two are nonmanipulated subject variables. As you'll see, word class is a within-subject variable. Evaluate this experiment from the point of view of confounds with the independent variables.

## The Method Section of the Schutz and Keislar Word-Class Study

*Subjects*

Children in the kindergarten-primary grades were tested in two different schools, one middle and one lower class, socio-economic class being defined in terms of the communities in which the schools were located. The lower class school, attended almost entirely by black pupils, had been earmarked as eligible for special poverty funds under ESEA. The middle class school, located in a suburban, middle class residential district, had practically a 100 per cent white population. In each school, the principal assigned one class at each of the three grade levels: kindergarten, first grade, and second grade. From each of these classes, ten children were randomly selected, providing 30 pupils from each of the two schools, or 60 subjects in all.

*Procedure*

In preparing the test instrument, a sample of seven one-syllable words for each word class was selected from the 500 most frequent words in the original Thorndike count. Care was taken so that no list contained common homonyms or minimal pairs. The nouns selected were *tree, chair, boy, leg, hat, fish, cup*; the verbs were *build, grow, drive, hit, run, swim, shoot*; and the function words were *if, down, front, or, and, on, of*. From this set the immediate memory task was constructed as a word analogue to the "digit span" memory test.

     The test consisted of eighteen items; for each item, a list of words in the same word class was read to the child following which he was to repeat the words in the list aloud. The first item consisted of a list of two function words, the second item of two nouns, and the third of two verbs. For items 4 through 6, the length of the list was increased by one for each word class, the order of the word class list was changed and the words within each list were rearranged. In this way, the subjects encountered first, lists of two words, then, lists of three words, and so on, until for items 16 through 18 they were given lists of seven words each, one list for each word class. Exactly the same format (order of lists and order of words within each list) was used for all subjects.

$E^1$ read each item, only once, at the rate of one word per second, and $S^1$ responded before the next item was read. An error was defined as any omission, addition, or incorrect order. After the child made an error in repeating a list, he was given no additional items of words in that class. Since each child made three and only three errors, hopefully there was little discouragement on the part of these young subjects. $S$'s score for any word class was the longest list of words of that class he could repeat without making an error. Each subject, therefore, was given three scores, one for each word class: the highest possible score was seven.

---

**A.** It's unlikely that there would be any subject bias with subjects that young. However, there could have been a little experimenter bias. The experimenters might have treated children from different social classes somewhat differently, and almost certainly they treated children of different ages in different ways. Both may have been unavoidable, given the nature of the variables. We will trust that Schutz and Keislar were careful to avoid saying the function words any differently from words in the other two classes.

The two nonmanipulated variables, age and social class, are obviously confounded with a larger number of other variables. Social class is certainly confounded with both school and neighborhood, since both factors were part of the operational definition. Moreover, it's almost perfectly confounded with race. Age is confounded with years in school, number of tests previously taken, amount of interaction with strangers, and many other factors of possible importance.

What of the manipulated within-subject variable, word class? There are some variable-connected confounds such as length of word, meaningfulness of word, frequency of word in speech, and frequency of word in written material. Any differences between the word classes might have been caused by any of these variables rather than by the fact that some words were nouns or verbs and others were function words. What about practice or fatigue effects? Schutz and Keislar used a sort of repeated-measures approach, although this one is somewhat unusual. Each child was tested on each word class several times, until reaching a number he or she couldn't recall without an error.

---

[1] $E$ and $S$ are abbreviations for experimenter and subject, respectively. They were used commonly until about 1974, when the style changed.

The authors changed the order of word classes each time they increased the number of words; this was an attempt to avoid confounding with practice and fatigue effects. However, they used the same order for all subjects, so their attempt may not have been completely successful. For example, if all the lists were four words long when the list of function words was unusually difficult, it would cause function words to look more difficult than the other words, even if this weren't true. The children would never get any more function words, even though they might have been able to recall more than four if the particular words chosen were easier.

As you'll see in Chapter 6, Schutz and Keislar recognized this problem and performed a second experiment in which they corrected it.

*Before-After Problems:*
*Practice and Fatigue Effects and*
*Novelty Effects. Solutions:*
*Repetition, Changing Designs*

With before-after variables practice and fatigue effects are always confounded in that the level of the dependent variable before the independent variable occurs is compared to the level after, which is necessarily later. In addition, there's one new source of confounding that I'll call the **novelty effect,** although it's sometimes called the **Hawthorne effect** after the studies in which it was first noted. In the Hawthorne studies, it was found that worker efficiency increased when the light was increased, when it was decreased, when music was added, when it was removed—the operations themselves didn't matter. The only thing that mattered was the change in routine. "Change in routine" or "novelty" was confounded with each of the independent variables.

In order to draw causal conclusions from a study using a before-after design, it's necessary to assume that any observed change in the dependent variable was *not* caused by practice or fatigue effects or novelty effects or by any of the other possible sources of confounding.

Recall the typing experiment of Yeatts and Brantley with the child with cerebral palsy. The before-after variable was what they called the "modification," consisting of giving the child a raisin every time she typed a correct word. Yeatts and Brantley would like to conclude that any changes in typing rate were caused by the treatment. In fact, there was an increase in typing rate during the modification. On the average, the child's rate prior to treatment was about 1 correct word per minute; during modification it increased to 2. Then, after the treatment stopped, it went back to about 1 again. Is it reasonable to conclude that the treatment caused the change?

The no-treatment intervals differed in several ways from the treatment intervals. Perhaps there were subject or experimenter bias effects, with the child figuring out what was expected of her or Yeatts and Brantley influencing her behavior in other ways during the treatment sessions. There were opportunities for variable-connected confounds. Not only was the child receiving raisins, but extra attention, information, and the like during the treatment. Of course, the treatment was confounded with practice and fatigue effects; the child may have become practiced and then fatigued. Moreover, she was probably receiving other medical treatments, varying in sickness, and experiencing other extraneous-variable confounds. Finally, there could have been novelty effects in that all the recording, attention, and other actions might have caused the increase in typing rate.

Any or all of those confounds might be responsible for the increase rather than the treatment itself. However, I'm inclined to assume with Yeatts and Brantley that her typing rate wouldn't have increased without the modification. Although there are several confounded variables, I'm almost willing to believe they're irrelevant. But I'm not very confident.

Experimenters use several techniques to make it less risky to draw causal conclusions from before-after variables. Probably the most obvious is to repeat the experiment on a number of different occasions. If the same pattern always occurs, it makes it less likely that the change is coincidental. A different possibility is to try to make coincidental changes appear less likely by lengthening the period of observation before the independent variable occurs. If there's no change over a long interval, it seems less likely that there will be a coincidental change just at the moment the treatment is added. We must still assume that the lack of change would continue, but it seems more reasonable.

If these measures are insufficient, the before-after variable can be changed to a between-subject variable by adding an explicit control group receiving no treatment throughout the experiment. Then all the rules for drawing conclusions about between-subject variables apply. Alternatively, the before-after variable can be changed to a within-subject variable by repeating the treatment/no treatment levels several times and observing whether there are corresponding changes in the dependent variable. In fact, that's what Yeatts and Brantley were approaching by adding the postmodification observations. In that case all the rules applying to within-subject variables are applicable.

In any case, with before-after variables, drawing causal conclusions is always a little risky. The key question is whether it appears likely that any observed changes in the dependent variable would have occurred even without the occurrence of the independent variable.

**Problems**

These problems will give you practice in evaluating designs with before-after variables.

**Q.** A doting parakeet owner, very familiar with his pet's normal patterns of behavior, brought home an exciting new bird toy and put it in the bird's cage. He observed that the parakeet became more active than usual, hopping quickly from perch to perch. He concluded that the toy caused the change. Was he correct? Why?

**A.** Since he was familiar with the bird's behavior it seems probable that something in the situation was responsible for the change in activity. The change might have been caused by something like a novelty effect or even by experimenter bias, since the experimenter presumably had some control over the bird's behavior. To evaluate the novelty effect, the owner might see how the parakeet reacts to being closely observed in the presence of other new objects. To eliminate experimenter bias, it might be sufficient to have a different person deliver the toy and observe the behavior.

**Q.** A Soviet official became concerned about the effects of microwaves on heart disease. Prior to building a new microwave installation, he had the rate of heart disease in the surrounding area measured. Then he built the station. After 10 years he measured the rate of heart disease again and found an increase from 2 cases per 100 to 18 cases per 100. He concluded that the microwaves had caused the increase. Was he correct? Why?

**A.** You may have had a little trouble saying just what the problems were here. I'd say that the worst problem was the likelihood of many confounded extraneous variables. There were many changes during that 10-year period that could have caused a change in heart disease rate. You can probably see other problems. This is an inappropriate variable to use as a before-after variable, because too little is known about the situation.

**Q.** A group of 10 neighbors were discussing local problems one evening. They were afraid that the amount of noise and vandalism in the neighborhood would increase after the new proposed youth center opened. They developed operational definitions for "noise" and "vandalism" and kept track for 6 months prior to the opening and again for 6 months after. They indeed found a noticeable increase in both variables after the opening and felt that the youth center had caused the increase. Were they correct? Why?

**A.** It's possible they were correct, but this demonstration is weak. The main problem is that the group could hardly be sure that no other changes had occurred at the same time as the youth center opening. If nothing else, the time of year would have changed, which might affect the amount of noise and vandalism. However, if the amount of noise and vandalism remained constant during the preceding 6 months and rose suddenly and remained high and constant for 6 months after the opening, they might be justified in believing that the youth center was to blame for the increase.

Now it's time to turn to your sample articles and try to discover any confounded variables and attempts to eliminate confounds. These questions should guide you. Go through the questions for *each* of your independent variables separately.

1. For each independent variable, is there any reason to believe that the experimenters allowed extraneous variables to be confounded?
2. For each independent variable, does it appear likely that it is contaminated by subject or experimenter bias?

Section A  Reading Method Sections for Confounds       133

3. For each independent variable, does it appear likely that there are any important variable-connected confounds?
4. For each between-subject variable:
   (a) How were the subjects assigned to groups?
   (b) Were the groups matched on any variables?
   (c) Does it appear that the subjects were assigned to groups in such a way that the groups might differ?
5. For each within-subject variable:
   (a) Did different subjects receive different orders of the levels of the independent variable? If so, what orders were used?
   (b) Did each subject receive each level more than once? If so, how many repetitions were there, and how were they ordered?
   (c) Does it appear that the levels were ordered in such a way that there might be confounds with practice and fatigue effects?
6. For each before-after variable:
   (a) Does it appear likely that the results were caused by practice or fatigue effects?
   (b) Does it appear likely that the results were caused by novelty effects?
   (c) Does it appear reasonable to believe that a change might have occurred in the dependent variable without the occurrence of the independent variable?

For any independent variables with any final "yes" answers, you should hesitate to draw causal conclusions. You may choose to believe that the relationship is causal, but the evidence is weak. You may draw causal conclusions for any independent variables in which your final answer to all relevant questions was "no."

**Extra Library Assignment.** Find examples of each of these six kinds of independent variable:

|  | Between Subject | Within Subject | Before-After |
| --- | --- | --- | --- |
| Causal conclusions okay |  |  |  |
| No causal conclusions |  |  |  |

Section B
Completing
Your
Experiment

Part 1  Completing Your Experimental Design

Now you must take steps to avoid possible confounds in your own experiment or to make them less damaging. While doing so, you might want to modify your operational definitions, add treatments, or otherwise change your experimental design. If you find at the end of this process that your experiment is still relatively uncomplicated, you might want to add another independent and/or dependent variable.

### Dealing with Subject Bias

Is it likely that your subjects will develop hypotheses about the purpose of your experiment and change their behavior accordingly? Few behaviors can't be affected by willful subjects, but there are many situations in which subject bias is unlikely to be a serious problem. In most between-subject designs it's difficult for the subjects to form any hypotheses about the experiment unless they know the levels of the independent variable of the other participants. Thus, if you have a between-subject variable, you can usually guard against subject bias just by ensuring that the subjects don't know what happens to the others until after the experiment is completed.

Within-subject variables give more opportunity for forming hypotheses and acting on them, but not all variables are troublesome. In situations requiring maximal performance from individuals under various conditions, most middle-class subjects will try to do the best they can under all conditions even if they recognize that they're expected to perform worse in some situations than others. If it doesn't seem likely that you'll have problems from subject bias, then you needn't do anything about it.

If subject bias does seem likely, there are several protective measures you can take. One is to change to a between-subject variable if that will do the job. A second is to add a placebo treatment which the subjects can't tell from the real thing. If the subjects can't tell what treatment is occurring, they can't manipulate their behavior in a systematic way.

A third approach is to influence directly the hypothesis likely to be formed by the subjects. You can tell all the subjects the true purpose of the experiment, or you can deceive them and tell them a false purpose. Obviously, the truth is ethically preferable. In either case, subject bias may still influence the results, but at least you have a better idea of the direction of the influence. You can also go one step further and include subject bias as a variable in your experiment, giving different subjects different hypotheses and observing the effects.

★ What, if anything, will you do about possible subject bias effects in your experiment?

### Dealing with Experimenter Bias

Is it likely that the experimenter will consciously or unconsciously influence the behavior of the subjects, either in agreement or disagreement with hypotheses about the outcome? Some situations are relatively unlikely to be unaffected by experimenter bias. In a between-subject design in which subjects from all groups perform during one session, the experimenter would have to work very hard to treat subjects from different groups in different ways. The only influence likely to occur might arise during scoring or recording the results, and double-checking could reduce that possibility. In a within-subject design in which the experimenter doesn't interact with the subject during different treatments, it again would be difficult to exert much influence. Thus, if you set up your experiment in either of these ways, you minimize the opportunities of the experimenter to bias the performance of the subjects.

If neither of these is possible, a different approach is to use experimenters who are "blind" to the level of the independent variable that occurs to record the dependent variable. This usually means having one experimenter measure or manipulate the independent variables and a different one measure the dependent variable.

If none of these are possible, the only remaining approach is to limit the conclusions to be drawn from the experiment to take account of the possibility of experimenter bias.

★ What, if anything, will you do about possible experimenter bias effects in your experiment?

## Dealing with Variable-Connected Confounds

### Nonmanipulated Independent Variables

If you have any nonmanipulated independent variables, answer these questions:

★ What important variables are confounded with that variable?
★ Which of these might be related to the dependent variable?
★ Do you want to try to eliminate them as confounded variables?

If you answered "yes" to the last question, the only approach you can take is to try to eliminate the variables by holding them constant or by matching. Procedures for doing this are discussed in the discussions of between-subject and within-subject variables.

If you answered "no" to the last question, you must modify the conclusions to be drawn from your study accordingly.

### Manipulated Independent Variables

If you have manipulated independent variables, answer these questions.

★ As you manipulate your variable, what other variables are you manipulating at the same time?
★ Are any of these other variables likely to cause changes in your dependent variable?

If you answered "no" to the last question, there's no problem. If you answered "yes," you probably should try to add one or more control treatments to your experiment in which you manipulate the other variables, but don't manipulate the real independent variable.

In drug experiments, the control treatments are relatively straightforward placebos. The placebos must be chosen carefully to look and perhaps taste like the real drug and to have no effects of their own. Of course, they must be administered in such a way that the subjects don't know what drug they're receiving.

In physiological operations, the control subjects often are given sham operations in which the experimenter anesthetizes, cuts, stitches, and allows the subjects to recover, but removes nothing. Sometimes they even remove an equal amount of tissue but from an area unlikely to affect the behavior of interest. Such different control treatments control for different variables.

In other areas, the problems also can be quite complicated. One problem area has to do with time manipulations. Suppose you wanted to manipulate "amount of study time allowed" in a between-subject experiment. One group is to be given 1 hour to study some material, and another group is to be given 2 hours. If you set up the experiment so that everyone starts studying at the same time, say 9 AM, and test them immediately after they finish, then time of testing is confounded with study time. If you set it up so that time of testing is the same, say 10 AM, then time of starting is confounded with study time. If you try to equate both, then one group has to wait an hour before being tested. To control for most of the possibilities we'd need quite a number of groups:

| Time Start | Time Stop | Time Test |
|---|---|---|
| 9:00 | 10:00 | 10:00 |
| 8:00 | 10:00 | 10:00 |
| 8:00 | 9:00 | 10:00 |
| 8:00 | 9:00 | 9:00 |
| 9:00 | 10:00 | 11:00 |
| 10:00 | 11:00 | 11:00 |

Depending on how the results came out, the findings might be easy to interpret or might be discouragingly difficult.

Another problem area has to do with word variables. Word frequency, that is, the frequency of occurrence of each word in speech or writing, is a variable of interest to many researchers. However, it turns out that word frequency is confounded with many other variables, including word length, pronounceability, spelling patterns, number of syllables, and so forth. We'd like to add control treatments that vary in all other factors but not in frequency. However, that can't be done; our language isn't structured that way. We can hold some variables constant or match on them, as with nonmanipulated variables, but we can't control all of them.

If you have a variable entangled with many other variables, such as time or word frequency, the easiest approach is to allow the confound to occur and recognize that your findings may have more than one interpretation.

Before dealing with variable-connected confounds in your own study, try some practice problems.

**Problems**

These problems should help you learn to control variable-connected confounds when you're manipulating an independent variable.

**Q.** Suppose an elementary school principal wanted to evaluate the interest level of a new group of readers. She selected a group of children in each grade and gave each a copy of the new reader appropriate for their grade. The children were asked to look through the new readers and rate how interesting they found them and how interesting they found their regular readers. What's wrong with this study? How should it be redesigned?

**A.** The new readers differed from the old ones not only in content, but also in that they were new and unfamiliar. They were probably also cleaner and more attractive. They might receive higher ratings for these reasons rather than because of their content. In order to perform the proper study, the teacher should use children who are unfamiliar with both readers and give them new copies of both to rate.

**Q.** Suppose a researcher for a telephone company wanted to determine how rate of presentation of numbers affected people's ability to remember them. He chose four groups of people. Each group heard 100 phone numbers over a telephone. After hearing each one, they had to say "thank you," break the connection, and then dial the number. Then they pressed another button and heard the next number. Errors in dialing the numbers were recorded. Each group heard their numbers at a different rate: 1, 2, 2.5, or 3 digits per second. What's wrong with this study? How should it be redesigned?

**A.** The groups differed not only in rate, but also in a number of other time variables: total time in the experiment, length of time between phone numbers, proportion of total time spent listening, and others. The number of errors might have been affected by these variables.

Several changes could be made. One possibility would be to present only one number to each subject, but then it would require an enormous number of subjects to collect many errors. Another approach would be to add control groups. Groups could be added which all take the same total time to perform the experiment but differ in rate of presentation and the length of time between phone numbers. Other groups could be added that all have the same amount of time between the start of one phone number and the start of the next but differ in rate of presentation and length of time after dialing with nothing to do. Still other groups could be added that don't differ in rate of presentation but do in total time in the experiment and length of time between phone numbers. By comparing all the groups, the effect of rate might be discoverable if the results are orderly.

**Q.** Suppose a comedienne was unsure about whether to add some sentimental material to her nightclub routine. Doing her regular act, she kept track of the length of applause after each of 10 performances and found it to be quite constant. Then she added the new material for one show and kept track of the length of applause again. What's wrong with this study? How should it be redesigned?

**A.** Adding the new material would increase the length of the act, even assuming that the performer was able to prevent the change from showing up in her general attitude and performance of the rest of the act. It would be difficult to redesign the study in such a way as to completely eliminate these problems, but repeating the new and old acts on several different occasions would help some. She could equate the length of the acts by leaving out different portions of the old material on each repetition or by adding some other new "control material" as filler. However, that would add other problems.

**Q.** Suppose a psychologist interested in verbal memory processes wanted to know which are easier to recall, nouns naming very easily pictured objects (such as *chair* or *pencil*) or nouns naming objects difficult to picture (such as *history* or *beauty*). She obtained two lists of words, one of each type, found two groups of people, and gave one list to each group. Would she be able to conclude that any differences between the lists were caused by the different imagery values of the words?
**A.** The words almost certainly differ in a number of other ways, including familiarity, length, number of syllables, and nature of spelling. Therefore, she wouldn't be able to draw that conclusion. She might be able to match the lists on some variables but certainly wouldn't be able to remove all the variable-connected confounds.

**Q.** Suppose a cigarette manufacturer wanted to compare the taste of his cigarette to the taste of other brands. He selected 50 smokers and gave each one a regular carton of his cigarettes and regular cartons of four competing brands. The people were told to smoke the cigarettes in a certain order, at certain times of day, and so on so that all the cigarettes were equated on these factors. After each cigarette, the smokers were to rate how tasty the cigarette seemed. What's wrong with this study? How should it be redesigned?
**A.** The cigarettes differed not only in taste, but also in brand, packaging, filter style, and the like. To perform the study properly, the manufacturer should remove all the brand identifications and make the cigarettes as nearly as possible equal in appearance.

**Q.** Suppose a physiologist wanted to investigate the effects of removing a certain portion of the brain of a frog on eye-tongue coordination. He obtained two groups of frogs and measured their coordination. Then he operated on both groups: anesthesia, hole cut in skull, and hole repaired. For one group he also cut out the appropriate portion of the brain, which happened to be near the center. Would he be able to conclude that any differences in eye-tongue coordination after the operation that weren't there prior to the operation were caused by the removal of that specific portion of the brain?
**A.** Not quite. He tried to add the appropriate control group, but he forgot one factor. In order to get to the relevant portion of the brain, he had to cut through other tissue. That might have caused a change. The control group should have had equivalent cuts made but no part of the brain removed.

★ What, if anything, will you do about variable-connected confounds in your experiment?

### Between-Subject Variables Only: Avoiding Extra Differences between Groups

Regardless of whether your between-subject independent variables are manipulated, you can draw clearer conclusions about their relationships with other variables if you minimize the differences between groups. It's important to control both subject variables and extraneous variables. There are a number of ways to do that.

*Holding Constant Possible Confounding Variables*

By choosing to use only certain subjects, you're already holding constant lots of subject variables. Thus the question becomes "Do you want to hold additional subject variables constant?" You may want to limit your subjects only to those who can reasonably be expected to perform some task. For example, if your experiment required color perception, it would be reasonable for you to exclude color-blind individuals. You also might want to hold constant some variable that you think might be confounded with your independent variable. For example, in an experiment using college major as an independent variable you might want to restrict your subjects to seniors, since class membership might be confounded with college major. Of course, any restrictions placed on the subjects restrict the size of the group to which you can generalize your findings.

★ Will you restrict the kinds of subjects in your experiment by holding additional subject variables constant? If so, what variables?

You also will hold constant many extraneous variables such as place of experiment and attitude of experimenter. One technique for holding variables constant is to have subjects at different levels of the independent variable all perform at once, in mixed groups. Thus the experimenter acts the same to everyone, and variables such as day of the week and room temperature are all held constant across all subjects. This technique also allows the experimenter to complete the experiment more quickly than if the subjects must be run individually. However, it often requires that the whole experiment be done with paper and pencil. If people receive different treatments, they usually must either be written or shown in pictures that are handed out. Otherwise, people hear each other's treatments, form hypotheses about the experiment, and influence their behavior accordingly. The dependent measure also must usually be silent.

Because of the convenience of running subjects in groups, experimenters often want to use groups even when they can't run mixed groups without encountering subject bias. Therefore, they run unmixed groups. However, you should recognize that running all the subjects at one level of the independent variable in one unmixed group virtually guarantees the addition of confounded variables. In an experiment varying, say, teaching method, if all the subjects receiving one teaching method were run in one group and at one time of day by an experimenter/teacher in one mood, and all the subjects receiving a different method encountered a different situation, then all these variables would be confounded. Of course, the problem is increased if the subjects assign themselves to the groups, since subject variables also become confounded. For these reasons it's usually unwise to have all the subjects at each level perform as a group at one time.

★ Would it be possible for you to perform your experiment with the subjects in mixed groups? If so, will you do it?

*Matching Groups on
Possible Confounding Variables*

A different way to prevent variables from being confounded is to match on that variable, that is, to make sure that each level of the independent variable has the same proportion of each level of the matched variable. For example, if you were comparing groups of male and female students, you could match the groups on grade-point average by selecting your subjects in such a way that each group had 20 percent of its members with 3.40–4.00 averages, 40 percent had 2.50–3.39 averages, and 40 percent had 1.00–2.49 averages. Then you'd be sure any differences that occurred between the males and females didn't have to do with grade-point average.

However, you should recognize that by matching on subject variables you're probably restricting the groups to which you can generalize the results. In this example, females usually have higher grade-point averages than males. By matching, the females with low averages actually will be somewhat overrepresented while the males with high averages will be somewhat underrepresented. The problem may not be serious in this case, but imagine the restrictions raised if an experimenter tried to compare Americans with Russians and attempted to match the groups on political beliefs!

★ Will you match the groups at different levels of your independent variable on some subject variable? If so, what variable?

It's also possible to match your groups on extraneous variables such as time of day or day of week. Again, to match you must make sure that each level of the independent variable has the same proportion of each level of the matched variable. For example, you might want to ensure that 60 percent of each level of the independent variable was run in the morning and 40 percent in the afternoon.

★ Will you match the groups at different levels of your independent variable on some extraneous variables? If so, what variables?

If you've decided to match your groups on a subject variable or another extraneous variable, you must decide how many levels to distinguish, how to operationally define each level, and what proportion of each level to include in the experiment. If you want to look at the relationship of the matched variable with the dependent variables, you can certainly do so, although you don't have to. If you choose to look at the relationship of a matched variable with the dependent variables, you'll actually be adding a new nonmanipulated independent variable to your experiment, combined factorially with the original variable.

★ If you're matching on some variable, do you want to treat it as an independent variable? That is, do you want to look at its relationship with the dependent variables?

*Assigning Subjects
Randomly to Treatment Groups*

If your independent variables are manipulated, you should avoid confounds with all other uncontrolled variables by assigning your subjects randomly to their treatment groups. By using random assignment you can't be *sure* that all other variables are distributed equally across the levels of the independent variable, but it's the best you can do. You can expect to avoid confounds with all uncontrolled subject variables and with other uncontrolled extraneous variables such as type of weather or mood of experimenter.

In almost all circumstances the easiest technique to use for assigning subjects randomly to treatment groups is *block randomization*. Suppose your independent variable has five levels. To assign subjects through block randomization, the first five subjects are assigned to these five levels in random order, the next five in a different random order, and so on. To obtain the random orders you could do one of two things. You could write the five levels on slips of paper, mix them up, and write down the order; each new mixture would be a new random block. Or you could look at Table C.2 in Appendix C, which contains random blocks for up to 16 levels. Start anywhere in the table, go through one sequence, and find the numbers 1 through 5. That order is your first random block. Then go to the next sequence and do the same thing.

*A Special Problem:
Subject Dropout*

If your between-subject experiment requires subjects to return for two or more sessions, there's a special problem: not all the subjects will return. If the subjects were assigned randomly to groups, the *loss* of subjects from the groups is probably not random. The remaining groups are to some degree self-selected. Unfortunately there's no cure for this problem.

Suppose four groups of randomly assigned subjects came to one session and each group learned a list of words under a different set of instructions. There were 25 subjects in each group, and all were supposed to return in 3 days to recall the lists; however, only 20 in each group return. There's a temptation to believe it doesn't matter, but that's not so. Suppose the people didn't come back specifically because of the instructions they received. The missing people might seriously change the average performance of their groups. Filling the empty spaces with new subjects won't help. They're just more people of the sort who *do* return.

Subject dropout is a problem that must be accepted. If your subject dropout is small, you needn't worry, but if it's large, your results will be difficult to interpret.

### Within-Subject Variables Only: Avoiding Extra Differences between Levels

Whether or not your variable is manipulated, you can draw clearer conclusions about its relationship with the dependent variables if you minimize coincidental differences between levels.

*Holding Constant Possible Confounding Variables*

By choosing to observe under limited conditions you can hold possible variables constant. For example, you may be performing an experiment in which you manipulate your greeting to your regular bus driver and observe his reactions. You may decide to limit your observations to days when the bus is uncrowded, thereby eliminating the possibility that "crowdedness" might be confounded with your greeting. The same approach can be used with nonmanipulated variables. For example, you might investigate the relationship between day of the week and number of questions asked in English class as a within-subject variable with your classmates as subjects. You could limit your observations to days when a test was not occurring within two class meetings, thereby preventing test closeness from being confounded with day of the week.

Of course, any limitation of this sort restricts generalization from your experimental findings.

★ What variables will you hold constant in your experiment?

*Matching on Possible Confounded Variables*

Rather than holding variables constant, you may choose to match on them, as with between-subject variables. With the bus driver example, you could match on crowdedness, making sure that you use equal proportions of crowded and uncrowded days with each type of greeting. With the English class example you could try to match on test closeness by observing an equal proportion of "test close" and "test far" days on each day of the week, if that's possible.

If you choose to match on some variable, you must decide how many levels to distinguish, how to operationally define each level, and what proportion of each level should be included. As with between-subject variables, matching results in the addition of a new independent variable. If you're interested, you can observe its relationship with the dependent variables.

★ Will you match the levels of your independent variable on some other variable? If so, what variable?

★ Will you look at its relationship with the dependent variables?

*Ordering the Treatment Levels*

If your variable is manipulated you should order your treatment levels in such a way as to distribute practice and fatigue effects equally across all the levels of the independent variable. One general approach is to use order as a between-subject independent variable, that is, to give various orders to different subjects. The other approach is the repeated-measures approach, which is to measure each subject a number of times at each level of the independent variable, varying the order of the levels on the repetitions. The two approaches also can be combined.

The repeated-measures approach is usually more efficient but isn't always possible. It can be used only when all the levels of the independent variable can be presented more than once to the same subject. If each level takes a lot of time or effort, repetitions may not be reasonable. For example, if each level required the subject to write a short story, it probably would not be reasonable to ask the subject to participate in each level more than once. Also, the levels of many independent variables can't sensibly be repeated at all. For example, if an experiment were designed to see how segments of four different poems affected mood ratings, it wouldn't make sense to present the segments more than once. On the other hand, in an experiment designed to see how amount of violence in movie segments affected mood ratings, repeated measures could be used if several different segments could be found for each level of violence.

★ If you wanted, could you reasonably use a repeated-measures approach with your variable?

If not, you should use order as a between-subject variable. If so, you may choose to use either approach or a combination of them. Regardless of which you choose you'll be adding a manipulated independent variable that could be thought of as order of levels or position of levels.

Before deciding which approach to use you should read through the procedure for each one.

*Using Order As a
Between-Subject Independent Variable*

There are two commonly used ways to distribute practice and fatigue effects over different levels of the independent variable by using order as a between-subject independent variable. One is to use some number of counterbalanced orders, and the other is to use a large number of random orders.

Counterbalanced orders are chosen so that each level of the independent variable appears equally often at each position. For example, an experiment might have a four-level independent variable with the levels called A, B, C, and D. One fourth of the subjects could receive one of four counterbalanced orders: BACD, DCAB, CDBA, and ABDC. In counterbalancing, each level appears exactly once in each position.

The other technique for distributing practice and fatigue effects is to use a large number of random orders, perhaps a different order for each subject. You can develop the random orders either by mixing small slips of paper or by using a table such as Table C.2, as explained in the discussion on assigning subjects randomly to treatment groups.

How can you choose between counterbalancing and many random orders? Counterbalancing gives more certain control over practice and fatigue effects. If you use relatively few subjects, it's probably a safer choice. Many random orders allows for a larger number of orders and makes it less likely that one "freak" order will distort the results of the experiment.

★ If you're using order as a between-subject independent variable, will you use counterbalancing or many random orders?

*Using the
Repeated-Measures Approach*

How many times will you repeat each level? The more repetitions, the more stable the result. However, you must consider the total time and effort required of each subject. If each repetition is very brief, such as pressing a button when a light appears, you may be able to get each subject to make thousands of responses. If you had 10 levels, you could easily have 100 repetitions of each level. However, if each repetition requires substantial time or effort, you may be able to have only 2 or 3 repetitions.

★ If you're using the repeated-measures approach, how many repetitions of each level will you have? How many total measures per subject does that make (Number of repetitions × Number of levels)?

Using the repeated-measures approach there are three ways to distribute practice and fatigue effects over each level of the independent variable. Again, suppose there were an independent variable with four levels, A, B, C, and D. Suppose further that the experimenter decided to repeat each level four times. Some possible sequences generated by the three techniques are shown below:

Complete randomization: DAAABDDBACDBBCCC
Block randomization: DABCDBACBDACABDC
Counterbalancing: BCADDACBBCADDACB

As you can see, complete randomization means that all 16 measures are completely randomized; block randomization means that each block of the 4 levels is randomized, but each block contains exactly 1 repetition of each level; and counterbalancing means that the second block of the 4 levels is a reversal of the first block and so on.

The advantage of complete randomization is that with large numbers of repetitions, the sequence is quite unpredictable. With counterbalancing the subjects may recognize the patterned nature of the events and respond to the pattern rather than to the individual events. Block randomization is intermediate, in that subjects may recognize the type of pattern, but it's less predictable than counterbalancing.

The advantage of counterbalancing is that it best distributes practice and fatigue effects. It's especially desirable when the number of repetitions is small. Block randomization does a somewhat less complete job. As you can see in the sample sequence, it happened that level C fell last in each block. Complete randomization is the worst of all in this respect. With only four repetitions, see in the sample sequence how badly the As and Cs are distributed.

Thus, if you have a variable in which it's likely that the subjects will recognize any sequences and you plan to have a large number of repetitions of each level (say 20), you should prefer complete randomization. If you plan only a small number of repetitions (say 4), and the patterns aren't likely to be noticed, you should prefer counterbalancing. Block randomization is intermediate.

★ If you're using the repeated-measures approach, will you use complete randomization, block randomization, or counterbalancing?

Regardless of your choice, you also may want to use order as a between-subject variable. For example, you could use a different complete randomization for each or have four different counterbalanced orders for different groups of subjects.

★ If you're using the repeated-measures approach, will you also use order as a between-subject variable?

If you haven't yet decided whether to use the repeated-measures approach or to use only order as a between-subject variable, go back now to the beginning of this

# Section B  Completing Your Experiment

discussion on within-subject variables and make your choice. Then answer all the questions relevant to your choice.

**Before-After Variables Only:
Making Causal Conclusions
Appear More Reasonable**

With before-after variables the general design is to measure the dependent variables for some "baseline" interval and then measure it again after the occurrence of the independent variable. A long baseline interval prior to the occurrence of the independent variable makes it appear more likely that the dependent variable changes only when the independent variable occurs. After the independent variable occurs, a long observation period also can make it appear more reasonable to believe that the dependent variable was actually influenced by the independent variable. Of course, the longer you observe, the more work it takes to perform the experiment.

★ For how long will you collect baseline data prior to the occurrence of the independent variable?

★ For how long will you collect data after the occurrence of the independent variable?

Another step you can take to make your demonstration more convincing is to add and remove your independent variable several times and observe corresponding changes in the dependent variable. Of course, that's not possible in all situations. It's very much like a within-subject variable using repeated measures, except that time effects aren't carefully distributed.

★ Can you repeat the application of your independent variable? If so, you probably should do so. How many times will you repeat the application of your independent variable?

If any other variables change at the same time as the independent variable, any causal conclusions will be weakened. With variables manipulated in the laboratory it's common to try to hold constant as many as possible while manipulating only the independent one. In the field, the moment for manipulating the independent variable should be one when other changes are unlikely.

★ If you're manipulating your before-after variable, can you be reasonably certain that no other changes will occur at the same moment you make your manipulation?

If not, you probably should change to a between-subject variable by adding a group of randomly designated "no treatment" subjects. If you do that you can draw clear-cut causal conclusions by following the recommendations applicable to all manipulated between-subject variables.

The last concern you have with before-after variables is novelty effects.

★ Can you be reasonably confident that your dependent variable won't change just because something unusual is happening?

If not, then it would be good for you to add some explicit control treatments to deal with the effects. The treatments should include other novel events that will cause the novelty effect to occur but will not cause other specific effects. You can determine the effect of your independent variable by comparing it to that of the novelty effect alone. By adding the control treatments, you'll probably change your design to either a between-subject or within-subject one.

★ What, if any, control treatments will you add to your experiment to control for other variables or for the novelty effect?

With any before-after variable the best single way of making the demonstration of causality appear convincing is to repeat the before-after observations a large number of times under widely varying circumstances, doing careful baseline and post-event observations each time. If the same relationship between the variables appears in many situations, the argument that the relationship is causal becomes more persuasive.

## Selecting the Number of Subjects Required for Your Experiment

As I mentioned in Chapter 2, there are statistical techniques beyond the scope of this book to assist you in making the decision about how many subjects to use. The following suggestions are some informal rules of thumb arising from common applications of those techniques.

For an experiment using *one between-subject variable* you'll need about 20 subjects for each level of the independent variable. For animal studies fewer subjects are sometimes needed when it's reasonable to believe that there will be little difference between individuals.

For *one within-subject variable with order as a between-subject variable* you'll need about 25 subjects. You should decide how many orders you'll use and then plan on assigning an equal number of subjects to each order. For example, if you're using four orders, you could plan to assign 6 or 7 subjects to each order. If you plan to use 30 random orders, you could assign just one subject to each order.

For *one within-subject variable with repeated measures* the number of subjects required decreases as the number of repetitions of each level increases. For only 2 or 3 repetitions you should have about 20 subjects. For large numbers of repetitions, like 20, you can use as few as 4 or 5 subjects in some areas—just enough to show that the results don't apply to only one person.

For *before-after designs* the number of subjects required decreases as your amount of control over the situation increases. If you're not certain that you can control most variables in the situation, it would be good to have 20 or so subjects. If you can control the situation very well, 2 or 3 subjects is enough if you think this is a large enough sample to allow your results to be generalized to a larger group.

When you combine *two between-subject variables* you must have a separate group of subjects for each combination of the levels of the two independent variables. Each subject then is measured on each dependent variable. For example, if three ethnic backgrounds were combined with two sexes, you'd need six groups of subjects, one for each ethnic-sex combination. In general, if one variable has A levels and the other has B, A × B groups are needed. The more levels you have, the fewer subjects needed in each group, but there should be no fewer than 10 subjects per group in any experiment. More would be better. Thus, for a 3 × 4 design, you'd need at least 120 subjects; 240 would be better.

When you combine *one between-subject and one within-subject variable* every subject must participate in all levels of the within-subject variable. However, there are as many groups of subjects as there are levels of the between-subject variable. To look at the differences between the groups, you should have about 20 subjects in each group. For example, if you wanted to compare children in three different age groups in their performance on four different problems, you'd need about 60 children. Each child would solve all four problem types. The order of problem types could be another between-subject variable or varied by repeated measures. In the first case, different children in each age group would receive different orders of problem types. With repeated measures, each child would receive several problems of each type, in varied orders.

When you combine *one between-subject and one before-after variable*, again, every subject must participate in all levels of the before-after variable, but there are different groups of subjects corresponding to the levels of the between-subject variable. For example, a bar-pressing experiment could compare rats in two different "experience" groups in their responses to the before-after variable of hunger. You could obtain or create a group of "experienced" rats and a group of "inexperienced" rats, collect baseline data on each group, make all the rats hungry, and measure them again. Then perhaps you could feed them and measure their bar-pressing rate once more. As with all between-subject variables, you should have about 20 subjects per level, or perhaps fewer with animal subjects.

When you combine *two within-subject variables* every subject must participate in all levels of both independent variables in all possible combinations. For example, a memory experiment could vary both time of day and list length as within-subject variables. If there were three levels of time of day and four list lengths, each subject would have to memorize at least 12 different lists, one of each length at each time of day. The order could be varied between subject or each of the 12 list-time combinations could be repeated with different lists. The number of subjects would be the same as for one within-subject variable, but the effort required for each subject would increase.

When you combine *one within-subject variable and one before-after variable* every subject must participate in all levels of both independent variables in all possible combinations. For example, some self-destructive psychotics could be given a treatment consisting of presenting them with a candy or cigarette every time they refrained from self-destructive behavior for 10 consecutive minutes. This could be combined with a within-subject variable of sometimes working in the doctor's office and sometimes in the patients' ward. Baseline data on self-destructive behavior would be collected in both situations; then the treatment would be added and some new measurements taken in both situations. To counteract practice and fatigue effects, the experiment might be designed so that one day the work started in the office and then moved to the ward and the next day the order was reversed. As with two within-subject variables, the number of subjects required doesn't increase in this combination, but there's an increase in the work per subject.

★ How many subjects does your experiment require?

### Summarizing Your Experimental Design

Your experimental design is now (finally!) complete. To summarize it for the benefit of others you should describe the following things:

**1.** Tell who your subjects are and how many there will be.

**2.** Describe the basic experimental situation: where the experiment is to be conducted, under what circumstances, and so on.

**3.** Name your independent variables.

**4.** Tell how many levels each independent variable has and either name or briefly describe the levels, perhaps briefly describing the operational definitions.

**5.** Tell whether each independent variable is varied between subject, within subject, or before-after. For between-subject variables tell how the subjects were assigned to each level. For within-subject variables describe any steps taken to avoid confounds with practice and fatigue effects.

**6.** If your experiment had more than one independent variable, tell how they were combined.

**7.** Name your dependent variables and perhaps briefly describe the operational definitions.

## Part 2  Performing Your Experiment

In performing the experiment you should have two important objectives. One is to minimize the discomfort of your experimental subjects. The second is to collect data that can be interpreted as clearly as possible; this implies that confounded variables should be minimized.

**Prepare for the Physical and Mental Comfort of the Subjects**

For animal subjects conditions during and outside the experimental situation should be as humane and comfortable as possible. Guidelines are available in *Rules for animals,* available from the American Psychological Association. For human subjects in the lab there should be comfortable places to wait and safe places for coats and parcels. Subjects should be met promptly and treated in a friendly, supportive manner throughout the experiment. If they're supposed to receive credit or pay for participation, it should be given promptly. Unnecessary discomforts such as bright lights or drafts should be eliminated. In the field the experiment should inconvenience the subjects as little as possible.

**Prepare for Obtaining Informed Consent**

Prior to participating in the experiment the subjects should be told as much as possible about what they'll be required to do and should be given a realistic opportunity to decline. Any unpleasant aspects of the study should be made clear. If they do decline to participate for any reason, you should be prepared to let them go without pressure. Special problems arise when the subjects are children, mental patients, prisoners, or the like. Guidelines are available in *Ethical principles in the conduct of research with human participants,* also available from the American Psychological Association.

The following constitutes a *good request for informed consent:* This is an experiment dealing with the relationship between some personal characteristics and attitudes. First I'll ask you to fill out a questionnaire about your attitudes toward organized religion and politics. Then I'll ask you some questions about your background, including questions about your education and your parents' education. Will you participate in this experiment?

Two examples of *bad requests* follow: (1) I'm going to ask you some questions about yourself in this experiment. Will you participate? (2) Your teacher told me that I could use your class in an experiment and that he expects you to participate. Will you?

### Prepare for Completely Explaining the Experiment As Soon As Possible

This may be in written or verbal form but should be in terms the subject can understand. It may include information about how the individual performed relative to others, but all such data should be confidential. If deception was used, great care should be taken to explain the deception and the necessity for its use, unless it's felt that the explanation would cause more harm than the deception.

After explaining the purpose of the experiment you may want to ask the subject not to discuss it with any other possible subjects.

*Good explanation of experiment:* You participated in a situation in which you heard a speech about brotherhood and were then asked questions about your feelings toward racism. Other people heard a different speech, which was pretty neutral in content, being about forestry. They answered the same questions you did. I wanted to see whether people who heard the brotherhood speech answered differently. Do you have any questions?

*Bad explanation:* This was a between-subject design with two levels of the independent variable, brotherhood and neutral. The dependent variable was feelings toward racism. I wanted to see if there was a causal relationship between the two variables.

*Bad explanation:* Any questions?

### Prepare Any Needed Instructions

These may be written or verbal. They should be as simple as possible and make very clear what kind of performance is expected. Examples should be included, and there should be ample opportunity for the subjects to ask questions. Try to anticipate common questions and include their answers in the instructions. This ensures that the instructions heard by different subjects will be more nearly equal.

*Good instructions:* During this experiment I'll ask you to learn 10 lists of 12 words each. The lists will be presented in this machine (show it), 1 word every 2 seconds, like this (demonstrate it). Your task will be to study each word as it appears. After you've seen all 12 words from 1 list, I'll ask you to say out loud as many of them as you can, in any order. You'll have 3 chances to learn each list. The words on the list will be in a different order each time. Then we'll go on to the next list. Do you understand? (Answer any questions.) Let's do a sample list of just 3 words so you can see how it works. Ready? Go. (Sample.) Any questions?

## Section B  Completing Your Experiment

*Bad instructions:* This machine will show you 10 lists of 12 words 3 times each. Your task will be to learn each list. Do you understand?

### Prepare for Data Collection

You may be able to arrange to have some data collection done by machines. If so, opportunities for experimenter bias are clearly reduced and you may save yourself a lot of work. Failing automatic data collection, you should prepare data-recording sheets thoughtfully to make it very easy to record the events of interest without effort or confusion. If possible, it's helpful to have two people recording the data so that errors can be detected. See Figure 4.1.

---

**Figure 4.1**
Sample data sheet
(hypothetical data)

Experiment name:  Instruction emphasis, spring of 1979

Subject number:  _16_

Type of instructions:  Standard   Emphasize speed  (Emphasize accuracy)

Order of lists:  ABCD  (BADC)  CDAB  DCBA

| List A | | List B | | List C | | List D | |
|---|---|---|---|---|---|---|---|
| Correct answer | Subject's answer | Correct answer | Subject's answer | Correct answer | Subject's answer | Correct answer | Subject's answer |
| up   | + | down | +  | down | + | up   | + |
| up   | + | up   | 0  | down | + | down | + |
| down | + | up   | 0  | down | + | up   | + |
| down | 0 | down | +  | up   | + | down | + |
| up   | + | down | +  | up   | + | down | + |
| down | + | up   | +  | up   | + | up   | + |

Number of errors:  _1_    _2_    _0_    _0_

Time to finish:  _3.24 sec._   _3.21 sec._   _3.16 sec._   _3.08 sec._

---

### Prepare Specialized Materials

For many experiments this process can be very time consuming. Machines to present materials may have to be located or built. Tests to measure skills, attitudes, traits, or other qualities may have to be found or developed. List of problems, words, pictures, stories, and the like may have to be developed, with appropriate techniques of presentation. Each area of research requires specialized knowledge of materials; whole chapters or even books could be written on each area. You should learn about them by reading articles or textbooks in the area or by asking knowledgeable people.

### Prepare the Experimental Situation Carefully to Avoid Unnecessary Confounds

You should arrange the situation to avoid interruptions insofar as possible, putting up "Do not disturb" signs or whatever seems appropriate. If you have groups of subjects, some will always be late. Put a sign on the door telling them what to do. In the field you have less control, but you can try to choose your situation carefully to avoid problems.

If something big does happen—for example, your machine breaks down, the lights go out, or a giant lawnmower keeps anyone from hearing—you should plan to replace those subjects. If something small happens—for example, a student walks in and out again during one session—you should report it as a minor problem in your experiment and keep in mind that it may have affected your results.

### Practice the Experiment

Go through all the steps without real subjects. Greet your subjects, get informed consent, give instructions, run your machines, collect responses, and give post-experiment feedback. Go through all the procedures until you're sure you can do it smoothly so that your performance isn't confounded with any of the variables and you don't embarrass yourself. If you can, have some friends serve as sample subjects.

### Now Do It!

Get your subjects. Use the same procedures throughout the experiment; don't change midway. Record the data as carefully as you can. Treat the subjects nicely.

This is a summary of the steps you took in completing your experimental design, preparing for your experiment, and performing it:

**1.** Deal with any possible subject or experimenter bias and with any variable-connected confounds as well as you can.
**2.** With between-subject variables, take steps to avoid extra differences between groups. With within-subject variables, take steps to avoid extra differences be-

tween levels. With before-after variables, take steps to make causal conclusions appear as reasonable as possible.

**3.** Decide how many subjects you'll need in your experiment.

**4.** Summarize your experimental design.

**5.** Prepare for the physical and mental comfort of your subjects, for obtaining informed consent, and for explaining the experiment afterward.

**6.** Prepare instructions, data sheets, any specialized materials, and the experimental situation.

**7.** Practice the experiment completely, and then perform it.

**Chapter Five
What Happens?**

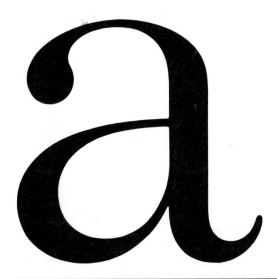

**Section A
Reading Results
Sections**

Logically enough, the function of the results section is to report the results of the experiment. Presumably, anyone who goes through the procedures described in the method section of an experiment will obtain the same results as those described in the results section. Taken together, the method and results sections describe the scientific facts contributed by the experiment. The introduction and discussion sections are historical and theoretical interpretations of those facts.

The results of an experiment consist of the observed relationship between the independent and dependent variables. That information can be communicated in the text of the results section, or it may be presented in tables or figures. Regardless of how the results are presented it's necessary for the reader to pay careful attention to the actual data and to interpret it meaningfully. The authors may not emphasize the aspects of most interest to the reader or may even make mistakes.

There are at least four general statistical approaches to presenting the results of an experiment, and each one requires somewhat different skills for complete understanding. I'll call these (1) the minimal-statistics approach, (2) the correlational approach, (3) the differences-between-frequencies approach, and (4) the differences-between-means approach. You'll read articles using each approach. In many published articles several approaches are used to emphasize different aspects of the results. There also are other somewhat less common approaches, but understanding the four common approaches will enable you to understand most papers.

### The Minimal-Statistics Approach

The first results section you'll read uses the minimal-statistics approach. It presents all the data from the experiment in a format that makes it relatively easy to see the relationships between the independent and dependent variables. The authors point out some of the relationships they consider most interesting, and the reader is free to study the results and see if anything else of interest is present.

The following is the results section of the Yeatts and Brantley study of the typing speed of a 7-year-old girl with cerebral palsy. Remember that the "modification" was to give a raisin every time she typed a correct word. As you read this brief section, try to answer this question:

★ What was the observed relationship between the "treatment" and the dependent variable? That is, what level of the dependent variable occurred at each level of the independent variable?

### The Results Section of the Yeatts and Brantley Typing Study

Results were plotted on a semi-log scale and graphed across days for comparison.

Word rates in Fig. 1 systematically decreased during initial baseline and with introduction of reinforcement both rates increased sharply, remaining above baseline level. During baseline 2, rates returned to initial baseline level.

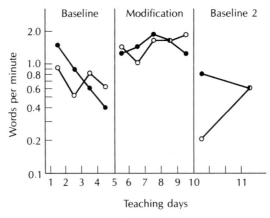

○ Rate of typing from dictation
● Rate of typing from a model

**Figure 1.** Word rates for typing from a model and from dictation.

What were the results? During the original baseline (no treatment) period, the rate of typing was about .7 words per min., perhaps showing a decline across the observation days. During treatment, the rate increased to about 1.5 words per min., about double. Then, after they stopped giving the raisins, the rate dropped to about .5 words per min. As long as the treatment continued, there was a temporary increase in the dependent variable. After it was discontinued, the rate declined perhaps to below its original level.

Yeatts and Brantley weren't particularly interested in the difference between dictation and written copy, and there's little interesting to be said about the relation between that variable and the dependent variable. It might be important that dictation showed such a severe drop after the raisins stopped, but it could also be an accident. With only 30 min. of measurement on just one subject it's not clear how reliable that finding is.

In this experiment the results were presented in a figure. This figure is customary in that it shows the levels of the independent variable on the horizontal axis and places the corresponding levels of the dependent variable relative to the vertical axis. The other independent variable is shown on a coded form within the figure. The same results also could have been displayed as in Table 5.1.

**Table 5.1.** Tabular Display of Word Rates for Typing from a Model and from Dictation

|  | Words per Minute | |
|---|---|---|
|  | From Dictation | From Written Model |
| Baseline day 1 | .9 | 1.5 |
| Baseline day 2 | .5 | .9 |
| Baseline day 3 | .8 | .6 |
| Baseline day 4 | .6 | .4 |
| Treatment day 1 | 1.5 | 1.4 |
| Treatment day 2 | 1.0 | 1.5 |
| Treatment day 3 | 1.7 | 1.8 |
| Treatment day 4 | 1.7 | 1.7 |
| Treatment day 5 | 1.8 | 1.4 |
| Posttreatment day 1 | .2 | .8 |
| Posttreatment day 2 | .6 | .6 |

Yeatts and Brantley's experiment had only one subject. Because relatively few data were collected, it was possible to present all the results. In many experiments more data are collected than can be presented in the report. Even if the data were presented there would be too much to comprehend just by looking. To facilitate comprehension and presentation some summarizing is needed. The summarizing is accomplished by using **descriptive statistics** such as correlation coefficients, frequencies, or means. The choice of descriptive statistics determines which of the remaining three approaches is used.

## The Correlational Approach

The second approach we'll discuss is the correlational approach. Rather than presenting all the data, the correlational approach presents only data summarizing the whole relationship between variables. It's not possible to know the precise level of the dependent variable at each level of the independent variable in many experiments using the correlational approach, but it is possible to know the nature of the whole relationship.

The concept of correlation was introduced in Chapter 2, with a brief mention of positive and negative relationships, regression lines, and correlation coefficients. Now it's time to look at some of these concepts in more detail. A good way to understand them is to understand a way of displaying experimental results called a **scatter graph.**

*Scatter Graphs*

The results from a hypothetical experiment investigating the relationship between hours of studying and amount of time needed to finish an objective exam are shown in Figure 5.1, a scatter graph.

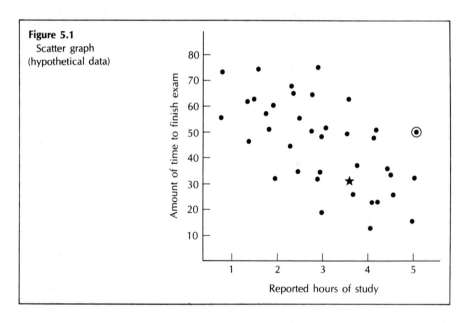

**Figure 5.1**
Scatter graph (hypothetical data)

In a scatter graph each axis displays the scores on one variable. Each dot usually represents one subject; the placement of the dot shows that subject's scores on both variables. For example, the person represented by the circled dot studied for about 5 hours and took about 50 min. to complete the exam. Can you tell how long the person represented by the star studied?

By looking at the whole graph you can learn a good deal about the relationship between the variables. In general, the longer the people studied, the less time they needed on the exam. Thus, this is a negative correlation. The scatter graph also shows a good deal of variability in the scores. Some of the people who studied for 3 hours finished the exam in as quickly as 20 min., while others took 75 min. In the scatter graphs shown in Figure 5.2, A shows a positive correlation, B shows a negative correlation, and C shows a nonmonotonic correlation. D shows two sets of scores that aren't correlated at all. High scores, medium scores, and low scores on one variable are completely unrelated to the other variable. Relationships such as those shown in A and B are called *linear,* because a straight line segment could be used to approximate the pattern. A curved line could be used to approximate the relationship shown in C. There is no relationship to approximate in D.

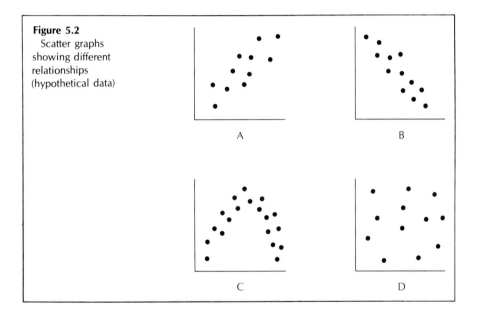

**Figure 5.2**
Scatter graphs showing different relationships (hypothetical data)

*Regression Lines*

Sometimes linear relationships are approximated by straight lines called **regression lines** or **lines of best fit.** They're formally calculated by using algebraic techniques, but you could approximate one by drawing a straight line through the middle of the dots in a scatter graph. Regression lines are summarized in articles with the usual algebraic equation for a straight line, $y = mx + b$. The symbol $m$ signifies the slope of the line, and $b$ is the point at which the regression line meets the vertical axis. Both $m$ and $b$ are calculated from experimental data and are given as numbers in an article, so that the equation for a regression line of the scatter graph in Figure 5.1 might be given as $y = -10x + 85$. The symbol $x$ signifies the score on the $x$ or horizontal axis, and $y$ is the score on the $y$ or vertical axis.

By filling the equation with any $x$ and solving it you can *predict* the corresponding $y$ on the regression line. For example, you can fill in the equation with $x = 2$ hours of study, and the corresponding $y$ is 65 min. needed to complete the exam. Regression lines are used for prediction. For a person who studied for 2 hours, the best prediction of time needed for the exam is 65 min.

**Problem**

This is a practice problem on applying a regression equation to make a prediction.

**Q.** Suppose an experiment showed the following relationship between people's bedtime on a 24-hour clock and the number of cups of coffee they drink after dinner:

Bedtime = 1.5 (number of cups of coffee) + 21.50 hours.

What bedtime, on a 24-hour clock, would you predict for a person who had two cups of after-dinner coffee? (Remember that the 24-hour clock starts over at 0 hours at midnight.)

**A.** You should have said Bedtime = 1.5 (2.00) + 21.50 hours = 3.00 + 21.50 = 24.50 hours, or 50 min. past midnight.

The regression equation summarizes the general relationship between two sets of scores. From the information in a regression equation, we'd expect that everybody with the same $x$ scores would have the same $y$ score; that is, everybody who studied for 2 hours would need exactly 65 min. to complete the exam. That's clearly not true. Some people took more and some took less than the time predicted by the regression equation. That scatter, or variability, isn't reflected by the equation. The two scatter graphs in Figure 5.3 would have almost the same regression line but are clearly different. **Correlation coefficients** express that variability.

**Figure 5.3**
Scatter graphs showing different amounts of variability (hypothetical data)

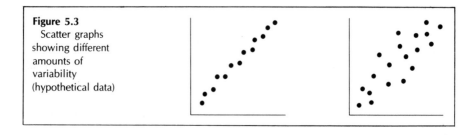

## Correlation Coefficients

As we discussed in Chapter 2, correlation coefficients express how much difference there is between the general relationship, as predicted by the regression equation, and the actual relationship between the two sets of scores. They usually vary between −1.00 and +1.00, with −1.00 meaning a perfect negative correlation, +1.00 meaning a perfect positive correlation, and .00 meaning no correlation at all.

By looking at a scatter graph you can tell how big the correlation coefficient would be if it were calculated. Figure 5.4 shows some samples.

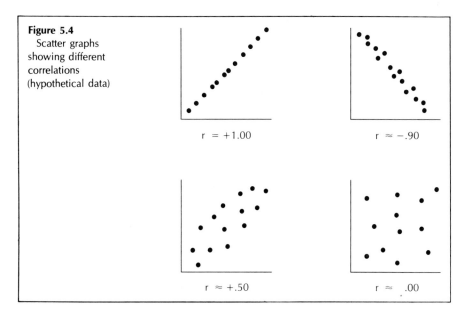

Figure 5.4
Scatter graphs showing different correlations (hypothetical data)

There are quite a number of correlation coefficients appropriate for different situations. The most commonly seen is the Pearson product-moment correlation coefficient, abbreviated $r$. Some others are Spearman's rho ($\rho$), Kendall's tau ($\tau$), the phi coefficient ($\phi$), the point biserial, the biserial, and the coefficient of concordance (K). If these statistics appear in an article it means the correlational approach is being used to describe the relationship between two variables.

The next results section you'll read is from the Sullivan and Skanes teacher rating study, whose introduction you read in Chapter 2. Recall that the variables were the grade received by students on the final exam in various courses and the rating given by the students to their teachers on a scale from 1 to 5. Sullivan and Skanes chose to calculate the mean exam grade and the mean teacher rating given by each *section* of students and then to correlate these two sets of scores. Since the students were in different courses, the courses were kept separate. Thus subject area may be thought of as another variable. However, Sullivan and Skanes weren't interested in it and don't present the results in such a way as to make it possible to tell the relationship between course area and student ratings or exam grade.

As you read this results section answer this question:

★ What was the relationship between the variables?

## The Results Section of the Sullivan and Skanes Teacher Evaluation Study

The correlations between student ratings and final examination marks are given in Table 1.

**Table 1**
Correlation between Instructor Rating
and Final Examination Achievement (December 1971)

| Subject | No. Sections | r |
|---|---|---|
| Biology 100F | 6 | −.2767 |
| Biology 1010 | 14 | .4157 |
| Chemistry 100F | 8 | .0785 |
| Chemistry 1000 | 6 | .5535 |
| Mathematics 100F | 8 | .4772 |
| Mathematics 1010 | 16 | .3379 |
| Mathematics 1150 | 9 | .3270 |
| Physics 1050 | 9 | .5720 |
| Psychology 1000 | 40 | .4009[a] |
| Science 115A | 14 | .5106[a] |
| Total | | + .394[b] |

[a] $p < .05$.
[b] $p < .01$.

Of the 10 correlations, it can readily be seen that 9 are positive and 8 are above +.32. The average correlation is +.39. A test for significance of a number of independent tests of the same hypothesis (Winer, 1971, p. 49) revealed a significant overall positive effect ($\chi^2 = 44.13$, $df = 20$, $p < .05$). In general, then, these results indicate that there is a modest, but significant, relationship between student evaluation of instruction and student achievement.

What were the results? Overall the students in sections receiving higher mean exam grades rated their instructors higher than students in sections receiving lower mean exam grades. The relationship between grades and teacher ratings was positive; the correlation coefficient was +.39.

It's apparent from the table of correlation coefficients that the strength of the relationship varied from one course to another. In Biology 100F, a course for students needing preparation for the regular college biology course, the relationship was even negative.

Could the overall relationship be accidental? If the experiment were repeated again somewhere else, is it likely that they would find no relationship or even an overall negative relationship? That's what that $\chi^2$ (chi-square) test is all about.

## Deciding Whether Results Are Accidental

Descriptive statistics such as the correlation coefficient and frequencies and means (we'll discuss them soon) summarize results. But how can you decide whether the results are trustworthy?

Ultimately, you must decide this question for yourself. If the same thing has happened a number of times, it seems more reliable than if it's happened only once. If the relationship between variables forms a "nice" pattern, it seems more trustworthy than if it appears to be scattered chaotically. There are other common-sense elements that can help make the decision as well.

To aid common sense in this decision, a whole family of statistics has been developed called **inferential statistics** or **hypothesis-testing statistics.** That's what $\chi^2$ in the Sullivan and Skanes study represents.

### Inferential Statistics

Inferential statistics evaluate the truthworthiness of results. There is a large variety of inferential statistics that are appropriate for different situations, but they all have the same general purpose: to determine how much trust can be placed in experimental results. Unfortunately they do that in a backward-sounding way.

Inferential statistics measure the probability, or likelihood, that the observed results are accidental. If the relationships between variables are accidental, they wouldn't be expected to occur again if someone repeated the experiment. If they're not accidental, they would be expected to occur over and over again.

If inferential statistics show that there's a high probability that a relationship is accidental, it's customary to conclude that the relationship *is* accidental and, hence, can be ignored as if there were no relationship at all. Such apparently accidental relationships are called **nonsignificant.** If there's a sufficiently *low* probability that a relationship is accidental, it's customary to conclude that the relationship is *not* accidental but real. These relationships are called **significant.**

Inferential statistics calculate the probability that a relationship is accidental. That probability is reported in the results section as *p;* we're told the approximate value of p. As p gets bigger, the statistical significance of the relationship being discussed gets smaller. If p is sufficiently small, say one chance in 100, or .01, the relationship is called significant. That means it's probably, though not certainly, real. If p is too large, say one chance in 5, or .20, the relationship is called nonsignificant. That means it's probably, though not certainly, accidental.[1]

In most papers approximations of p are reported rather than exact probabilities. It's common to report that p is less than or greater than some quantity rather than giving a precise figure. The statement $p < .01$ means that the probability that the relationship is accidental is smaller than .01 or one chance in 100; that difference probably would be called significant. The statement $p > .20$ means that p is greater than 1 chance in 5; that difference would almost certainly be called nonsignificant.

Many behavioral scientists have adopted $p = .05$ as a more or less official cutoff point. Relationships with $p < .05$ are usually called significant; relationships with $p > .05$ are usually called nonsignificant. If no p value is reported but a statement is made about a relationship being significant, it's safe to assume that the .05 level is being used unless you're told otherwise in the article.

In reading results sections it's usually easy to tell that inferential statistics are being employed: a p value is almost always given. However, it's not always easy to tell just what relationships are being discussed. Although a thorough knowledge of statistics is really needed, this book should give you a good start. In the case of the Sullivan and Skanes data, the $\chi^2$ test has been used to evaluate the overall pattern of results and to show that it's statistically significant at the .01 level. We've also shown in the table that the positive correlations for the psychology and science classes are significant at the .05 level when those classes are considered individually. None of the other classes considered alone has a significant relationship.

*Interpreting
Nonsignificant Relationships*

What exactly does it mean when a relationship is nonsignificant? In the Sullivan and Skanes results in the physics class, the correlation between the students' grades and the ratings they gave their teachers was + .57, but it was nonsignificant. In this case it seems likely that it failed to reach significance because there were only nine scores considered in the calculation. We may conclude properly that this experiment hasn't demonstrated a relationship between the two variables in the physics class, but we can't really conclude that the variables are *un*related in that class.

One important property of inferential statistics that you must learn is that they never allow us to conclude that variables are *not* related. They can tell us that the

---

[1] In statistical terminology the hypothesis that the differences are accidental is called the **null hypothesis.** Thus inferential or hypothesis-testing statistics measure the probability that the null hypothesis is true. If that probability is sufficiently small, we *reject* the null hypothesis.

probability that the relationship was accidental is rather large, but they can't tell us that the relationship *was* accidental. A related issue is that a sloppily performed experiment or one with few measures or one with a great deal of variability in the scores can camouflage real relationships. Thus it could be (as seems likely) that there's a real relationship between grades and ratings in the physics classes, but this experiment just wasn't sensitive enough to discover it, because there were so few classes.

The correct conclusion to draw from a nonsignificant effect is that no relationship between variables was *demonstrated*. You can't conclude that no relationship *exists*.

*Interpreting
Significant Relationships*

In the best of cases, statistical significance can mean only that the probability is relatively small that a relationship is accidental. We're not told that the relationship is definitely real. When *p* is less than .05 we usually conclude that the relationship is real; however, we should keep in mind that as often as 1 in 20 times relationships like that occur strictly by accident.

Even that caution isn't enough if improper statistics have been used, as happens all too often. We'll discuss a few kinds of errors in this chapter, but you should take a complete course in statistics if you want to be able to evaluate the appropriateness of the statistics reported in articles.

## The Differences-Between-Frequencies Approach

The third approach for presenting results counts the number of similar events and groups them together, calling them frequencies. To look at the results you look at the frequencies of the dependent variable to see how they change depending on the level of the independent variable. After counting, it's usual to translate the counts, or frequencies, into proportions or percentages.

Look now at the results section of the Loftus and Zanni question-wording experiment. Recall that the independent variables had to do with the questions asked about a film all the subjects had just seen. "Leadingness" had to do with whether the questions contained the word *a* or the word *the;* it was varied between subject. The other independent variable was whether the object asked about was really in the film; it was varied within subject, with three questions of each type. The dependent variable was the subject's response: yes, no, or I don't know. The treatment of the dependent variable in the text may be a little confusing because the authors wrote about the responses as if they were three separate dependent variables. As you read the article, try to answer these questions:

168    Chapter 5  What Happens?

**1.** What was the relationship between presence of object and the number of "yes" answers? "no" answers? "I don't know" answers?

**2.** What was the relationship between leadingness and the number of each kind of response?

**3.** How likely is it that the reported results were accidental?

## The Results Section of the Loftus and Zanni Question-Wording Study

Table 1 presents the percentage of "yes, "no," and "I don't know" responses for both the "the" and "a" subjects. Whether an item was actually present or not, subjects interrogated with *a* were twice as likely to respond "I don't know." Subjects interrogated with *the* tended to commit themselves to a "yes" or "no" response.

**Table 1**
Percentage of "Yes," "No," and "I don't know"
Responses to Items That Were Present and Not Present in the Film

|              | Present |      | Not Present |      |
|--------------|---------|------|-------------|------|
| Response     | "the"   | "a"  | "the"       | "a"  |
| Yes          | 17      | 20   | 15          | 7    |
| No           | 60      | 29   | 72          | 55   |
| I don't know | 23      | 51   | 13          | 38   |

Another aspect of these data is worthy of mention. First, when a subject is queried about an item that was not present in the film, "yes" responses are particularly interesting. A "yes" response indicated a subject reported that he saw something that was not, in fact, present. Using the indefinite article resulted in false "yes" responses 7% of the time. With the definite article, however, false "yes" responses occurred 15% of the time—over twice as often.

To test statistically for the difference between interrogation with *a* and *the*, a single score for each subject was generated. "Yes" responses were assigned a value of $+1$, "I don't know" responses were assigned a value of 0, and "No" responses were assigned a value of $-1$. A subject's mean score, then, reflected his confidence that the items were present. The difference between the "confidence scores" for the *a* and *the* questions was significant by a Mann-Whitney U test, $z = 2.98$, $p < .01$.

To test for the difference between items which were and were not present, two mean scores for each subject were generated, one for items which were present and the other for items which were not. Again, "yes" responses received a value of $+1$, "I don't know" responses a value of 0, and "No" responses a value of $-1$. As before, a subject's mean score for a particular type of item expressed his confidence that those items were present. Two Wilcoxon matched-pairs signed-ranks tests revealed that both the subjects who had been interrogated with *a* and the subjects who had been interrogated with *the* were more confident about items that had been present, $z = 3.97$, $p < .001$, and $z = 2.52$, $p < .01$, respectively.

What were the results? When the subjects were asked about objects that really were present in the film they answered yes more often than for objects that weren't present, and they answered no less often. They also answered I don't know more often for objects that were present.

Subjects asked *the* questions tended to respond I don't know less than subjects receiving *a* questions. They responded no considerably more often and, overall, responded yes more often. However, that latter result was true only for objects that were *not* present. When the objects being asked about weren't present, subjects asked questions phrased with *the* answered yes 15 percent of the time, while subjects asked *a* questions answered yes only 7 percent of the time. Loftus and Zanni point out that that's more than twice as often. However, it's still a relatively small change in the absolute number of yes answers. It really represents a change from 10 to 22 yes answers out of 150 possible.

How likely is it that the reported results were accidental? Three statistical comparisons were reported, each one showing that the difference in the dependent variables between the levels of the independent variable is statistically significant. That's another way of saying that the relationship between the variables is significant. However, only very careful reading will tell you exactly what is being compared with what.

The first comparison, using the Mann-Whitney U, tells us that there was a statistically significant difference between the answers of the subjects with *a* questions and *the* questions. It doesn't tell us the nature of the difference; we learned that by looking at the table. The inferential statistic tells us only that the whole pattern of observed differences between *a* and *the* questions probably didn't occur by accident.

A second comparison, using a Wilcoxon test, is limited to the subjects with *a* questions. It tells us that there was a statistically significant difference between their answers about present and not present items. The third comparison, also using a Wilcoxon test, tells us that the same thing is true for subjects with *the* questions.

*None* of these comparisons tells us specifically that there's a statistically significant difference between the 7 and 15 percent of yes answers for not present objects depending on whether *a* or *the* questions were asked. Such a comparison could have been made, but for some reason Loftus and Zanni chose not to make it. Only very careful reading would reveal that omission.

From these data alone I would doubt seriously that *the* questions cause a reliable increase in yes answers to questions about objects that weren't present. However, Loftus and Zanni actually repeated the whole experiment, using different subjects, a different film, and different questions. Their replication showed the same pattern of results as the original experiment: When subjects were asked about objects that weren't present, 6 percent of the subjects asked a questions responded yes, compared to 20 percent of the subjects asked *the* questions. Thus it appears that the difference is indeed reliable.

### The Differences-Between-Means Approach

The **mean** is a descriptive statistic that represents the "average" or central tendency of a set of scores. It's the total of all the scores divided by the number of scores. In experimental reports the mean is sometimes abbreviated $M$, $\bar{X}$, or $\mu$ (Greek letter mu) without explanation. If many experimental subjects are measured, or if a few subjects are measured many times, it's useful to summarize the data by reporting the mean rather than by reporting all the individual measurements. It's common to describe the mean level of the dependent variable at each level of the independent variable.

Now read the results section of the Cross and Davis marijuana study. Remember that the independent variable was reported frequency of marijuana use and the dependent variable was adjustment, as measured by a standardized test, the ISB. In this article $F$, $t$, and $\chi^2$ are inferential statistics, $n$ or $N$ stands for the number of measurements in that group, and $df$ and $MS$ are statistics necessary for calculating other statistics. In many results sections you'll see statistics you don't understand. Of course, it would be better if you did, but just grasp as much as you can. As you read, try to answer these questions:

1. What's the nature of the association between the independent and dependent variables?
2. How likely is it that the reported results were accidental?

### The Results Section of the Cross and Davis Marijuana Study

Table 1 presents mean ISB scores by drug use category and an analysis of variance based on these means. The $F$ ratio is nonsignificant and shows no difference between the user groups. As a check on the possibility that the very heaviest drug users ($n = 12$) in the sample may have higher scores, students who had used LSD 20 or more times (most of them smoked marijuana at least once daily) were extracted and compared with the rest of the sample. Their ISB mean was 147.2 versus 135.4 for the remainder of the sample ($t = 2.32$, $df = 176$, $p < .025$), so that heavy drug users had significantly higher maladjustment scores.

## Table 1
### ISB Adjustment Scores by Categories

| Category | M | Source | df | MS | F |
|---|---|---|---|---|---|
| Adamant nonuser | 133.5 | Between | 4 | 376.05 | 1.29 |
| Nonuser | 133.8 | Within | 173 | 291.95 | |
| Taster | 139.6 | | | | |
| Recreational user | 139.6 | | | | |
| Regular user | 139.4 | | | | |

One question on the inventory asked the students if they had ever had any psychological counseling or psychotherapy. Five nonusers and 16 users ($\chi^2 = 3.87, p < .05$) answered affirmatively.

What were the results of this study? The first two sentences and the table report the important findings. The people in the five groups at different levels of marijuana use filled out the ISB, on which a low score is supposed to indicate better adjustment. The mean scores for the groups are presented in the second column of the table. It appears from the means that the two groups who never used marijuana obtained somewhat lower scores in the ISB than the other two groups. However, the F test, or analysis of variance, showed that the means of the groups were not significantly different from each other. The probability that the differences are accidental wasn't reported here, but we can assume that it was greater than .05. From this we can conclude that the mean ISB scores of the five groups haven't been shown to differ. However, we should keep in mind the possibility that maladjustment really is associated with marijuana use but that this experiment wasn't well enough controlled to detect the association.

Having reported these major findings, Cross and Davis continued to make other comparisons. Such previously unplanned comparisons, made after the data have been collected, are sometimes called **post hoc,** or after the fact, comparisons. As we'll discuss shortly, post hoc comparisons aren't quite as trustworthy as comparisons planned before the data were collected, but they're often of interest anyway. First, Cross and Davis looked at the ISB scores of 12 very heavy drug users and found that the mean of those scores was significantly higher than that of the rest of the subjects. The t test showed that the probability that that difference was accidental was less than .025. Second, they looked at one question on the test and found that the proportion of marijuana users who reported having had some sort of professional counseling was significantly higher than the proportion of nonusers. Rather than reporting means (which would make no sense at all), they reported frequencies. These frequencies are easier to understand if we translate them to proportions. This takes a little effort, but by looking back at the method section we see that 100 subjects were users and 78 were nonusers. Sixteen of the users had professional counseling versus only 5 nonusers, which is 16 percent versus about 6.4 percent. The $X^2$ test showed that the probability of such a difference occurring by accident was less than .05.

The facts demonstrated by the experiment can be summarized as follows: (1) For these 178 University of Connecticut psychology students, ISB-measured adjustment scores weren't significantly associated with reported frequency of marijuana use, at least up to moderate levels. (2) The 12 students who reported that they used drugs very heavily had significantly higher maladjustment scores than the other students. (3) A higher proportion of the students who reported that they have used drugs than of students who reported otherwise said that they had had professional counseling. However, some additional caution is warranted by the last two comparisons because they were post hoc.

### Post Hoc Comparisons Require Extra Caution

Post hoc comparisons are comparisons not planned before the experiment but performed afterward because they looked interesting. Cross and Davis chose to compare the 12 heavy users to the others *because* they had particularly high scores on the ISB. They singled out the counseling question *because* it showed a difference in an interesting direction. The reason why post hoc comparisons are troublesome is that some differences do occur just by accident. When experimenters test a relationship just because it looks like it might be significant, they might pick out just such an accidental effect.

All the inferential statistics you've learned about so far are proper for comparisons planned *before* the data were collected. To perform post hoc tests properly, different statistics must be used. Some common ones are called Scheffe's test, Dunnett's test, and the Newman-Keuls test. If inappropriate statistics are used, post hoc analyses aren't as trustworthy as comparisons planned before the experiment. The statistics used by Cross and Davis were $t$ and $X^2$, which aren't really proper. Therefore, the conclusions about heavy users being maladjusted and users having more counseling shouldn't be trusted as much as if they'd been planned. These differences are more likely to be accidental than indicated by the $t$ and $X^2$ tests.

From all these facts one reader might reasonably conclude that frequency of marijuana use isn't associated with adjustment; another might conclude that marijuana users are more likely to be seriously disturbed. These questions of interpretation are taken up more fully in Chapter 6, which deals with the discussion section. For now we'll concentrate on discovering the facts demonstrated by an experiment.

### Variability

From the information given by Cross and Davis we can't really tell how the individual subjects scored on the ISB. All of them might have been very close to the reported mean scores, or some could have been very high or very low. We have no idea how much the scores spread out around the average score. To express that kind of information there's a different kind of descriptive statistic called a measure of **variability.**

If all the scores are close to the mean, then variability is low; if the scores are spread out, with some very high and low scores, then variability is higher. Figure 5.5 shows the number of males and females who got different numbers of parking tickets. You can see that the means for the two groups would be about the same, but the variability is greater for the males than for females.

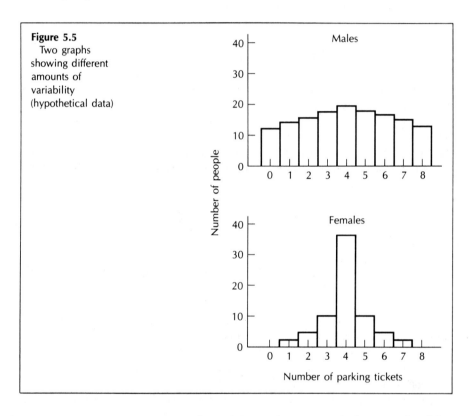

**Figure 5.5** Two graphs showing different amounts of variability (hypothetical data)

A very simple measure of variability is the **range,** which is just the difference between the highest and lowest scores. In the parking ticket figures, the range for the males was 8 tickets, but was only 6 for the females. If Cross and Davis had told us that the range of ISB scores was only 5 points, we'd know that all the subjects were quite close. However, the range isn't used very often in scientific writing. The most commonly used measures of variability are the **standard deviation** (sometimes abbreviated $s$, $S$, $SD$, or with a Greek letter sigma ($\sigma$) and its square, the **variance** (abbreviated $s^2$, $S^2$, Var, or $\sigma^2$).

To interpret the standard deviation it's useful to remember a few numbers.

The interval defined by going one standard deviation higher and one lower than the mean contains about 68 percent of the scores in the group. For example, with a mean of 136 points on the ISB, if the standard deviation were 2 points, it would mean that about 68 percent of the subjects scored between 134 and 138 points, that is 136±2. The other 34 percent of the subjects had either less than 134 or more than 138. If you go two standard deviations in either direction about 95 percent of the scores are included within that interval. More than 99 percent of the scores are included in the interval three standard deviations on either side of the mean. Thus, if the standard deviation in Cross and Davis's study were 5 points, we would know that about 99 percent of the subjects scored between 121 and 151 points, that is, 136±(3 × 5 points).

The variance is the square of the standard deviation, so you can translate the variance into standard deviations by taking the square root. For example, if a set of scores had a mean of 100 and a variance of 25, the standard deviation would be 5. Then you'd know that about 68 percent of the scores lay between 95 and 105 and about 99 percent between 85 and 115.

In fact, in Cross and Davis's experiment the mean square within term can be thought of as a variance score. It was 291.95, and the square root of that is 17.09. Thus we can estimate the standard deviation of the ISB scores at about 17 points and know that about 95 percent of the people scored between about 103 and 171 on the ISB.

### Problems

The problems will give you practice in interpreting standard deviations and variances.

**Q.** Suppose an experiment reported that the mean time required to solve a problem was 12.47 sec. and the standard deviation was 2.10 sec. What interval of scores would include about 95 percent of the solution times?

**A.** To encompass about 95 percent of the scores you must go two standard deviations in either direction from the mean, that is, 12.47 ± 4.20 sec. So the answer is that about 95 percent of the people solved the problem in between 8.27 and 16.67 sec.

**Q.** If an experiment reported that the mean number of correct answers on a test was 84.2 and the variance was 16 correct answers, what interval of scores would include about 99 percent of the scores?

**A.** The mean is 84.2, and the standard deviation is the square root of 16, or 4. To encompass about 99 percent of the scores you must go three standard deviations in either direction from the mean, that is, 84.2 ± 12 points. So the answer is that about 99 percent of the scores lie between 72.2 and 96.2 points.

When there's a lot of variability, group averages are less trustworthy than if there's less variability. That means that information about variability can help you decide whether differences between groups are trustworthy. Figure 5.6 should make this connection a bit clearer.

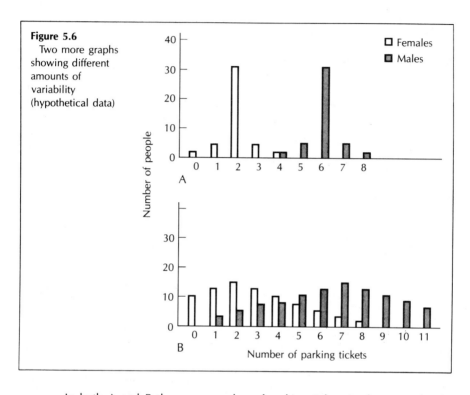

Figure 5.6
Two more graphs showing different amounts of variability (hypothetical data)

In both A and B the mean number of parking tickets is about two for the males and six for the females. However, that difference seems more trustworthy in A, where there's less variability. Inferential statistics are based on measures of variability, so it's not surprising that the information is related.

Now let's look at the results section of the Wallace et al. teaching methods study, which presents standard deviations as estimates of variability. Recall that there were three teaching methods used in training counseling students how to help others with decision-making problems. The dependent variable is the total score each student received when being rated by some judges during a trial interview.

As you read this section, try to answer these questions:

**1.** What is the relation between the teaching methods and the amount learned?

**2.** What do the measures of variability make you think about the trustworthiness of the results?

3. What do the inferential statistics make you think about the trustworthiness of the results?

## The Results Section of the Wallace et al. Teaching-Method Study

Data from the spring and summer terms were treated separately, as an original study and a replication. Thus, two one-factor (teaching method) analyses of variance were conducted on the students' counseling performance scores. Significant omnibus $F$s were followed up with Tukey's wholly significant difference technique (Games, 1971).

Table 1 depicts the students' counseling performance scores. A significant main effect was found in the spring term, $F(2, 21) = 5.44$ $p < .02$, and in the summer term, $F(2, 27) = 5.84, p < .01$.

**Table 1**
Summary Statistics on Counseling Performance of Students Exposed to Each Teaching Method

| Term | Teaching Method | $n$ | $\bar{X}$ | SD |
|---|---|---|---|---|
| Spring | 1 | 8 | 41.62 | 16.74 |
|  | 2 | 8 | 45.25 | 14.08 |
|  | 3 | 8 | 63.00 | 9.96 |
| Summer | 1 | 10 | 52.40 | 14.67 |
|  | 2 | 10 | 48.90 | 11.20 |
|  | 3 | 10 | 65.90 | 8.57 |

Results of the post hoc follow-up analysis are shown in Table 2. In both spring and summer significant differences occurred between those students taught by Method 3 and those taught by each of the other two methods. No significant differences were found, however, between Methods 1 and 2.

**Table 2**
Summary of Post Hoc Comparisons of Teaching Methods on Counseling Performance

| Term | Comparison | Difference |
|---|---|---|
| Spring | 1 vs. 2 | 3.62 |
|  | 1 vs. 3 | 21.37[a] |
|  | 2 vs. 3 | 17.75[a] |
| Summer | 1 vs. 2 | 3.50 |
|  | 1 vs. 3 | 13.50[a] |
|  | 2 vs. 3 | 17.00[a] |

[a]$p < .05$.

What were the results of the experiment? The means in Table 1 suggest that Method 3 consistently led to higher scores than either other method. The standard deviations show that there was a fair amount of variability in the scores given to the different students. In fact, for those given Method 1 in the spring, the standard deviation suggests that to include 99 percent of the students' scores we'd need an interval from $-8.60$ to $+91.84$ points. Since the possible range is only from 0 to 88 that's clearly a statistical anomaly, but it *is* plain that the scores vary quite a bit from one student to another.

Despite the differences between the students within each group, the inferential statistics show that some of the groups did differ significantly from each other. The first two $F$ tests are planned comparisons, showing that during each session at least two of the three groups differed significantly from each other. The results shown in Table 2 are post hoc comparisons telling exactly which groups differed. They use a statistical test appropriate for post hoc comparisons, the Tukey test. They suggest that in both spring and summer Method 3 was significantly higher than Methods 1 and 2, and Methods 1 and 2 weren't different from each other. It seems likely that their results are trustworthy, and that if the experiment were repeated, Method 3 would again be superior to the other two methods which would be unlikely to differ from each other.

**Main Effects**

Warren et al. mention that the "main effects" are statistically significant. A main effect refers to the relationship between one independent and one dependent variable. In Warren et al. that's all there were, so the term isn't particularly useful. However, in studies with factorial designs it's often helpful to look individually at the relationship between each independent and dependent variable. Each such relationship can be called a **main effect** of that independent variable. For readers interested in only one of the independent variables, separating the effects can make the results more useful. For any reader, separating the results can make it easier to understand the whole pattern.

Now let's look at the results of a study that had more than one independent and dependent variable, the Lutkus mirror image drawing study. Recall that there were four groups of subjects, two who tried to trace a zigzag figure and two with a diamond figure. One group with each figure was given no special instruction. Each subject was given 10 trials to perform the task. The other group was told to try to image the figure to be traced and was then given 10 trials to perform the task. There also were two dependent variables, time to complete the figure, and errors. Thus there are six possible main effects we could look at, the main effects of each of the three independent variables on each of the two dependent variables. As you read this section there will be several things you don't understand. For now, try to answer just these questions:

1. What's the relationship between instructions and time to complete the task, that is, what's the main effect of instructions on time?

2. What's the main effect of instructions on errors?

3. The numbers aren't given, but can you figure out what the main effect of practice (that is, trials) on time must have been?

## Excerpts from the Results Section of the Lutkus Mirror Image Drawing Study

Analyses of variance (Figure by Condition by Trials) were calculated for both the time and error measures. The image instructed groups were significantly faster [$F(1, 60) = 9.94, p < .01$] and made fewer errors [$F(1, 60) = 7.34, p < .01$]. As would be expected with a motor task, the practice effect was highly significant [$F(9, 540) = 27.5 \, p < .001$]. Practice interacted with the type of Condition in that the Imagery groups began with much faster times than the Standard groups and then the gap was narrowed as the Standard groups improved over trials [$F(9, 540) = 3.54, p < .001$]. Table 1 presents mean times and errors for all four groups, averaged over the 10 trials.

Table 1
Mean Tracing Times (Seconds) and Errors

|  |  | Times | | Errors | |
| --- | --- | --- | --- | --- | --- |
|  |  | Standard | Image | Standard | Image |
| Zig-zag | Mean | 51.9 | 30.2 | 14.08 | 12.32 |
|  | SD | 30.3 | 18.7 | 6.72 | 6.32 |
| Diamond | Mean | 76.9 | 37.6 | 22.16 | 10.72 |
|  | SD | 62.8 | 28.3 | 15.12 | 7.84 |

The error data showed that the diamond benefited most from the imagery condition, i.e., the Figure by Condition interaction was significant [$F(1, 60) = 4.01, p < .05$].

---

What was the effect of instructions on the two dependent variables? The group means show that the image instructions allowed people to complete their task more quickly and with fewer errors. These are the main effects of instructions on time and errors. The second sentence of the text says that these two main effects were tested and found to be statistically significant. The probability is less than .01 (1/100) that the time difference between the two groups was accidental. The same is true for the differences in errors.

What was the effect of practice on time? Lutkus didn't present any descriptive statistics, but he does tell us that the practice effect was highly significant. It seems logical that performance must have improved from the first to the tenth trial, that is, the people must have gotten faster. That would be the main effect of practice on time. Notice that when main effects are described only the independent and dependent variable involved are mentioned. To look at main effects you must average across all the other variables and ignore them.

### Problems

For some practice in recognizing and verbally describing main effects, answer the following questions. To describe a main effect, describe the level the dependent variable takes at each level of the independent variable without mentioning the other variables at all.

Q. Suppose a consumer research group performed an experiment on washing machines, varying two independent variables. They used either 4, 6, or 8 oz. of detergent and set the washer for either 6 or 10 min. of washing time. They washed six loads of white wash, as identical as possible, and used a light meter to measure the brightness of each load. The results are shown in Table 5.2.

**Table 5.2.** Percent Reflected Light (Hypothetical Data)

| Washing Time | Amount of Detergent | | |
| --- | --- | --- | --- |
| | 4 oz. | 6 oz. | 8 oz. |
| 6 min. | 30 | 40 | 50 |
| 10 min. | 40 | 50 | 60 |

Describe verbally the main effect of amount of detergent on reflectance. Then describe the main effect of washing time.

A. The main effect of amount of detergent could have been described something like this: 8 oz. of detergent gave the most reflectance, 6 oz. gave less, and 4 oz. gave least. The main effect of washing time was that 10 min. washing time gave more reflectance than 6 min.

Q. Suppose an experiment were performed in which 120 different bank personnel directors were given a description of a fictitious job applicant and asked to rate the likelihood that they would hire such a person on a scale from 1 to 7 with 7 meaning "definitely hire." The descriptions were identical except for two variables.

One independent variable was sex of the fictitious applicant, with the personnel directors receiving applications from either a Harold or Harriett Ferguson. The other independent variable was the reported hobby of the applicant: sky-diving, bird-watching, pottery-making, and coin-collecting. Fifteen directors received each of the resulting eight descriptions. The results are as shown in Figure 5.7. Describe the main effects of each of the independent variables.

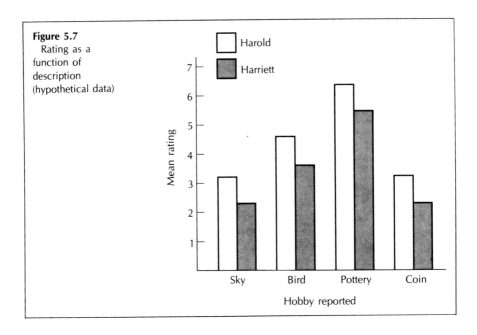

**Figure 5.7** Rating as a function of description (hypothetical data)

**A.** The main effect of sex was that Harold received higher ratings than Harriett. The main effect of hobby was that the pottery-makers received the highest ratings, followed by the bird-watchers. The sky-divers and coin-collectors were lowest and apparently didn't differ.

In the Lutkus experiment, what was the main effect of type of figure on time? Lutkus doesn't mention it, but by looking at the table we can see that, overall, the diamond took more time to complete than the zigzag. What about the main effect of figure on errors? If we average across, it's clear overall that there were more errors on the diamond than on the zigzag. But that was true only with the standard instructions. In fact, with the image instructions there were slightly fewer errors on the diamond. The effect of type of figure changed, depending on which level of instructions we look at. That kind of relationship between variables is called an **interaction.**

### Interactions

Interactions occur when the relationship between one independent variable and the dependent variable are influenced by another independent variable.

It's customary to call interactions by the names of the independent variables involved. In the Lutkus experiment there was an interaction between instructions and type of figure. It would be customary to call that an "instructions by type of figure" interaction. In Lutkus's words, "The error data showed that the diamond benefited most from the imagery condition, i.e., the Figure by Condition interaction was significant . . ." (page 389). Lutkus also mentioned that "Practice interacted with the type of Condition in that the Imagery groups began with much faster times than the Standard groups, and then the gap was narrowed as the Standard groups improved over trials . . ." (page 389). That could be called a "practice by instructions" interaction. The effects of practice were different depending on which level of instructions we look at.

Sometimes interactions occur with main effects in the same set of data, as in Lutkus; sometimes main effects occur alone; and sometimes interactions occur alone. It's useful to be able to recognize and describe main effects and interactions as they occur. To describe a main effect, the average level taken by the dependent variable at each level of the independent variable should be described. For example, look at the set of hypothetical data in Figure 5.8. The main effect of problem type is that insight problems take the longest to solve, word puzzle problems take the next longest, and arithmetic problems take the shortest. The main effect of age is that, as age increases, time to solution decreases.

There's also an interaction between age and problem type; the differences between the problems change as we look at different ages. Sometimes it's possible to describe an interaction by summarizing an obvious pattern. In this case we could just

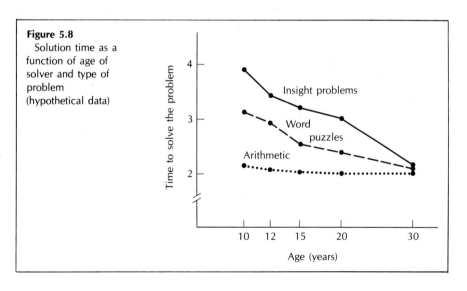

**Figure 5.8**
Solution time as a function of age of solver and type of problem (hypothetical data)

say that as age increases the differences between the problem types decrease. In other cases no obvious pattern exists. When that happens, one approach that always works is arbitrarily to choose one independent variable and go through each level in turn. For each level, describe the relationship between the *other* independent variable and the dependent variable. In this case we might arbitrarily choose age. Starting at 10 years, we can say that at 10 years there's a large difference between the problem types, with insight problems being most difficult, followed by word puzzles, followed by arithmetic problems. At age 12, the order is the same, but the differences are smaller. And so on until age 30, when we say that there's no difference between the problem types.

Alternatively, we could arbitrarily choose problem type and say something like this: For insight problems, there's a big decrease in time required to solve as people get older. For word problems there's also a decrease, but less dramatic. And for arithmetic problems there's little or no decrease. Usually one description is easier or theoretically more interesting than the other, and in a published article only that one is given. In describing interactions yourself, you might want to try both possible descriptions to see which one seems better.

### Problems

These problems should help you learn to recognize and describe main effects and interactions.

For each set describe any main effects of each independent variable and any interaction between them. If no effect is present, you should say so.

**Q.** See Figure 5.9.

**A.** The main effect of time of day is that more food gets eaten at 9 PM than at

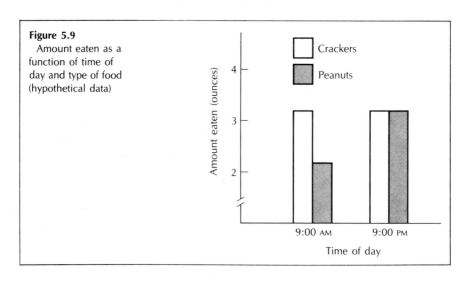

**Figure 5.9**
Amount eaten as a function of time of day and type of food (hypothetical data)

9 AM. The main effect of type of food is that, overall, more crackers get eaten than peanuts. The interaction could be described in two ways. One is to say that at 9 AM more crackers get eaten than peanuts, while at 9 PM there's no difference. The other description would be that for crackers time of day doesn't affect the amount eaten but for peanuts more are eaten at 9 PM than at 9 AM.

**Q.** See Table 5.3.

Table 5.3. Proportion of People Observed Singing "The National Anthem" at Football Games (Hypothetical Data)

|  | College Games | Professional Games |
| --- | --- | --- |
| Lead by band | .90 | .80 |
| Lead by singer | .85 | .75 |

**A.** The main effect of type of game is that more people sing at college games than at pro games. The main effect of type of leader is that more people sing when they are led by a band than when they're led by a singer. There's no interaction between the two variables.

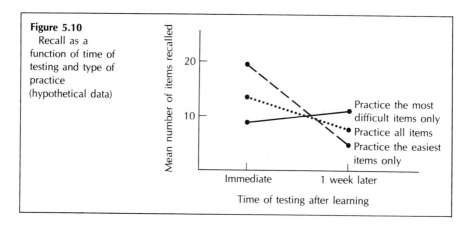

Figure 5.10 Recall as a function of time of testing and type of practice (hypothetical data)

**Q.** See Figure 5.10.

**A.** The main effect of time of testing is that less is recalled after 1 week than is recalled immediately. The main effect of type of practice is unclear; it seems likely that there is none. However, there is an interaction. One way to describe it would be to say that on immediate testing the technique of practicing only the easiest items led to the highest recall, followed by practicing all the items, and the technique of practicing only the most difficult items led to the worst recall. When recall was tested 1 week later, the order was reversed. You may have described the interaction from the other point of view.

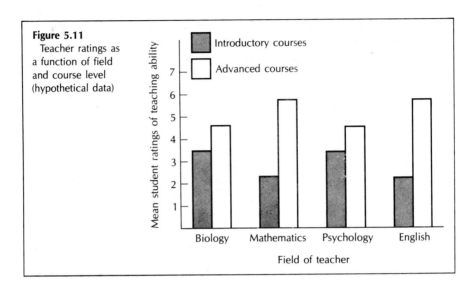

Figure 5.11
Teacher ratings as a function of field and course level (hypothetical data)

**Q.** See Figure 5.11.
**A.** There is apparently no main effect of the field of the teacher. However, there is a main effect of how advanced the courses were, with the advanced courses getting higher teacher ratings than the introductory courses. There also was a field by advancement interaction, such that there was a big difference between the advanced courses and introductory courses in English and mathematics but a small difference in biology and psychology.

**Q.** See Figure 5.12.
**A.** Overall, the main effect of score on general aptitude test is that people with scores of 70 and below and 130 and above got the lowest scores on the conventionality test. People with scores between 71 and 84 and between 116 and 129 got somewhat higher scores. And people with scores on the general aptitude test between

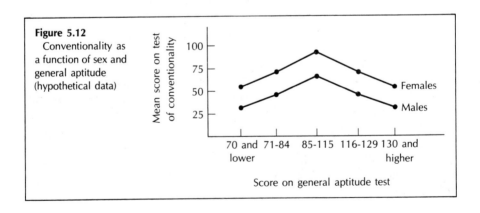

Figure 5.12
Conventionality as a function of sex and general aptitude (hypothetical data)

85 and 115 got the highest scores of all on the conventionality test. The main effect of sex was that females scored higher on the conventionality test than males. There was no interaction between the variables; the effect of sex was the same at all scores of the general aptitude test.

Q. See Table 5.4.

Table 5.4. Percent of Drivers Running Yellow Lights (Hypothetical Data)

| Type of Vehicle | Age of Driver | | |
|---|---|---|---|
| | Younger Than 25 | 25–45 | Older Than 45 |
| Conventional car | 75 | 50 | 25 |
| Sports car | 40 | 40 | 50 |
| Van | 50 | 60 | 5 |
| Pickup | 75 | 20 | 20 |

A. The main effect of age of driver was that the younger drivers most often ran the yellow light, followed by the drivers aged 25 to 45, and the older drivers went through the yellow light least frequently. The main effect of type of vehicle was that people driving conventional cars ran the yellow light most often, followed by people driving sports cars. The drivers of vans and pickups went through least often and, overall, didn't differ from each other. An interaction between age of driver and type of vehicle is present, but is entirely too unpleasant to describe verbally.

Recall now the McDaniel and Vestal source credibility experiment, which varied issue relevance and source credibility of a written handout and measured recall of the information as a dependent variable. As you read the section, you'll notice that in the last paragraph they treat "time of measurement" as a third independent variable and discuss its interactions with the other variables. Some of this may be confusing, but try to understand as much as you can. As you read, try to answer this question:

★ Describe the main effects and interaction of the two main independent variables of interest.

## The Results Section of the McDaniel and Vestal Source Credibility Study

Each subject's retention of communication content was assessed by scores obtained on the six-item questionnaire, the same questions 1 week later and six new questions also 1 week later.

As predicted, variations in levels of source credibility and issue relevance differentially influence communication content retention. Analysis of variance on retention scores (source credibility by issue relevance by retention assessment method) indicates that source credibility and issue relevance interact, $F (1, 40) = 6.04, p < .025$. Figure 1 demonstrates this interaction where scores for the groups are averaged across the three retention assessment methods. A paired-comparisons analysis indicates that the high source credibility/high issue relevance group retained significantly more content than the high source credibility/low issue relevance group, $t (40) = 2.09, p < .05$, and the low source credibility/high issue relevance group, $t (40) = 2.55, p < .02$. Mean scores for the high source credibility/high issue relevance, high source credibility/low issue relevance, low source credibility/high issue relevance and low source credibility/low issue relevance are, respectively, 3.00, 2.18, 2.00, and 2.54.

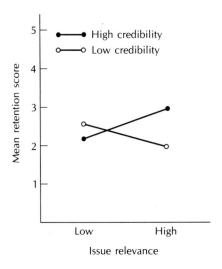

**Figure 1.** Mean retention scores for the four groups.

The analysis of variance also indicates that retention of communication content varies as a function of the time and method of measurement, $F (2, 80) = 34.45, p < .001$. A paired comparisons analysis reveals that immediate retention scores are higher than long-term retention measured by the same questions, $t (80) = 4.31, p < .001$, and long-term retention measured by the new items, $t (80) = 8.33, p < .001$. The difference between identical item and new item long-term retention scores, $t (80) = 4.03, p < .001$, indicates that long-term retention is facilitated on items previously answered. Means for immediate long-term identical and long-term new item retention scores are, respectively, 3.34, 2.41, and 1.54. The nonsignificant interactions between retention assessment methods and source credibility, issue relevance, and source credibility by issue relevance indicate that questionnaire item rehearsal was not confounded with long-term retention.

To summarize, the main finding of interest was the source credibility by issue relevance interaction, which showed that people recalled more from a high credibility source when the issue was highly relevant but recalled more from a low credibility source when the issue was less relevant. The other findings are less important.

### Interactions Occurring When There Are More Than Two Independent Variables[2]

As you've probably noticed, in both the Lutkus mirror image drawing experiment and in the McDaniel and Vestal source credibility experiment there were actually three independent variables. We've discussed the main effects and a number of interactions in each study. In fact, whenever there are more than two independent variables combined factorially in one experiment, each one can have a main effect, and any two can interact with each other. When interactions occur between two independent variables they're called **two-way interactions.**

*Two-Way Interactions*

Discovering and describing interactions between two variables when there are three or more independent variables in an experiment is difficult for many people. However, it gets easier when you recognize that the rule is the same: There is an interaction present between two variables if the relationship between one independent variable and the dependent variable changes at different levels of another independent variable. If there are three independent variables in an experiment, there can be as many as three two-way interactions: A × B, A × C, and B × C. Look at the sample problem in Figure 5.13.

The three independent variables are sex of speaker, sex of listener, and whether the speaker was telling the truth or lying. We'll go through each of the possible effects one at a time.

· *Main effect of sex of speaker:* None. Overall the male and female speakers look into the listeners' eyes about equally.

*Main effect of sex of listener:* None.

*Main effect of lying:* Speakers look directly into listeners' eyes more when lying than when telling the truth.

*Sex of speaker by sex of listener interaction:* When males speak to males they look directly at them less than when they speak to females. The opposite is true for females.

---

[2] This section is designed to present more detailed information on the concept of interactions. Readers not requiring this information may skip to the section on interpreting main effects and interactions.

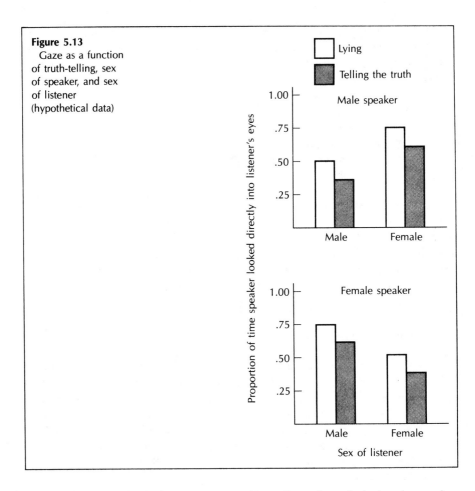

**Figure 5.13**
Gaze as a function of truth-telling, sex of speaker, and sex of listener (hypothetical data)

*Sex of speaker by lying interaction:* None. Regardless of whether the speaker is male or female, they look directly at the listener more when they lie.

*Sex of listener by lying interaction:* None. Regardless of whether the listener was male or female they were looked at more by lying speakers.

In each case the variables not under consideration were ignored; the dependent variable was averaged across the ignored variables.

### Problems

These problems will help you to recognize main effects and two-way interactions in sets of data with three independent variables.

For each set of data describe on your answer sheet any main effects or two-way interactions that are present. If an effect isn't present, write "none" on your answer sheet.

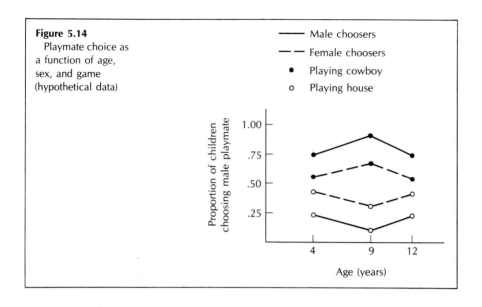

**Figure 5.14**
Playmate choice as a function of age, sex, and game (hypothetical data)

**Q.** See Figure 5.14.

Age:

Sex of chooser:

Game:

Age × sex:

Age × game:

Sex × game:

**A.** There were no effects of age, sex, or age by sex. The main effect of game was that the children tended to choose male playmates more for cowboys than for playing house. The age by game interaction was that with cowboys the preference for males was highest at age 9 and somewhat lower at 4 and 12, whereas with house the preference for males was lowest at age 9 and somewhat higher at 4 and 12. The sex by game interaction was that males showed more preference than females for males when playing cowboys and less preference when playing house.

**Q.** See Table 5.5.

Social class:

Grade:

Word class:

Social class × grade:

Social class × word class:

Grade × word class:

Table 5.5 Mean Length of Maximum List
That Could Be Recalled by Each Child (Hypothetical Data)

| | Word Class | | |
|---|---|---|---|
| | Function Words | Nouns | Verbs |
| Lower-class Children | | | |
| Kindergarten | 3.0 | 3.5 | 3.5 |
| Grade 1 | 3.5 | 4.0 | 4.0 |
| Grade 2 | 4.0 | 4.5 | 4.5 |
| Middle-class children | | | |
| Kindergarten | 4.0 | 4.0 | 4.0 |
| Grade 1 | 4.5 | 4.5 | 4.5 |
| Grade 2 | 5.0 | 5.0 | 5.0 |

**A.** There were main effects of all three independent variables. The middle-class children recalled longer lists than the lower-class children. As grade increased, so did length of list recalled. And overall more nouns and verbs were recalled than function words. There was no social class by grade interaction and no grade by word class interaction, but there was a social class by word class interaction. The middle-class children recalled the same number of each type of word, but the lower-class children recalled fewer function words than verbs or nouns.

You may have noticed that the last practice problem had a strong resemblance to the Schutz and Keislar word class/social class/age experiment. In fact that practice problem was an idealization of their actual results. Recall that in their introduction section they predicted that the social class by word class interaction would occur. They stated it this way: "Hypothesis 1. Deficiency in word recall of lower-class children in the kindergarten-primary grades is greater for function words than for nouns and verbs in relation to the performance of middle class children" (page 14). While reading this results section, notice that all the effects are as described in the last practice problem except, of course, they are less regular.

## The Results Section of the Schutz and Keislar Word-Class Study

The means and SD's for each grade by school are presented in Table 1. The data were analyzed using a 3 × 2 × 3 factorial design (grade, social class, and word class) with one repeated factor. The results showed, as might be expected, significant differences between the social classes, $F(1, 54) = 7.54$, among the grades, $F(2, 54) = 10.34$, and word classes, $F(2, 108) = 10.84$. The interaction predicted by Hypothesis 1 was found to be significant, $F(2, 108) = 6.68, p < .01$. The triple interaction predicted by Hypothesis 2, however, was not significant.

## Table 1
### Means and Standard Deviations on Immediate Memory Test by Word Class, Grade, and Socio-economic Level[a]

| Socio-economic Level | Grade Level | Word Class | | | | | |
|---|---|---|---|---|---|---|---|
| | | Function Words | | Nouns | | Verbs | |
| | | Mean | S.D. | Mean | S.D. | Mean | S.D. |
| Low | Kindergarten | 3.1 | .88 | 3.6 | .70 | 3.4 | .84 |
| | Grade 1 | 3.7 | .48 | 4.5 | .53 | 4.4 | .70 |
| | 2 | 3.7 | .95 | 4.9 | .92 | 4.2 | 1.20 |
| Middle | Kindergarten | 3.9 | .57 | 3.9 | .57 | 3.9 | .57 |
| | Grade 1 | 4.3 | .67 | 4.9 | 1.29 | 4.7 | .68 |
| | 2 | 5.1 | .88 | 5.0 | 1.25 | 4.0 | .47 |

[a] Score is the number of words in longest list repeated in order without error.

### Three-Way Interactions

The last sentence of the Schutz and Keislar results section mentions a triple interaction. Such things do indeed exist and are sometimes called **three-way interactions** because they involve three independent variables.

When a three-way interaction occurs, all three variables together determine the level taken by the dependent variable. The two-way interaction between any two of the variables changes, depending on the level of the third independent variable. In these fictitious data there is only a three-way interaction; there are no main effects or two-way interactions. It would be called an anxiety level by problem difficulty by intelligence interaction. The results are unlikely, but pure three-way interactions are rare. See Table 5.6.

### Table 5.6. Mean Time in Sec. to Solution (Hypothetical Data)

| | High Anxious People | | Low Anxious People | |
|---|---|---|---|---|
| | High IQ | Low IQ | High IQ | Low IQ |
| Tricky problems | 120 | 40 | 40 | 120 |
| Obvious problems | 40 | 120 | 120 | 40 |

To describe the three-way interaction you must describe the whole pattern of results. One good way to do that is to start with one level of one independent variable and describe the interaction between the other two variables and then go to the next level and do the same thing. In this case we might describe the three-way interaction this way: Among high anxious people, those with high IQs solve obvious problems faster than tricky problems, but those with low IQs solve obvious problems slower.

Among low anxious people, those with high IQs solve the tricky problems faster, but those with low IQs solve the obvious problems faster. That's quite a mouthful, but there's no way to shorten it much. It's just a complicated relationship.

Three-way interactions can occur in combination with any number of main effects or two-way interactions. As you've no doubt guessed, Schutz and Keislar predicted a three-way interaction in their second hypothesis. It was phrased this way: "Hypothesis 2. At the kindergarten-primary level, the difference between the social classes for recall of function words, relative to recall of nouns and verbs, is larger for children in the higher grades than for children in the lower grades" (page 14). This prediction along with all their other predictions would have resulted in findings something like those in Table 5.7. Their actual results are reproduced in Table 5.8.

Table 5.7. Mean Length of Maximum List That Could Be Recalled by Each Child (Hypothetical Data)

|  | Word Class | | |
| --- | --- | --- | --- |
|  | Function Words | Nouns | Verbs |
| Lower-class children | | | |
| Kindergarten | 3.0 | 3.5 | 3.5 |
| Grade 1 | 3.1[a] | 4.0 | 4.0 |
| Grade 2 | 3.2[a] | 4.5 | 4.5 |
| Middle-class children | | | |
| Kindergarten | 4.0 | 4.0 | 4.0 |
| Grade 1 | 4.5 | 4.5 | 4.5 |
| Grade 2 | 5.0 | 5.0 | 5.0 |

[a]These are the scores most important to the predicted three-way interaction.

Table 5.8. Schutz and Keislar's Actual Results

|  | Word Class | | |
| --- | --- | --- | --- |
|  | Function Words | Nouns | Verbs |
| Lower-class children | | | |
| Kindergarten | 3.1 | 3.6 | 3.4 |
| Grade 1 | 3.7 | 4.5 | 4.4 |
| Grade 2 | 3.7 | 4.9 | 4.2 |
| Middle-class children | | | |
| Kindergarten | 3.9 | 3.9 | 3.9 |
| Grade 1 | 4.3 | 4.9 | 4.7 |
| Grade 2 | 5.1 | 5.0 | 4.0 |

You can see the results were in the direction predicted by the three-way interaction, but rather weak. At second grade the difference between the lower- and middle-class children in number of function words recalled was greater than at first

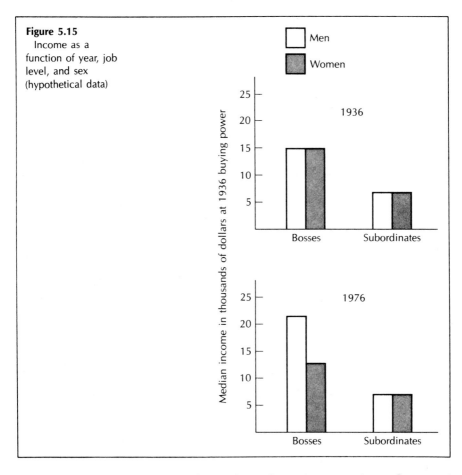

**Figure 5.15** Income as a function of year, job level, and sex (hypothetical data)

grade or kindergarten. However, the analysis of variance, an inferential statistical technique, must have failed to show a significant social class by grade by word class interaction because Schutz and Keislar reported that the triple interaction was nonsignificant.

Just as there can be three-way interactions there also can be four-way interactions or even more. When a four-way interaction occurs, the three-way interaction among any three of the variables changes, depending on the level of the fourth independent variable. However, few experiments involve four-way interactions. They're just too complicated to be understood in a useful way by most people.

### Problems

These problems should help you to understand the concept of three-way interactions. Each set of data shows a three-way interaction along with other effects. Try to verbally describe just the three-way interaction for each set of data.

**Q.** See Figure 5.15.
**A.** The sex by job level by year interaction could be described in this way. In 1936 there was no difference between men and women at the different job levels, but

in 1976 an interaction between sex and job level became apparent such that females who were bosses earned less than males, but there still was no difference between the sexes at the lower job levels.

**Q.** See Figure 5.16.

**A.** The sex by year in school by athletic ability interaction might be described by saying that for women there's no interaction between athletic ability and year in school but for men there is an interaction such that freshman men show more sensitivity if they are high in athletic ability but senior men show less.

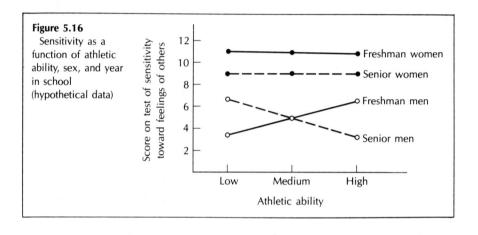

**Figure 5.16** Sensitivity as a function of athletic ability, sex, and year in school (hypothetical data)

## Interpreting Main Effects and Interactions

Now you know how to recognize and describe main effects and interactions, but what do they mean?

When an independent variable is associated with a dependent variable it means that predictions can be made from one variable to the other. If an experiment has two independent variables that don't interact, they can be thought of as completely separate. You can make predictions from each one about the dependent variable. If you put both together, you just add the predictions together. For example, suppose an experiment showed that one kind of study technique led to higher recall than another and that increased time spent studying also increased recall and that there was no interaction. A person could select the superior study technique and expect improved recall. Or recall could be increased by increasing study time. If someone did both, they'd get both benefits added together. As a second example, suppose an experiment showed that people who exercise a lot tend to live longer and that people who smoke tend to have shorter lives and that the variables don't interact. If we have two people, one who exercises a lot and the other who doesn't, we can predict which one will live longer. If we have two different people, one who smokes and the other who doesn't, again we can make a prediction based on that information alone. If we know both facts about a person, we can add them together and get a more accurate prediction than from either one considered separately.

But what does it mean if two independent variables interact? It means that the nature of the association of one variable with the dependent variable depends on the level of the *other* independent variable. To make predictions about the dependent variable we *must* know the levels of both independent variables. For example, suppose the effects of a certain drug interacted with the sex of the recipients: For males injections of the drug cause a decrease in heart disease but for females there's no effect. Does the drug decrease heart disease? The only completely correct answer is that it depends on the sex of the recipient. When two variables interact you can't fully describe the effect of one variable without qualifying your description in terms of the other.

If three or more variables interact, you must extend this line of thinking. In a three-way interaction you can't describe the two-way interaction between any two of the variables without qualifying your description in terms of the remaining variable.

### Interpreting the Dependent Variable

All the studies in this book had dependent variables that are relatively easy to interpret. In some cases high numbers mean better performance, as with number correct or speed of response. In other cases high numbers mean poorer performance, as with number of errors, trials to criterion, or reaction times. A little thought is sufficient to interpret such variables.

However, in many experiments the dependent variables are less straightforward, although they can be interpreted with some effort. For example, one common dependent variable is difference scores. An experiment on the effects of a particular drug on number of eye movements might display its results as in Figure 5.17. With

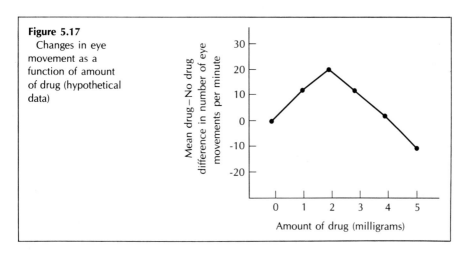

**Figure 5.17**
 Changes in eye movement as a function of amount of drug (hypothetical data)

difference scores there's no information given about the actual level of the dependent variable at each level of the independent variable. Only the difference between some of the levels is presented.

It can be further complicated by converting the difference scores to ratios, as in this present case the results could be presented as *percentage* of eye movements with the drug as compared to without the drug. To read the figures as given you read the level on the graph as the difference in number of eye movements between that dosage and no drug at all. At 0 drug, of course, there's no difference. At a dosage of 2 mg there was an average of 20 more eye movements per minute with the drug as compared to no drug. At a dosage of 5 mg there were 10 fewer eye movements per minute than with no drug.

As another example of a relatively complicated dependent variable we'll consider cumulative responses. An experiment using a particular treatment to reduce walking behavior in pigeons might display its results as in Figure 5.18. To read this figure you must recognize that each time the pigeon takes a step the graph goes up one notch. When it gets to the top (as happened at about 22 min.) it resets at the bottom and starts again. When the pigeon stands still the graph remains at the same level, horizontal. Thus we can see that after treatment began the pigeon's walking behavior decreased considerably. When treatment stopped, the walking behavior remained low for a period but then started again.

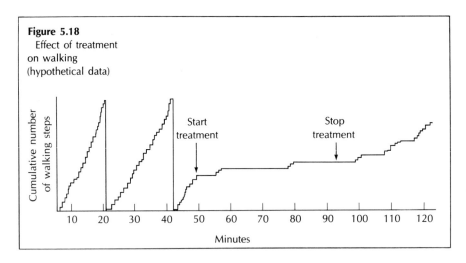

**Figure 5.18**
Effect of treatment on walking (hypothetical data)

Thus with many variables, some study and care is needed to make them interpretable. In still other experiments you may not be able to interpret the dependent variable without learning a good deal more about the research area. If the dependent variable in an experiment were the number of alpha waves showing in an individual's brain waves, it might be displayed as the actual tracing of the machine that records the brain waves, an electroencephalograph. To interpret the results you'd have to learn how to read electroencephalograms. Many research areas have specialized dependent variables. If you try to read an article that contains a variable like that, you'll have to search for information in textbooks about that research area or ask someone for help.

Clearly, you must understand the nature of the dependent variable before you can understand anything about the results, so this is your first step in reading a results section.

**A Word on Factor Analysis**

You may find you'd like to read an article that uses **factor analysis.** Factor analysis is a statistical technique developed out of the correlational approach. An article using factor analysis will almost certainly have some tables of correlation coefficients.

The most common application of factor analysis is in test construction. A test-maker develops, say, 100 test items and gives them to a group of subjects. Of those 100 items a group of perhaps 10 items might be answered in one way by some subjects and in a different way by other subjects. Another group of items might show a different pattern, and so on for all the items. Factor analysis is used to discover these item groups by correlating the answers given by each subject with all the other answers. In an article using factor analysis the items are likely to be displayed in the discovered groups, which are called *factors*. Correlation coefficients may be displayed for each item and called *factor loadings*. Those correlations tell how closely that item correlates with all the other items in the group in terms of the answers given to that item by the subjects.

Obviously, to fully understand an article using factor analysis you should know much more, but this amount of information should be enough to prevent any gross misunderstanding.

**Summary**

You've seen four approaches to describing results. In all four approaches the important information to be conveyed is the relationship between the independent and dependent variables. That information can be given in text, in tables, or in figures. The first step is to make certain you understand the nature of the dependent variable. If you fail to do that, you can't possibly understand what the experiment shows.

In the minimal statistics approach all the data are simply presented and the authors emphasize the important aspects in the text.

In the correlational approach the relationships between sets of data are measured and reported by using correlation coefficients. Inferential statistics may be used to measure the probability that the relationships are accidental.

In the differences-between-frequencies approach the data are summarized by counting similar events and perhaps translating them into percentages or proportions.

Inferential statistics may be used to measure the probability that the distribution of events is accidental.

In the differences-between-means approach the data are summarized by calculating means and measures of variability. Inferential statistics may be used to measure the probability that differences between means are accidental.

In experiments with two or more independent variables in a factorial design the results are often discussed in terms of main effects and interactions. Inferential statistics may be used to measure the probability that these effects are accidental. This is most common with the differences-between-means approach, but discussion of main effects and interactions also is meaningful when the other approaches are used.

These approaches are often used in combination, and there are other approaches as well.

The goal of this chapter has been to give enough information for an intuitive understanding of results sections of many experiments. It has included enough basic concepts to allow the interpretation of many experiments not using highly specialized presentation methods or statistical techniques. However, many kinds of commonly used statistics have been omitted, and not enough information has been given for you to understand discussions of statistical fine points that often arise in results and discussion sections. Moreover, this chapter has not equipped you to detect experimenter errors in choosing or interpreting statistics although such errors certainly occur. One or more courses in statistics are required to give you more complete understanding.

Now it's time to turn to your own sample article and try to answer these questions:

**1.** Exactly *what* are the dependent variables? How do you interpret a high or low score?
**2.** What are the relationships between the independent and dependent variables? Your phrasing of this answer depends on which approach has been used.
**3.** Are there any inferential statistics? Which differences or relationships are statistically significant?

**Extra Library Assignment.** Find examples of articles using each of the four approaches: minimal statistics, correlational, differences between frequencies, and differences between means.

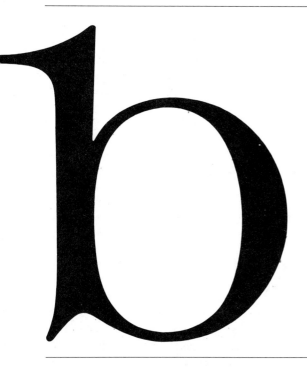

**Section B
Organizing and
Analyzing Your
Results**

Having completed your experiment, you now have a collection of data. To understand it fully you must first decide what general approach to use: minimal statistics, correlational, or one of the two differences-between-levels approaches. After making this decision you'll organize your data and study the pattern of results carefully to discover what relationships between variables are present. If you're using statistics, you should determine which descriptive statistics are appropriate to summarize your data. Finally, you may want to use some inferential statistics to determine which relationships or differences are statistically significant. We'll discuss each of these processes in this section.

The decisions you've already made about the nature of your variables, your number of subjects, your experimental design, and other factors largely determine the kind of statistical procedures that are appropriate for your experiment. In planning future experiments, after you're more familiar with these statistical techniques, you should keep them in mind when designing the experiment. Experienced experimenters already know what statistical procedures they'll be employing even before they run the first subject. (As a beginning experimenter, you couldn't do everything at once, so I left this element until now.)

### Decide Which Statistical Approaches to Use

For each independent and dependent variable pair in your experiment you must decide whether to use primarily the minimal-statistics approach, the correlational approach, or one of the two differences-between-levels approaches. You'll decide later whether to use frequencies or means. If you have more than one independent or dependent variable you may use different approaches for different pairs of variables if you like.

If you have relatively few data and you believe they show a convincing pattern without using any statistics, you may elect the minimal-statistics approach. That's the common approach in the physical sciences, and it is used frequently in physiological research and before-after designs in the behavioral sciences. The advantage is that no data are lost or disguised through summarizing. Disadvantages arise when there are many data to be presented. It may be confusing and is certain to be expensive.

★ List here any independent/dependent variable pairs for which you'll use the minimal-statistics approach.

If you listed all your variables, you can skip to the step on organizing your data. If you have some variables left, you must decide whether to use one of the differences-between-levels approaches or the correlational approach. This decision depends largely on what you want to emphasize. If you want to emphasize the whole relationship between an independent and dependent variable, you should choose the correlational approach. That choice is often most appropriate when your independent variable is nonmanipulated and not selected. The correlational approach emphasizes naturally occurring patterns of relationships between measured variables.

If instead you want to emphasize the differences between the levels of the independent variable, you should choose one of those approaches. That choice is often most appropriate for manipulated or selected independent variables.

★ For each remaining pair of independent and dependent variables decide whether to use the correlational approach or one of the two differences-between-levels approaches.

### Organize Your Results

Regardless of which approach you've chosen you should organize your raw data in such a way that the relationship between each independent and dependent variable is as clear as possible. You may have collected your data in an organized way, perhaps recording them all on one or two well-organized data sheets or having a machine automatically draw a graph showing the relationship clearly. If your data

## Section B  Organizing and Analyzing Your Results

aren't already well organized, you should make up one or more tables showing the relationships among all the independent and dependent variables.

### The Miminal-Statistics Approach

You should make one table for each dependent variable with a title at the top to remind you which dependent variable is shown. In a study using a before-after variable with two subjects you may well have decided to use the minimal-statistics approach. If you had three pretreatment measures, four during-treatment measures, and two posttreatment measures, the organized data might look like Table 5.9. Each space in the table is filled with the score or level of each subject on the dependent variable. Notice how the table aligns all the similar measurements and makes it relatively easy to compare them.

**Table 5.9.** Data Sheet for the Minimal-Statistics Approach

| Number of Errors | | |
|---|---|---|
| Measures | Subject 1 | Subject 2 |
| Pretreatment<br>1<br>2<br>3<br>During treatment<br>1<br>2<br>3<br>4<br>Posttreatment<br>1<br>2 | | |

### The Differences-Between-Levels Approaches

Again, you should make one table for each dependent variable, labeled appropriately. The table should be organized in such a way that measurements in all the similar combinations of independent variables are grouped together. In a 2 × 3 between-subjects design with 20 subjects in each group, the organized data might look like Table 5.10. Each subject's score on the dependent variable would fill the spaces in the table. Leave space at the bottom of the columns for calculations.

Table 5.10. Data Sheet for the Difference-Between-Levels Approach—Example 1

| | Number of Glances ||||||
| --- | --- | --- | --- | --- | --- | --- |
| | Males ||| Females |||
| | Age (years) ||||||
| Subject | 10–19 | 20–29 | 30–39 | 10–19 | 20–29 | 30–39 |
| 1 | | | | | | |
| 2 | | | | | | |
| 3 | | | | | | |
| 4 | | | | | | |
| 5 | | | | | | |
| 6 | | | | | | |
| 7 | | | | | | |
| 8 | | | | | | |
| 9 | | | | | | |
| 10 | | | | | | |
| 11 | | | | | | |
| 12 | | | | | | |
| 13 | | | | | | |
| 14 | | | | | | |
| 15 | | | | | | |
| 16 | | | | | | |
| 17 | | | | | | |
| 18 | | | | | | |
| 19 | | | | | | |
| 20 | | | | | | |
| | | | | | | |

In a design with three levels of a within-subject variable on which each of 10 subjects was measured 5 times on each level, the data might look like Table 5.11.

Table 5.11. Data Sheet for the Difference-Between-Levels Approach—Example 2

| | Estimated, Line Length ||||||||||||||| 
| --- | --- | --- | --- | --- | --- | --- | --- | --- | --- | --- | --- | --- | --- | --- | --- |
| | Amount of Light |||||||||||||||
| | Dark ||||| | Medium ||||| | Light |||||
| | Measurement Number |||||||||||||||
| Subject | 1 | 2 | 3 | 4 | 5 | | 1 | 2 | 3 | 4 | 5 | | 1 | 2 | 3 | 4 | 5 |
| 1 | | | | | | | | | | | | | | | | | |
| 2 | | | | | | | | | | | | | | | | | |
| 3 | | | | | | | | | | | | | | | | | |
| 4 | | | | | | | | | | | | | | | | | |
| 5 | | | | | | | | | | | | | | | | | |
| 6 | | | | | | | | | | | | | | | | | |
| 7 | | | | | | | | | | | | | | | | | |
| 8 | | | | | | | | | | | | | | | | | |
| 9 | | | | | | | | | | | | | | | | | |
| 10 | | | | | | | | | | | | | | | | | |

Each subject's score on the dependent variable at each time of measurement fills the spaces in the table. To make it easier to use statistics to summarize and analyze your results, leave space at the ends of rows and columns.

*The Correlational Approach*

In the correlational approach the distinction between independent and dependent variables is less important, so the organization is somewhat different. In a study using three variables where each of 40 subjects was measured just once on each variable, you might have decided to use the correlational approach to emphasize the pattern of relationships among the variables. Table 5.12 shows one way you could organize your data.

**Table 5.12.** Data Sheet for the Correlational Approach

| Education, Sex, and Response | | | |
|---|---|---|---|
| Subject | Years of Education | Sex | Response to Questionnaire |
| 1 | | | |
| 2 | | | |
| 3 | | | |
| 4 | | | |
| 5 | | | |
| • | | | |
| • | | | |
| • | | | |
| 37 | | | |
| 38 | | | |
| 39 | | | |
| 40 | | | |
| | | | |

Each subject's level on each variable would fill the spaces in the table. For some variables that would be a number; for nominal variables it might be a word naming the level such as freshman or male or yes. It might help you to see the relationships between the variables if you ordered the subjects in some way, perhaps grouping all the males and females together or ordering one of the variables from lowest to highest. However, that's not necessary. You'll simplify later calculations if you leave space between the columns and to the right and some space beneath the columns.

*All Approaches*

When you organize your data you needn't get it all onto one sheet, but it should all be organized in an easy-to-understand way. Put a title at the top of each table to remind you of the data and purpose of your experiment; you might want to look back at your results sometime in the future. Also be sure you identify the dependent measure recorded in the table. Label all the rows and columns well enough so that you'll be able to look back at your table and know what the numbers mean. Don't just label things "Level 1" and "Level 2"; give them useful names.

When you copy your data onto the table be especially careful to avoid mistakes. Any errors you make will enter into any statistics you use and, more importantly, will make your results false. Carefully copy and proofread the table for errors before continuing.

★ Now organize your raw data.

### Look at Your Results

After organizing the data it should be relatively easy for you to study them for relationships between the independent and dependent variables. Statistics are a useful tool, but they're not a substitute for an intuitive understanding of your results. Look at your data. Are groups of high and low numbers on the dependent variable related to differences in the independent variables?

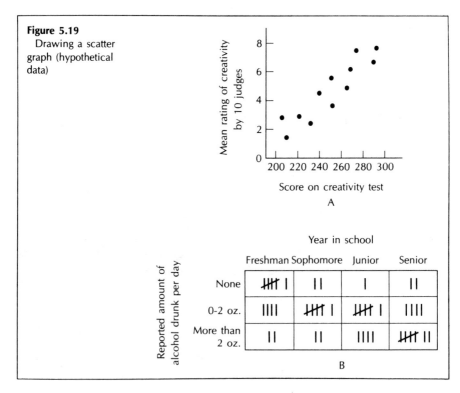

**Figure 5.19** Drawing a scatter graph (hypothetical data)

*Drawing a Scatter Graph*

If you're using the correlational approach, it may be useful to draw a scatter graph to show the relationships between your variables.

Draw a graph with two axes. On the horizontal axis arrange all the levels of an independent variable; on the vertical axis arrange all the levels of a dependent variable. On the graph place one dot or X relative to each pair of measurements; that often means one symbol for each subject. Some examples are shown in Figure 5.19. If both variables have a large number of ordered levels, as is true with the variables in A, you should be able to look at your scatter graph and guess whether there's a relationship between the two variables, as we discussed in Section A of this chapter. Be especially careful to notice whether the relationship is more or less in a straight line or if it curves. If there's a pronounced curve, that can be quite interesting in its own right, but you should remember it because there are statistical implications.

If one or both variables have relatively few levels, as in B, many pairs of scores would lead to dots at exactly the same point. You can keep better track by drawing boxes for each possible pair and placing symbols in the appropriate boxes. If your scatter graph has much overlap, as in B, you can organize your data further by counting the number of symbols in each box. Calculating meaningful proportions or percentages can make various relationships much easier to see. The data from B are redisplayed in Table 5.13 in their new organizations. Notice that although they all show the same data, the relationships emphasized are different. You may want to try several different organization techniques to see which one appears more meaningful to you.

**Table 5.13.** Organizational Techniques (Hypothetical Data)

| Reported Amount of Alcohol Drunk per Day | Freshmen | Sophomores | Juniors | Seniors | Total |
|---|---|---|---|---|---|
| | Number of students in each group | | | | |
| None | 6 | 2 | 1 | 2 | 11 |
| 0–2 oz. | 4 | 6 | 6 | 4 | 20 |
| More than 2 oz. | 2 | 2 | 4 | 7 | 15 |
| Total | 12 | 10 | 11 | 13 | 46 |
| | Proportion of each class reporting various levels of alcohol use | | | | |
| None | .50 | .20 | .09 | .15 | |
| 0–2 oz. | .33 | .60 | .55 | .31 | |
| More than 2 oz. | .17 | .20 | .36 | .54 | |
| Total | 1.00 | 1.00 | 1.00 | 1.00 | |
| | Percentage of each alcohol use group from each class | | | | |
| None | 55 | 18 | 9 | 18 | 100 |
| 0–2 oz. | 20 | 30 | 30 | 20 | 100 |
| More than 2 oz. | 13 | 13 | 27 | 47 | 100 |

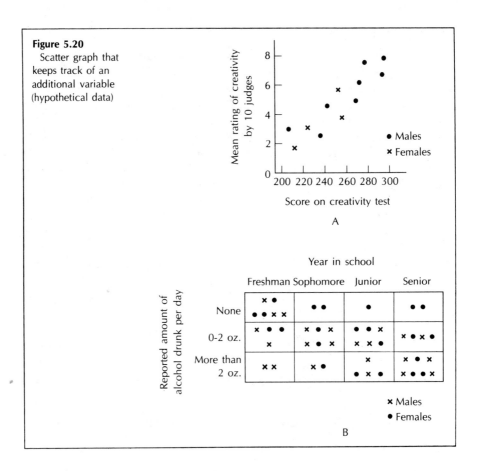

Figure 5.20
Scatter graph that keeps track of an additional variable (hypothetical data)

You also can use a scatter graph to keep track of an additional variable by using different symbols corresponding to the levels of that variable, as shown in Figure 5.20. The scatter graph for creativity shows that there's a generally positive correlation between the test scores and creativity ratings *and* that females generally score lower on both measures than males. The scatter graph for alcoholic consumption shows that reported alcoholic consumption generally increases with year in school and that, as a group, females report less drinking than males.

### Drawing a Frequency Distribution

If you're using one of the differences-between-levels approaches, a different kind of visual display may be useful: a **frequency distribution.**

Arrange all the levels of one dependent variable on the horizontal axis of a graph, leaving plenty of space between the levels. The vertical axis shows the number, or frequency, of scores at each level of the dependent variable. Therefore, you should

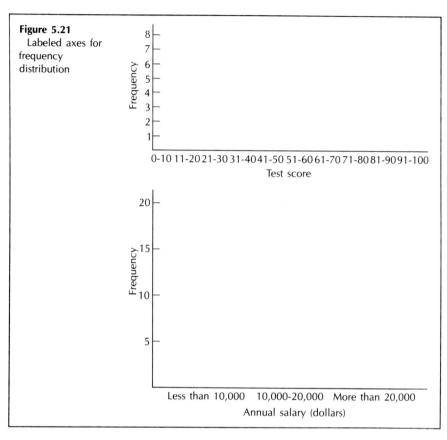

**Figure 5.21** Labeled axes for frequency distribution

number up the vertical axis to whatever level seems reasonable. Some unfilled graphs for frequency distributions are shown in Figure 5.21.

To see the relationship between that dependent variable and one independent variable, choose a symbol like X or 0 for each level of that independent variable. Start with one level and place one symbol for each event above the appropriate level of the dependent variable. Pile the scores up to show the frequency at each level, as shown in Figure 5.22. Then do the same thing for another level of the independent variable, placing these symbols just to one side of the others, as shown in Figure 5.23.

After placing all the symbols, look at the distributions you've created to get an intuitive feeling for differences between the different levels of the independent variable. In the study method data, for example, it appears that the scores for Study Method 1 are higher than the scores for Method 2. In the income data it appears that the college group had higher incomes than the noncollege group.

As with the correlational approach, it may be helpful in some instances to further organize your data by making a table showing the frequencies and perhaps calculating proportions or percentages to emphasize interesting relationships. This was done for the income data with the results shown in Table 5.14.

**Table 5.14.** Displays of Differences between Frequencies (Hypothetical Data)

| Annual Salary (Dollars) | College | Noncollege | Total |
|---|---|---|---|
| | Number of executives in each group | | |
| Less than 10,000 | 5 | 6 | 11 |
| 10,000–20,000 | 4 | 16 | 20 |
| More than 20,000 | 20 | 10 | 30 |
| Total | 29 | 32 | 61 |
| | Proportion of executives at each salary level with college and noncollege educations | | |
| Less than 10,000 | .45 | .55 | 1.00 |
| 10,000–20,000 | .20 | .80 | 1.00 |
| More than 20,000 | .67 | .33 | 1.00 |
| | Percentage of college and noncollege executives at each salary level | | |
| Less than 10,000 | 17.2 | 18.8 | |
| 10,000–20,000 | 13.7 | 50.0 | |
| More than 20,000 | 69.0 | 31.2 | |

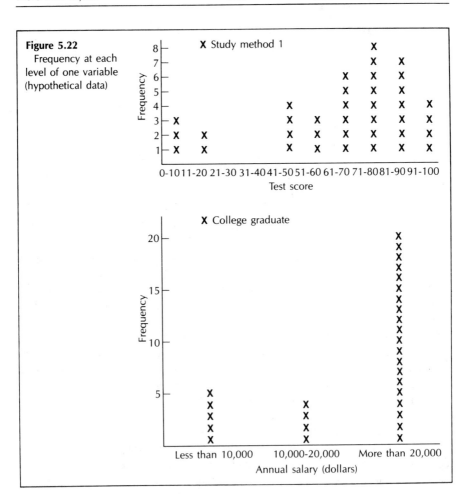

**Figure 5.22** Frequency at each level of one variable (hypothetical data)

★ Now look at your results, using a scatter graph or frequency distribution if either appears helpful. You also might use a frequency table transformed into proportions or percentages to help clarify the relationships.

**Determine the Scale
of Measurement of Your Variables**

Different summary techniques, display formats, and statistical analyses are appropriate for data measured in different ways. When you "measure" a variable, you assign numbers to the levels in some way. We'll discuss in detail three possible tech-

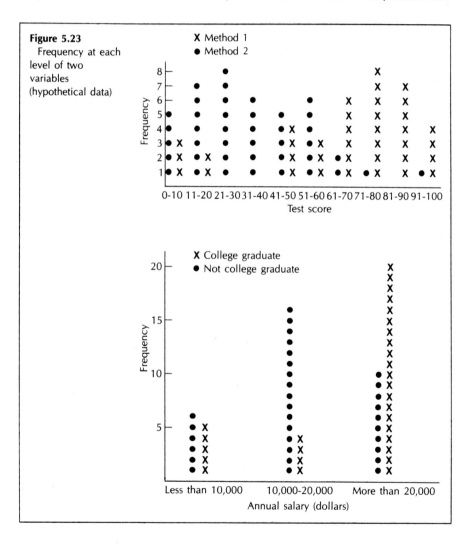

**Figure 5.23**
Frequency at each level of two variables (hypothetical data)

niques for assigning numbers, which lead to three different **scales of measurement: nominal, ordinal,** and **interval.** You must determine the scale of measurement of each of your variables in order to determine what display formats are appropriate and what descriptive and inferential statistics can be applied usefully to your data.

### Variables Measured on a Nominal Scale

In Chapter 2 we discussed nominal variables. These are some examples: occupation, marital status, position on a football team, response of agreement or disagreement, and buying or not buying a product. Numbers can be assigned to the levels of a nominal variable quite arbitrarily. For example, for marital status, single = 1, married = 2, divorced = 3, and widowed = 4. These numbers are meaningless; arithmetic manipulation of them makes no sense at all. If you have a group consisting of two single and two divorced people, it makes no sense to say that 1 + 1 + 3 + 3 = 8; 8 ÷ 4 = 2, and therefore the mean score = married! Moreover, the order of the levels is meaningless. It would be just as sensible to say divorced = 1, widowed = 2, married = 3, and single = 4.

### Variables Measured on an Ordinal Scale

Ordinal variables can't be arranged arbitrarily; their order is meaningful. These are some examples: class standing, highest academic degree earned, and problem difficulty (in some definitions). In class standing one student is first, another is second, and so on; they can't be rearranged in assigning numbers to them. In highest academic degree earned the order from high school to bachelor's to master's to doctorate is clear. If a teacher made up six problems knowing that one was the easiest, another somewhat harder, and so on, problem difficulty would be an ordinal variable.

Although the order is meaningful in an ordinal scale, it's not clear how much difference there is between levels. In class standing there may be a small difference between the two students ranked 1 and 2 and a large difference between the ones ranked 4 and 5. However, the difference in the numbers assigned is the same in both cases: The difference is one. In problem difficulty there may be a big difference between Problems 1 and 2 and very little real difference between 5 and 6. Thus the assigned numbers don't necessarily reflect the interval between levels.

With ordinal variables it makes sense to say that different levels have more or less of the variable being measured. However, many common arithmetic manipulations still lack meaning. Suppose the teacher made up three pairs of problems: 1 and 6, 2 and 5, and 3 and 4. Each pair has the same "mean rank" of 3.5. However, all the pairs aren't of the same real mean difficulty. The reason is that the numbers aren't really good reflections of the differences in difficulty of the problems.

*Variables Measured on an
Interval Scale (or Stronger)*

Interval variables accurately reflect the size of the intervals between levels. Temperature is measured on an interval scale; there's as much difference between 5 and 10 degrees as between 40 and 45 degrees.

Many variables in the behavioral sciences are assumed to be interval variables, although the assumptions aren't necessarily true. For example, a 5-point rating scale might be assumed to have equal intervals between the points, although it may not be true that the interval between "disagree very much" and "disagree mildly" is equal to the interval between "disagree mildly" and "neither agree nor disagree." Similarly, the number of errors on a test is usually treated as an interval measure of test difficulty, although the difficulty difference between 1 and 2 errors might be different from the difficulty difference between 9 and 10 errors. In general most rating scales and test scores are assumed to have interval scales. Although there's some disagreement, many behavioral scientists feel that these assumptions do little harm unless the size of the intervals differs grossly. For example, few researchers would treat this "liking" scale as interval: 1 = dislike violently, 2 = neutral feelings, and 3 = like at least a little. It would be treated as an ordinal scale.

With interval variables most common arithmetic manipulations are meaningful. If one person solves 8 problems and another solves 10, it seems reasonable to say that the average number of problems solved was 9.

You may have heard of an even stronger scale of measurement called a **ratio scale.** Ratio scales have a true zero point; some examples are physical length, weight, and time. For our purposes there's no need to distinguish between interval and ratio scales; we'll group them all together as interval scales.

To decide which scale of measurement is used for a particular variable, determine how numbers were assigned to the levels of the variable. If there are no numbers, or if the numbers are completely arbitrary, the scale is nominal. If numbers reflect the order, but probably don't reflect the size of the difference between the levels, the scale is ordinal. If you assume that the numbers do reflect the size of the interval, you may assume that the scale is interval.

★ For each of your independent variables, state whether you assume it's measured on a nominal, ordinal, or interval scale. Then do the same for your dependent variables.

Do you have any nominal variables with only two levels? If so, you should answer one other question: Are the levels of the variable representative of particular points on one underlying continuum such as rich/poor, difficult/easy, or agree/disagree or are they really discrete, that is, separate such as male/female, monkey/rat, or story instructions/picture instructions? The answer to this question affects the display techniques and statistics that are appropriate for you.

★ For any nominal variables with only two levels, state whether it's likely to be really discrete or whether there's an underlying continuum.

For any variables you assumed to be measured on an interval scale you should now answer one more question: Are the scores on that variable likely to be distributed in a way radically different from a normal, bell-shaped distribution? That is, are most of the scores near the center, with relatively few scores extremely high or low, or are almost all the scores near one extreme or the other? If you drew a frequency distribution, looking at it now would be helpful. If you didn't, you might consider drawing one now, with the variable in question on the horizontal axis.

If one of your variables were the number of published papers of a group of 20 graduate students, the distribution would be far from normal. You might find that 15 students had 0 publications, 4 students had 1, and 1 student had 12. Application of statistics such as the mean to distributions of that kind gives misleading results. In that example the mean number of publications would be .80, or nearly one publication per student. However, that number would be far from representative of average performance. A different kind of statistics should be used, as will be discussed shortly.

★ For any variables you assumed to be measured on an interval scale, state whether its distribution is likely to differ radically from a normal, bell-shaped curve.

**One New Skill:
How to Read Flow Charts**

Soon you'll be deciding which of several descriptive statistics, inferential statistics, and display formats are appropriate for your data. This book presents that decision-making information in the form of flow charts. A flow chart is a series of operations you go through one at a time, following arrows from one operation to the next. In the flow charts in this book there are four kinds of operations. One kind is a direct instruction such as "Use Pearson $r$." That's pretty straightforward. After performing the operation you go on to the next one. A second kind is a question such as "Does the variable have an interval scale?" If you answer a question with a "yes," you should follow the "yes" arrow; if you answer "no," you should follow the "no" arrow. The remaining two operations are "Start" and "Stop." They're pretty obvious.

Here's an example of a flow chart for determining the scale of measurement of a variable. Step through it, one operation at a time, to be sure you understand how flow charts work. First step through it with the variable "number on the back of racing cars." Then you should step through it with one or two of your variables to make sure you come up with the same answer as before. See Figure 5.24.

You should have started with "Start" and then named "number on the back of racing cars." The answer to the first question is "yes," since the levels do have numbers, so follow the arrow to the next question. The answer to that question also is

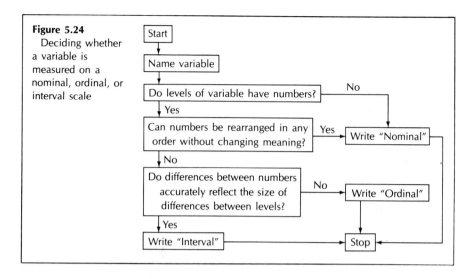

Figure 5.24
Deciding whether a variable is measured on a nominal, ordinal, or interval scale

"yes," since the assignment of numbers to racing cars is completely arbitrary, so following the arrow you should write "Nominal" and then go to "Stop."

We'll be using this tool throughout the rest of this chapter. In the next step you'll be deciding which, if any, descriptive and inferential statistics to use. The statistics we've discussed most—the mean, standard deviation, analysis of variance, and Pearson r—are all what are called **parametric statistics.** As you'll see from working through the flow charts, they should be used only with variables assumed to have interval scales of measurement, and that do not deviate radically from a normal distribution. Other variables should be summarized and analyzed with **nonparametric statistics,** of which we've mentioned a few: the mode, median, $X^2$ test, Wilcoxon test, and Mann-Whitney U. As the final step in each of several flow charts, you're instructed to use a certain statistic. Instructions for how to calculate a few of the most common are included in Appendix C of this book. However, if you have another statistics book with which you're familiar, you should use it because its instruction is probably more complete. For the statistics not included in this book there are many excellent explanatory texts; I won't venture to recommend any above others. There are also a number of good "cookbook" types of statistics books without explanation. One is by Bruning and Kintz, the *Computational Handbook of Statistics,* published in 1977 by Scott, Foresman. Another is by Linton and Gallo, *The Practical Statistician: Simplified Handbook of Statistics,* published in 1975 by Brooks/Cole. If you use these or any other cookbook approach, keep in mind that you really need a complete knowledge of statistics to understand the meaning of your calculations fully and to have greater flexibility in the procedures you can use.

### Summarize Your Data

*The Minimal-Statistics Approach*

For any variables on which you plan to use the minimal-statistics approach you won't summarize the data at all.

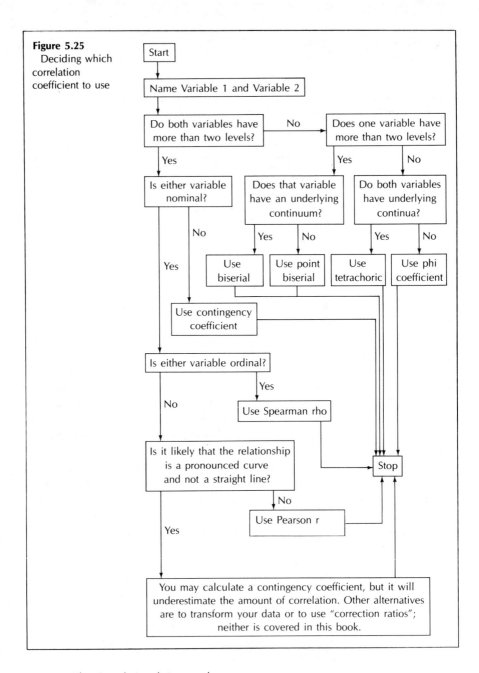

**Figure 5.25** Deciding which correlation coefficient to use

### The Correlational Approach

If you're using the correlational approach, the descriptive statistic you want is a correlation coefficient. The correlation coefficient you calculate is determined by the nature of the two variables and of the relationship between them.

For any pair of variables for which you plan to use the correlational approach step through the flow chart shown in Figure 5.25 to determine which correlation coefficient is appropriate for your data.

★ Which correlation coefficient will you use?
★ Depending on your choice, now calculate the appropriate statistic. Instructions on calculation of the Pearson r are included in Appendix C. For the other statistics you must refer to other books. A book by Siegel called *Non-parametric statistics* and published in 1956 by McGraw-Hill is especially helpful for most of these calculations.

### The Differences-Between-Levels Approaches

If you're using either of these approaches, the descriptive statistics you're most likely to want are a measure of the average, or central tendency, of the dependent variable at each level of each independent variable and a measure of variability. The statistics you calculate are determined by the scale of measurement of the dependent variable.

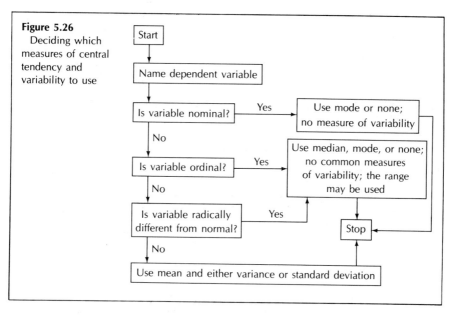

**Figure 5.26** Deciding which measures of central tendency and variability to use

For any pair of variables for which you plan to use one of the differences-between-levels approaches, step through this flow chart to determine which descriptive statistics are appropriate for your data. As you can see, you'll use the mean only when the dependent variable is measured on an interval scale and isn't likely to be distributed very differently from normal. Thus these are the only circumstances in which you can use the differences-between-means approach. When your dependent variable doesn't meet those assumptions, you must use the differences-between-frequencies approach. See Figure 5.26.

★ Which measures of central tendency will you use for each variable? Will you also calculate a measure of variability?

★ Depending on your choice, now calculate the appropriate statistics for each level of your independent variables. Instructions on how to calculate the mode, median, range, mean, variance, and standard deviation are all included in Appendix C of this book.

### Display Your Results

Whether or not you've summarized your data by using descriptive statistics, you must decide how to display it so that you can tell others what you've found. There are three basic formats you can use: sentences in text, tables, or figures. Text is the cheapest and easiest one if you have relatively little data, perhaps just three or four means or a few correlation coefficients. Figures make it easy to see relationships between one dependent variable and possibly two or three independent variables. However, in published papers they're expensive, so publishers tend to discourage their use. Tables can be used to show the relationships among any number of variables and are relatively efficient and easy to publish. However, it's harder for the reader to see the relationships in a table than in a figure.

In one experiment you may want to use more than one presentation technique. For example, if you have two dependent variables, you may want to present the data relevant to one of them in written form and the other in a figure. Whenever you present data in a table or figure, you must also describe verbally the general pattern. However, any set of numbers should ultimately be presented only once; presenting them twice is inefficient.

If you have relatively few data, you should use a written presentation. Figures aren't really good for displaying more than three independent variables or more than one dependent variable. If you have more than that, you should plan to display the results in a table or subdivide your results so that you can display them in more than one figure. You should use figures only if you believe their use will make your results easier to understand.

*Presenting the Results in Text*

Form coherent sentences that make it clear what numbers go with what. For example, six means in a 3 × 2 design could be reported as "The treatment means were males = 10, 9, and 8 errors for image, story, and standard instructions, respectively; females = 11, 9 and 7 errors."

*Presenting the Results in Tables*

An example of a slightly complicated table is shown in Table 5.15. A list of the important elements follows.

Table 5.15. Means and Standard Deviations for Trials to Criterion, Number of Errors, and Total Recall for All Situations and Anxiety Groups (Hypothetical Data)

| | Anxiety Level | | | | | | | |
|---|---|---|---|---|---|---|---|---|
| | High ($n = 17$) | | Medium ($n = 20$) | | Low ($n = 18$) | | Mean | |
| | Situation | | | | | | | |
| | Test | Neutral | Test | Neutral | Test | Neutral | Test | Neutral |
| Trials to criterion | 10.0 | 8.9 | 9.5 | 8.7[b] | 9.0 | 8.6 | 9.5 | 8.7 |
| | (1.2) | (2.0) | (1.1) | (1.5) | (2.3) | (.9) | | |
| Number of errors[a] | 27.3 | 16.2 | 24.2 | 15.1[b] | 20.9 | 14.8 | 24.1 | 15.3 |
| | (4.6) | (6.2) | (4.3) | (4.7) | (5.1) | (5.0) | | |
| Total recall | 90.0 | 95.4 | 91.3 | 97.0[b] | 95.6 | 99.0 | 92.3 | 97.2 |
| | (10.6) | (12.3) | (11.1) | (14.2) | (9.2) | (10.6) | | |

[a] Errors were defined as omissions only. Misspellings and synonyms were counted as correct.
[b] One medium anxiety subject was not measured in the neutral situation. Thus $n = 19$.

1. The shape of the table should be rectangular. You need not draw in all the possible horizontal and vertical lines, but the table should be aligned in such a way that all lines *could* be drawn.

2. The rows and columns should be arranged in a logical order.

3. The column and row labels should be as clear as possible within space limitations and should be placed directly in line with the row or columns for which they're meant.

4. At minimum the title should identify all the dependent variables displayed in the table. Other information may be included as well. A title that says "Experimental results" is insufficient.

5. Footnotes should be used for lengthy comments about parts of the table.

*Presenting the Results in Figures*

There are three kinds of figures commonly used in the behavioral sciences: bar graphs, line graphs, and scatter graphs. (See Figure 5.27.) All three have one independent variable on the horizontal axis and one dependent variable on the vertical axis. One or two independent variables also may be presented within the figure in a coded form. To decide which kind of figure is appropriate for your data, step through the flow chart shown in Figure 5.28.

★ Which kind of figure, if any, will you use to display your results?

Some examples of relatively complicated graphs are shown in Figure 5.27 to guide you in the construction of your own figures. As with tables, there are some conventions relevant to the construction of figures.

**Figure 5.27**
Examples of complicated graphs (hypothetical data)

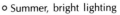

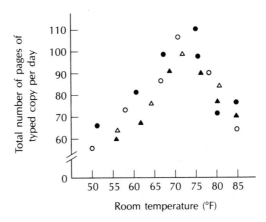

A. Effects of room temperature, season, and lighting on typing performance of 15 typists in insurance company typing pool (hypothetical data)

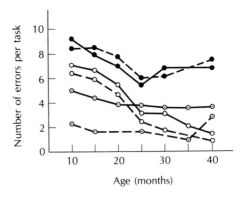

B. Relationships between age and type of task for two rats (hypothetical data)

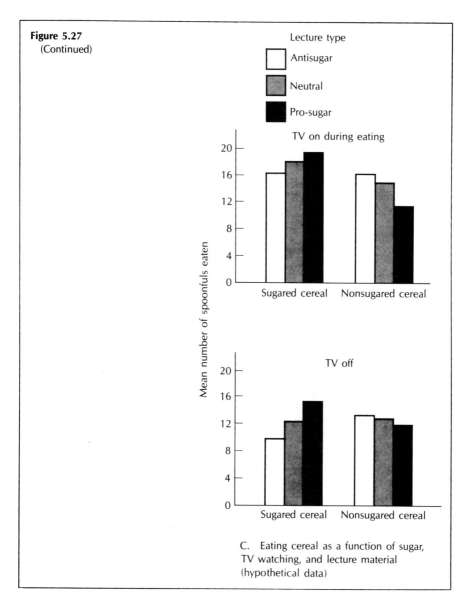

**Figure 5.27** (Continued)

C. Eating cereal as a function of sugar, TV watching, and lecture material (hypothetical data)

1. The shape of the figure should be generally rectangular.
2. The independent variable should be on the horizontal axis, and the dependent variable should be on the vertical axis.
3. The levels on both axes should be clearly labeled. The arrangement should have low numbers at the bottom of the vertical scale and on the left of the horizontal scale.
4. Whenever possible the vertical scale should be arranged so that a zero

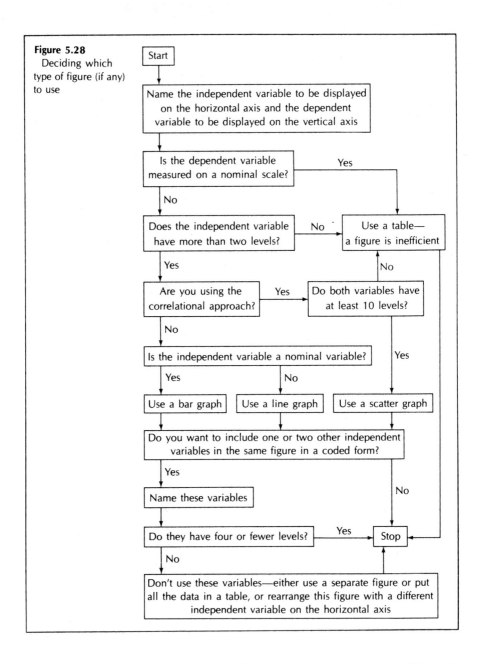

**Figure 5.28** Deciding which type of figure (if any) to use

point falls at the bottom. If that's not possible, it's customary to make a small break in the vertical axis.

5. Both axes should be clearly labeled.
6. The levels of any additional independent variables should be clearly labeled.
7. There should be an explanatory title beneath the figure.

★ Now make up sentences, tables, or figures to display your results.

### Think Again about Your Results

Now that you've summarized your results, look at them again and think about them. What's the answer to the question you were asking in your experiment? What are the associations between your independent and dependent variables? Are correlations positive or negative? Do there appear to be main effects or interactions?

These are your results. They're the facts demonstrated by your experiment. We'll discuss their interpretation in the next chapter.

### Use Inferential Statistics

Unless you're using the minimal-statistics approach you should probably use some inferential statistics to measure the probability that your pattern of results is accidental. The type of inferential statistic you use depends on the nature of your experiment and of your variables.

*Minimal-Statistics Approach*

No inferential statistics are used.

*Correlational Approach*

You may want to test the statistical significance of your correlation coefficient. If it's statistically significant, that means that the relationship you found between the two variables probably is not accidental. If someone were to repeat your experiment, he or she could reasonably expect to find a relationship in the same direction. If the relationship isn't statistically significant, then the relationship between your two variables apparently isn't different from .00. If someone were to repeat your experiment, the correlation might be close to zero or even in the opposite direction from yours.

*Differences-Between-
Levels Approaches*

You can use a variety of parametric or nonparametric statistics to tell you the probability that the differences are accidental between levels of the independent variable you observed. If you compare two levels of an independent variable and find that the frequencies or means of the dependent variable differ significantly, it means that if someone were to repeat the experiment, he or she would probably find a difference in the same direction.

Sometimes you may be comparing several means or frequencies at one time. If you do so and find a significant difference, it means that at least two of the levels differ significantly, but not necessarily more than that. Thus, if you had the results shown in Table 5.16, and inferential statistics showed that the frequencies differed significantly, it might only mean that Level 1 is significantly different from Level 6. It's possible that none of the other differences are significant.

Table 5.16. Results (Hypothetical Data)

|  | Level of Independent Variable | | | | | |
| --- | --- | --- | --- | --- | --- | --- |
|  | 1 | 2 | 3 | 4 | 5 | 6 |
| Frequency or mean | 10 | 12 | 14 | 11 | 16 | 20 |

You can use a technique called **analysis of variance** to test the statistical significance of main effects and interactions separately if you have a design with two or more independent variables combined factorially and a dependent variable of the right kind. Sometimes that helps you to understand the results better.

If your variables are not combined factorially and you want to look at the significant differences in more detail, you can use one of several post hoc comparisons after completing an analysis of variance. In this book I mention only the Newman-Keuls test, but some others include the Duncan's multiple-range test, the Tukey test, and Scheffe's test. All are discussed in most intermediate statistics books.

*Decide Which*
*Inferential Statistics to Use*

The flow chart in Figure 5.29 and Table 5.17 are designed to help you choose among some of the most commonly used inferential statistics. This list is by no means exhaustive.

Each dependent variable must be analyzed separately. If you have more than one dependent variable you should step through the flow chart once for each one. Sometimes you may want to step through the flow chart separately for each independent variable as well.

★ What inferential statistics will you use for which variables?
★ Depending on your decision, calculate the appropriate inferential statistics now and use them to decide which of your results are likely to recur if someone were to try to repeat your experiment. Instructions on the calculation of $X^2$, both t tests, the analyses of variance listed, and the Newman-Keuls test are included in Appendix C.

# Section B  Organizing and Analyzing Your Results

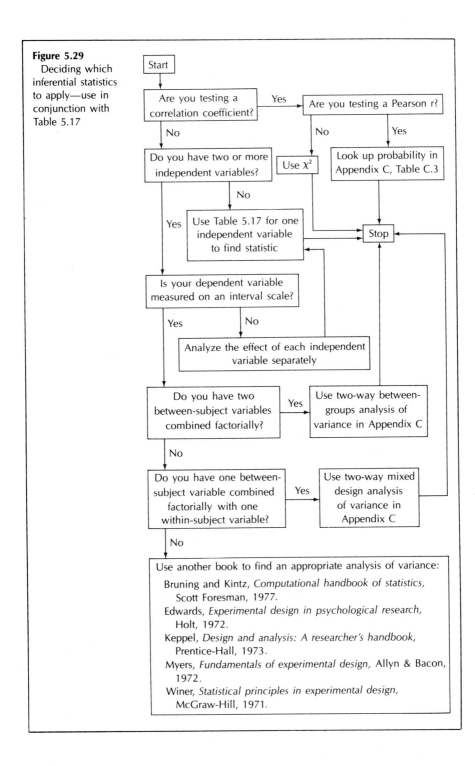

**Figure 5.29**
Deciding which inferential statistics to apply—use in conjunction with Table 5.17

**Table 5.17.** Inferential Statistics to Use with One Independent Variable (Hypothetical Data)

| Scale of Measurement of Dependent Variable | Characteristics of Independent Variable | | | |
| --- | --- | --- | --- | --- |
| | Only Two Levels | | More Than Two Levels | |
| | Between Subject | Within Subject | Between Subject | Within Subject |
| Nominal | $\chi^{2a}$ | McNemar test | $\chi^{2a}$ | Cochran Q |
| Ordinal | Mann-Whitney U test | Sign test | Kruskal-Wallis one-way analysis of variance | Friedman two-way analysis of variance by ranks |
| Interval | Independent-measures $t$ test[a] | Related-measures $t$ test[a] | One-way between-groups analysis of variance,[a] perhaps followed by Newman-Keuls[a] | One-way within-subjects analysis of variance,[a] perhaps followed by Newman-Keuls[a] |

[a]Calculational procedures given in Appendix C. The other statistics can be found in Seigel's *Non-parametric statistics,* McGraw-Hill, 1956, as well as in many other books.

This is a summary of the steps you took with your experimental results:

    **1.** Decide which statistical approaches to use for each independent and dependent variable pair.
    **2.** Organize the results, and look at them.
    **3.** Determine the scale of measurement of the variables.
    **4.** Summarize the data, if appropriate, by using descriptive statistics.
    **5.** Display the results so that others can understand them.
    **6.** Think again about the results.
    **7.** Use inferential statistics to calculate the probability that your pattern of results is accidental.

Chapter Six
What Do the Results Mean?

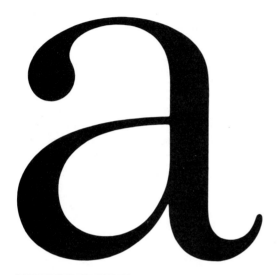

**Section A
Reading
Discussion
Sections**

The purpose of discussion sections is to present the authors' conclusions about the events reported in the method and results sections. These conclusions are normally related to the issues raised in the introduction section. If the experiment was designed to discover relationships among variables, the conclusions consist of descriptions of the demonstrated relationships. If specific questions were raised in the introduction, the conclusions consist of answers to the questions. If the experiment was designed to test a prediction from a theory, the conclusions consist of statements about whether the predictions were supported.

Many authors state conclusions in the form of explicit predictions or causal relationships between variables. Others are less explicit. They assume that the relationships are obvious to the reader and go beyond them to more subtle implications. Depending on their interests, some authors emphasize practical implications of the events while others emphasize theoretical implications. Some authors are content simply to point out the relationship between the observed events and the issues raised in the introduction section.

You need not agree with the authors' conclusions or interpretations. It sometimes happens that your own interests are different from the researchers', leading you to look for different practical or theoretical implications. On some occasions it happens that authors draw conclusions not really warranted by their experiments. In effect, you may want to write your own discussion section, reinterpreting the experimental observations in light of your own interests and views about the strengths and shortcomings of the experiment.

In reading discussion sections you may have several tasks. You must understand the conclusions described or assumed by the authors. You should decide whether you believe these conclusions are warranted. Sometimes you may want to reinterpret the experiment, looking for implications that interest you.

The discussion section from the Yeatts and Brantley typing experiment with the child with cerebral palsy was only one sentence long. The question raised in the introduction section of that paper concerned whether this kind of fine motor performance could be improved. Recall that the facts were that during the four sessions when the child received a raisin for each correct word she typed, her typing rate increased from about 1 to about 2 words per minute. Based on these facts, Yeatts and Brantley concluded the following: "These data clearly demonstrate that consistent improvement in fine motor performance, although variable, of this cerebral palsied child could be made under carefully controlled and reinforced conditions in spite of cerebral palsy and doubts of hospital personnel" (page 198).

### Recognize the Conclusions Drawn by the Authors

The most direct conclusions drawn from an experiment consist of a statement of the relationship between variables. Yeatts and Brantley concluded that their independent variable, the raisin treatment, caused a change in the dependent variable, typing rate.

### Evaluate the Conclusions

Do you think that conclusion is warranted? You've already learned many skills necessary for critically evaluating authors' conclusions.

*Can the Results Be Explained on the Basis of Confounded Variables?*

As we discussed in Chapter 4, the increase in typing rate could have been caused by novelty effects, by subject or experimenter bias, or by a number of variable-connected or extraneous-variable confounds. Personally, I'm willing to assume that these confounds don't matter much in this experiment.

*Are the Results
Trustworthy, or Likely
to Be Due to Chance Factors?*

In Chapter 5 you learned that inferential statistics can help you decide whether results are trustworthy, but you must always depend on your best judgment for the final decision. In this case there were no inferential statistics. Because of the nice overall pattern of the increase in typing rate when the treatment started and the decrease when it stopped, I'm willing to believe that the child's typing rate didn't just accidentally increase at the moment the treatment was added. Again, I could be wrong.

*Do the Conclusions Follow
Logically from the Results?*

This is a question we haven't considered in detail. Suppose Yeatts and Brantley had said something like "This treatment is superior to other techniques for increasing typing speed" or "Young children are more likely to benefit from such treatment than older children." These conclusions would be completely unwarranted by the experiment, because the necessary comparisons weren't performed. To draw conclusions about untested comparisons is quite inappropriate, but occasionally it does happen in published articles.

Another kind of difficulty that can arise involves drawing conclusions about levels of the independent variable that weren't included in the experiment. If Yeatts and Brantley had suggested that their treatment would have been more effective if they had continued longer, that conclusion wouldn't be based on their data. In many cases, however, it seems quite appropriate to generalize to other levels. For example, if an experiment produced the results shown in Figure 6.1 it would seem reasonable to assume that at 20 min. of study time about 60 percent of the material would be recalled. It might even be reasonable to guess at the amount likely to be recalled at 40 min. of study time, but that would be riskier.

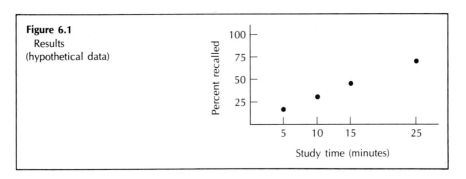

**Figure 6.1**
Results
(hypothetical data)

A third possible problem is that conclusions may go too far afield from the operational definitions used in the experiment. Yeatts and Brantley would be out of line to suggest that their treatment caused improvement in the child's language abilities. They measured typing speed; to use that as an operational definition of language ability would be unreasonable.

Yet another possible problem is that a conclusion might be drawn that involves a hidden assumption of causality that isn't warranted. For example, an experiment might demonstrate that students in schools with remedial reading programs generally perform better than students from schools without such programs. It's tempting to conclude that all schools should have remedial reading programs, but that involves a hidden assumption of causality: the program *caused* improvement. Since the experiment involved a nonmanipulated variable, the causal conclusion isn't warranted, and the suggestion that practical action be taken isn't based logically on the experimental findings.

There are other logical errors that can arise in drawing conclusions from experiments in addition to those I've mentioned. It's important for you to question the logical relation of any conclusion with the experiment on which it's supposedly based.

**Problems**

These problems give practice in recognizing conclusions that don't follow logically from the experiment. In each case criticize the conclusion that is stated (that is, try to tell what's wrong with it) and give a more correct one.

Q. A personality test was given to a group of mature doctors rated "eminent" by their colleagues and to a group of mature doctors all of whom had lost a malpractice suit. There were large differences on several test scales between the groups. Conclusion: This test could be given to medical school applicants to accurately predict who will later be eminent and who will be likely to be found guilty of malpractice.

A. This conclusion is unwarranted because no comparison was made that showed any personality differences between the two types of doctors when they were applicants to medical school. A more appropriate conclusion is that these types of doctor differ in their personalities in reliable ways. If you know that a mature doctor is eminent, you could make predictions about his or her personality based on these results.

Q. An experiment was performed in which three randomly assigned groups of students each took a math test. Before giving the test, the experimenter looked at the thermometer in the room (which the students couldn't see). To one group she said, "My goodness it's hot in here . . . 80 degrees." To another group, she said, "My goodness it's cool in here . . . 60 degrees." She said nothing about the temperature to the third group. In fact, the temperature was 70 degrees in all cases. The results showed that the "hot" subjects performed worst, the "cold" subjects next, and the normal subjects best. Conclusion: High temperatures cause poor performance, cool temperatures are somewhat less damaging, and normal room temperatures cause the best performance.

**A.** The conclusion is unwarranted because room temperature wasn't varied. The operational definition of the independent variable doesn't match the conclusion drawn about it. A more appropriate conclusion might be that *beliefs* about room temperature cause differences in performance.

**Q.** A total of 100 men and 100 women were stopped on the street and asked whether they read their horoscopes. In both groups 50 people said yes. They were asked whether they thought the horoscopes were accurate; 45 of the women said yes but only 15 of the men. Conclusion: Horoscopes are more accurate for women than for men.

**A.** This conclusion is unwarranted because the accuracy of the horoscopes wasn't measured. The operational definition of the dependent variable doesn't match the conclusion drawn about it. A more appropriate conclusion would be that more women than men say that horoscopes are accurate.

**Q.** Three groups of 20 families were interviewed and asked how "satisifed" they were with their lives. In one group the combined take-home pay of the family members was between $50,001 and $60,000. In the second group it was between $60,001 and $70,000. In the third group it was between $70,001 and $80,000. Overall there was no difference in satisfaction between the groups. Conclusion: Income is unrelated to satisfaction.

**A.** This conclusion is warranted for the three income groups measured, but it's not appropriate to go beyond the groups actually compared. If families with incomes of $5,000 or $2,000,000 had been included, differences might well have been found. A more appropriate conclusion is more limited: Between $50,000 and $80,000 income is unrelated to satisfaction.

**Q.** A group of high anxious people and a group of low anxious people all took tests on which they were required to make creative responses in order to solve a series of abstract problems. The high anxious people solved significantly fewer problems. Conclusion: The higher a person's anxiety, the fewer problems of this sort he'll be able to solve.

**A.** Again, this conclusion is warranted for the two groups compared, but it's unwarranted to go beyond these levels. If people with medium anxiety had been included, it's actually likely that they would have solved more problems than either group. A more appropriate conclusion is that high anxious people solve fewer problems of this type than low anxious people.

**Q.** Sixty baby rabbits were all separated from their mothers and raised in isolation until they were mature. Then they were mated and had babies of their own. Forty-five (75 percent) of them paid no attention to their offspring. Conclusion: Being deprived of their own mothers generally causes rabbits to pay no attention to their young.

**A.** This conclusion is unwarranted because the deprived rabbits weren't compared to anything. How would nondeprived rabbits behave? We have no idea. The only conclusion that can be drawn is that when rabbits are separated from their mothers about 75 percent pay no attention to their young.

**Q.** A group of 20 students memorizes a new list of words for each of 40 consecutive days. They do this by going through a deck of cards containing the words until they can recite the whole list out loud without looking at the cards. The time required to learn each list is measured. During the course of the experiment 10 of the students are given a new experimental drug on even-numbered days; the other 10 are given the drug on odd-numbered days. The results showed that all the students required about twice as long to learn the list on drugged days. Conclusion: The drug causes a temporary intellectual impairment.

**A.** The conclusion is unwarranted because it's not clear that the effects of the drug are intellectual. It might affect the students' motivation, speech, vision, attention, or even coordination. A more appropriate conclusion is that the drug caused their learning to slow down.

**Q.** A group of workers was given a mood test every working day for a month. The results showed that people were least happy on Mondays and had a steady increase until Friday. Conclusion: The people got happier because they were looking forward to the weekend.

**A.** This experiment had nothing to do with discovering the cause of the increase in happiness. No conclusion can be drawn other than that there was an increase. This information could be used to predict happiness levels on various days.

**Q.** A group of doctors collected information on the blood cholesterol levels of their patients for a number of years. They found that individuals with high blood cholesterol levels were much more likely to have either heart attacks or strokes than individuals with lower levels. Conclusion: Cholesterol in the diets of patients similar to these patients should be decreased.

**A.** No evidence has been gathered that a decrease in dietary cholesterol would be beneficial. This conclusion involves at least one hidden causal conclusion: The high cholesterol levels caused the illnesses. Since cholesterol level was a non-manipulated variable, this experiment doesn't allow causal conclusions to be drawn. Since there's no evidence that cholesterol level increased the rate of illness, it's not appropriate to conclude that decreasing the cholesterol would decrease illness. A more appropriate conclusion is that individuals with high blood cholesterol levels can be expected to have a higher rate of heart disease and strokes.

It seems to me that Yeatts and Brantley's conclusion follows logically from their experiment. Since I'm willing to discount the confounded variables *and* to believe that their results weren't accidental *and* to think their conclusion is logical, I'm willing to accept their conclusion. You must make your own decision.

We'll look next at the discussion section from the Cross and Davis marijuana experiment. The original question asked in the introduction was "Is frequency of marijuana use associated with adjustment?" The experiment demonstrated these facts: (1) For 178 University of Connecticut psychology student volunteers, adjustment (as measured by the Incomplete Sentences Blank, or ISB) wasn't significantly different for groups of students reporting different levels of marijuana use, up to moderate dosage levels. (2) The 12 students who reported very heavy drug use had significantly higher maladjustment scores than the other students. (3) A higher proportion of students reporting drug use than of students reporting no drug use said that they had professional counseling at some time.

As you read this section, try to answer these questions:

**1.** What conclusions were drawn by Cross and Davis about the association between frequency of marijuana use and adjustment?
**2.** Are their conclusions warranted?

## The Discussion Section of the Cross and Davis Marijuana Study

The results presented in Table 1 suggest little or no relationship between marijuana use and adjustment, as shown by ISB scores, across the whole range of college student users and nonusers. A post hoc analysis, however, of the heaviest drug users, who were heavy marijuana smokers, showed that they admitted to more maladjustment than did the sample as a whole. Given the high frequency of marijuana use on college campuses, these results are not surprising. Any behavior that is so commonplace (56% in this study and a recent Gallup Poll says 42% nationwide) is unlikely to be greatly maladaptive even though individual users have become seriously disturbed (Keeler, 1968). Furthermore, even though moderate usage is not associated with maladjustment, it is plausible that very heavy users are in considerable psychological distress.

Regarding the ISB, it should be noted that strict application of Rotter's (Rotter & Rafferty, 1950) suggested cutoff of 135 would classify over half this sample ($M = 136.19$) as maladjusted. The college population has changed a great deal in the last 20 years and one of the major changes has been a willingness to state thoughts and faults directly. This social frankness leads to ISB responses which are classifiable in a pathological direction. The ISB is still a sensitive instrument with clearly demonstrated validity (Murstein, 1965) and its usefulness for comparative purposes is not impaired.

As an index of maladjustment, some mental health professionals would point to the fact that more drug users had been in counseling or psychotherapy than nonusers. Others might defend the idea that psychological assistance does not admit to maladjustment any more than it does to an introspective attitude and curiosity about one's thoughts and behavior. We believe, however, that since the type of counseling or therapy more readily available to these students was relatively short-term, crisis-oriented treatment from the student mental health center, it is tenable to regard this finding as an index of maladjustment. This is still not conclusive evidence, however, for only a small number of either users or nonusers had been in treatment. If student drug users are more introspective and creative than nonusers as Keniston (1968, 1969) believes, this introspectiveness might also be associated with having sought psychotherapy. Seeking psychological assistance is a multidetermined phenomenon, just as maladjustment is.

*Conclusions*

Marijuana use in the college population is not in and of itself associated with maladjustment. Heavy drug use, however, is related to maladjustment. One might surmise that drug education programs on the college level should focus on the desire for changing usage among heavy users, rather than trying to alter the behavior of the modal, occasional user.

---

What conclusions were drawn by Cross and Davis about the association between frequency of marijuana use and adjustment? Based on their findings they concluded that there was no association, with the added suggestion that perhaps heavy drug use is related to maladjustment. On that basis they tentatively suggested that it would be desirable to change the use patterns of heavy drug users.

Are their conclusions warranted? The experiment was designed to compare the adjustment of individuals who use marijuana in different amounts. The direction of the results agrees with their conclusions.

With respect to their main conclusion, it's appropriate that they should say that no association was demonstrated between the two variables. The fact that both variables were nonmanipulated prevents us from drawing causal conclusions, but statements about associations are warranted. However, as we discussed in Chapter 5, there's a statistical problem in that a lack of relationship can't be supported statistically. It could be that a relationship really exists but wasn't detected by this experiment.

Their suggestion that heavy drug use is associated with maladjustment suffers from two problems. One is that it's based on questionable post hoc statistics, as discussed in Chapter 5. Another is that we should keep clearly in mind that this group of heavy marijuana users also used many other drugs heavily. The other drugs might be responsible for any maladjustment.

Their additional suggestion that the use patterns of heavy marijuana users should be changed suffers from these problems and from another one: It implies a hidden assumption of a causal relationship between heavy drug use and maladjustment. Only if marijuana use *caused* the maladjustment would a reduction in marijuana use cause a decrease in maladjustment. Since marijuana use was a nonmanipulated variable this experiment allows no causal conclusions to be drawn. Thus this conclusion doesn't follow logically from the experiment.

In addition to these difficulties all the conclusions suffer from questionable operational definitions. It's clear that Cross and Davis were aware of this problem, because they devoted much of the discussion section to arguing that their definitions were adequate.

Were Cross and Davis actually measuring "frequency of marijuana use" and "adjustment?" The first was operationally defined as responses to a written questionnaire and the second, as scores on the Incomplete Sentences Blank.

A high score on the drug questionnaire could indicate that the student uses marijuana very frequently; that's what Cross and Davis assumed. However, a high score could mean instead that the student wants to call attention to himself or herself or wants to invalidate the experiment or made a mistake in filling out the questionnaire. Similarly, a low score might come from a student who really doesn't use marijuana often, as the authors assumed, or from one who uses it frequently but is afraid to say so or who wants to invalidate the experiment. Thus the experiment may not tell us about "frequency of marijuana use," although it certainly tells us about "*reported* frequency of marijuana use."

The major operational definition of adjustment in this study was the score received on the Rotter Incomplete Sentences Blank, a test designed to measure maladjustment. What does a high score on that test mean? It could mean that the student is maladjusted, as assumed by Cross and Davis. But they point out that maladjustment isn't the only explanation of a high score on that test; social frankness also causes high scores. That could mean that a student with a high ISB score isn't necessarily maladjusted but is more willing to report symptoms than other students. Cross and Davis recognized that possibility but didn't believe that it mattered in their experiment. They present a reference that presumably gives evidence that the ISB correlated with adjustment in 1965. If you knew more about the ISB you might be able to think of other interpretations for high and low scores on the ISB.

There was also a post hoc kind of operational definition of adjustment: the report of the student about having professional counseling. What does an affirmative response mean? Cross and Davis assumed that it meant the student had sought counseling because of maladjustment. However, it could also mean that the student had counseling not because of maladjustment but because of boredom, curiosity, or some other reason. It could even mean that the student has really had no counseling but decided to say the opposite for some reason. (Of course, you could say that *that's* maladjusted.)

Thus one possible interpretation of the facts demonstrated by the study is that, indeed, maladjustment isn't associated with marijuana use up to moderate levels but heavy drug use is associated with maladjustment. Another possible interpretation is that heavy drug use isn't associated with maladjustment but rather that heavy users are more willing to report symptoms experienced equally by students at all levels of marijuana use. Another possible interpretation, based on the counseling information, is that drug use is associated with a tendency toward introspection, or curiosity about mental processes. Yet another possible interpretation is that there is no relation between *reported* drug use and adjustment but there may be an association between real drug use and adjustment. You can devise other interpretations as well.

In summary, then, it's not clear that Cross and Davis's conclusions are warranted by their experiment. Their operational definitions aren't entirely convincing, so it's not clear that the experiment really tells us about marijuana use and adjustment. The finding of no relationship between the variables can't be supported statistically and could have occurred just because there was too much variability in the experiment. The two statistically significant differences were based on post hoc comparisons and are therefore a little suspicious; they might be accidental. The suggestion that it might be desirable to change the use patterns of heavy drug users suffers both statistically and because it involves a hidden assumption of causality.

From the results of this experiment I'm willing to conclude that if any relationship at all exists between frequency of marijuana use and adjustment, then the correlation is negative. There's little in this experiment to suggest that marijuana users are *better* adjusted than nonusers. I believe it's inappropriate to suggest on the basis of these results that it's desirable to change the marijuana use patterns of heavy drug users. However, it may be true that heavy marijuana use is a good predictor of maladjustment.

### Practical Implications
### of Experimental Results

Neither Yeatts and Brantley nor Cross and Davis chose to emphasize practical implications of their findings, but an interested reader could do so. The most important thing to remember when deriving practical implications is that they require generalization from the experimental subjects, situation, and operational definitions to more practical situations.

Someone reading the Yeatts and Brantley cerebral palsy article might be interested in the possibility that hospital personnel could use their treatment to help children with cerebral palsy improve their typing skills. Additional generalization might suggest that children afflicted with other diseases also might be helped or that improvements could be made in other fine motor skills. Without direct testing it's not clear that such generalization would be warranted. However, it does seem promising.

In the case of the Cross and Davis marijuana study one practical implication might be that therapists shouldn't expect people who say they use marijuana a lot to be very different in adjustment from other people. You can probably think of other implications. However, be careful not to start assuming that this experiment demonstrated something about the causal effects of marijuana use on adjustment. For example, it might be tempting to suggest on the basis of this experiment that no one should use marijuana in an attempt to improve their adjustment. But the experiment didn't show that smoking marijuana causes no improvement in adjustment, so that practical implication would be unwarranted.

We'll look next at the discussion section from the Wallace et al. teaching methods study. Recall that they used three different methods to teach counselor trainees how to help students with decision-making. Method 1 was the traditional lecture and reading, Method 2 added a film, and Method 3 added the film plus a session in which the trainees were supervised and criticized during a trial counseling session.

Their original question was a practical one: "Do the more complicated and expensive methods cause sufficiently more learning to warrant their inclusion in regular courses?" Their results showed that for their 54 beginning graduate students, the ones taught by Method 3 were judged to be significantly better counselors on the basis of two tape-recorded counseling sessions.

As you read this section try to answer these questions:

1. What conclusions were drawn by Wallace et al. about the relationship between the teaching methods and the performance of their counselor trainees?
2. Do you agree with their conclusions?
3. What practical implications do Wallace et al. suggest for their findings?
4. Do you believe that such generalization is warranted?

**The Discussion Section
of the Wallace et al.
Teaching-Method Study**

The traditional approach to teaching decision-making counseling (lectures and assigned readings), even when fortified by a filmed instructional model, is apparently inferior to a teaching method incorporating some of the features of microcounseling. Students who role played decision-making counseling in a small group and who received videotaped and supervisory feedback on their performance displayed substantially better decision-making counseling skills than did those taught by the other two methods alone. These results are in accord with other studies that have shown microcounseling techniques can be implemented effectively in the teaching of complex counselor behaviors (Bellucci, 1972; Ivey, 1971).

The experimental teaching method employed in this study differed somewhat from the traditional microcounseling paradigm. Students were treated in small groups rather than individually, and furthermore, they each practiced only part of rather than the complete decision-making model among themselves before being posttested on the entire process with a coached client. Whereas the original microcounseling paradigm might require extensive time commitments on the part of an instructor, the modification employed here can be incorporated into an existing academic course in little more than a couple of class periods.

What conclusion did Wallace et al. draw? They concluded that microcounseling caused better learning than either other method.

Is that conclusion warranted? It does appear to follow logically from the results. Moreover, it seems unlikely that the results were accidental. The same pattern of results occurred twice, in two different semesters; both times inferential statistics showed significant differences between the methods. However, some problems may be caused by confounded variables, as we discussed in Chapter 4. Wallace et al. don't mention anywhere that their results might have been caused by subject or experimenter bias effects. Nor do they mention the possible confounds arising from the fact that the trainees receiving each method were in different class sections. Either they were unaware of those problems or felt them to be too unimportant to merit discussion. I'll assume that the conclusion is warranted, but with reservations.

What practical implications do Wallace et al. suggest for their findings? They implied that microcounseling techniques like these would be effective in any course designed to train counselors.

Is that generalization warranted? As we discussed in Chapter 3, I think it's probably safe to generalize to other students and situations. If I were teaching a course to counselor trainees, this study is persuasive enough for me to try to include microcounseling. If the technique were very expensive or the consequences of failure very serious, the possible confounds would cause me to hesitate. However, their technique appears reasonably simple, and no one will be hurt even if microcounseling isn't a very effective teaching tool.

Next we'll look at the discussion section of the Loftus and Zanni question-wording study, in which practical implications are discussed at greater length. The purpose of that experiment was to determine how a change from *a* to *the* in the wording of a question affected the answer to the question. Recall that there were actually two experiments, of which you read only one. The results of both experiments are reproduced in Table 6.1.

**Table 6.1.** Percentage of "Yes," "No," and "I Don't Know" Responses to Items That Were Present and Not Present in the Film

| Response | Present | | Not Present | |
|---|---|---|---|---|
| | "the" | "a" | "the" | "a" |
| Experiment I | | | | |
| Yes | 17 | 20 | 17 | 7 |
| No | 60 | 29 | 72 | 55 |
| I don't know | 23 | 51 | 13 | 38 |
| Experiment II | | | | |
| Yes | 18 | 15 | 20 | 6 |
| No | 62 | 28 | 69 | 56 |
| I don't know | 20 | 57 | 11 | 38 |

In their discussion section Loftus and Zanni state their conclusions and discuss both theoretical and practical implications of their findings. We'll discuss the theoretical aspects of the study shortly. For now, read the section and try to answer these questions:

1. What conclusions were drawn by Loftus and Zanni?
2. Are these conclusions warranted?
3. What practical implications do they suggest?
4. Are these generalizations warranted?

**The Discussion Section of the Loftus and Zanni Question-Wording Study**

The research reported here required subjects to view a film of a traffic accident and then to answer questions about the film. A major finding was that the questions containing an indefinite article led to many more "I don't know" responses. The phrase "a broken headlight" could refer to any of a number of broken headlights which a subject might have seen. Since it is impossible to inspect all of the headlights carefully in the time allowed for viewing the accident, it is impossible to be sure that no headlight was broken. In this case, then, the subject must deal with uncertainty about whether a broken headlight actually existed at all, and a larger number of "uncertain" or "don't know" responses result. On the other hand, "Did you see the broken headlight?" more strongly implies the existence of a specific broken headlight, and the subject need not deal with the uncertainty about whether the broken headlight existed. This finding may also indicate that it is easier to be confident that you have not seen some specific item than to be confident that you have not seen any instance of a general class of items.

A second result is that questions containing a definite article resulted in a greater number of false recognitions ("recognition" of events that had never occurred). At least two explanations for this finding are possible. One is that the definite article produces a bias favoring a "yes" or "no" response; in other words, *the* changes a subject's criteria for how much objective evidence he needs to say "yes" or "no." The other is that the definite article leads a subject to infer that the object was in fact present, causing for some a reconstruction in their original memory for the event. While the present data cannot differentiate between response bias and reconstructive memory explanations, a recent study (Loftus & Palmer, 1974) does not indicate that questions asked subsequent to an event can cause a reconstruction in one's memory of that event. In that study subjects viewed films of automobile accidents and then answered questions about events occurring in the films. The question "About how fast were the cars going when they smashed into each other?" elicited a higher estimate of speed than "About how fast were the cars going when they hit each other?" Furthermore, on a retest 1 week later, those subjects who received the verb *smashed* were more likely to say "yes" to the question "Did you see any broken glass?" even though broken glass did not exist in the accident.

The implication of these results for courtroom examinations, police interrogations, and accident investigations is fairly clear cut. The main aim of interrogations conducted by attorneys before the court, for example, is to provide information about events which have actually taken place. Different forms of questions can be consciously used to elicit desired answers from a witness, and also to create a desired influence upon the jury. In the present research, the indefinite article elicited more false responses. Questions which either by form or content suggest to the witness the answer desired or "lead" him to that desired answer are called "leading questions" in the courtroom, and the existence of rules for excluding them (e.g., Supreme Court Reporter, 1973) is a definite recognition of their power of suggestion. While an attorney can seemingly easily "sense" when to object to a leading question asked by another attorney, the definition of leading is a long way from being precise. Any complete definition must eventually consider the subtle suggestibility that individual words can carry with them.

---

What conclusions were drawn by Loftus and Zanni? One conclusion was that phrasing the question with *the* caused many fewer "I don't know" responses. A second conclusion was that phrasing a question with *the* caused more false recognitions. The remainder of the material in the first two paragraphs involves possible theoretical explanations for that pattern of results; the last paragraph discusses practical implications.

Were Loftus and Zanni's conclusions warranted? I think so. Looking at the results, one finds that the conclusions logically follow from them. There seems to be no reason to believe that the results were caused by confounded variables. The decrease in "I don't know" responses was quite strong in both experiments, so it seems unlikely that it's accidental. Moreover, that result was supported by inferential statistics. The increase in "yes" responses about objects which weren't present is somewhat weaker and wasn't supported by inferential statistics. However, the pattern was the same in both experiments, so it seems that that effect was also real.

What practical implications do Loftus and Zanni suggest? They suggest that the same pattern of results would be likely to occur in "real-life" courtroom or interrogation situations with witnesses to real events.

Are these generalizations warranted? As we discussed in Chapter 3, I have some doubts. A graduate student who has watched a film knowing that questions will be asked about it may not respond the same way to leading questions as a witness to a real accident or crime. If I were a police officer trying to learn the truth about real events, this study is strong enough to influence me to use *a* questions rather than *the*. However, I'm not sure that it would really make any difference in the answers given by real witnesses.

## Theoretical Implications of Experimental Results

In this discussion section as in many others the authors speculated about various possible theoretical implications of their results.

To account for the decrease in "I don't know" answers caused by *the* questions Loftus and Zanni suggested that phrasing the question with *the* actually changed the meaning of the question. The reason why people answer "I don't know" when asked about *an* object is that they're answering the question "Was an object present?" But when people are asked about *the* object, they assume it was present and answer the question "Did you *see* the object?"

To account for the increase in "yes" responses to questions about objects that weren't there Loftus and Zanni offered two theories. One, the response-bias theory, suggests that phrasing a question with *the* makes a person more willing to make a decision based on less objective information. Since the person is more willing to say "yes" or "no" he or she is more likely to make mistakes. The other theory, the memory reconstruction theory, suggests that phrasing a question with *the* sometimes makes the person believe he or she really saw the object.

### *Purposes of Theories*

Why do authors include such theoretical speculation in their experiments? Why bother to have theories at all? Why not just perform experiments to learn each thing we need to know?

One reason has to do with efficiency. There are so many facts that we must organize them in some way. Theories organize knowledge. They make it possible to think in terms of larger, general patterns. Once the knowledge is organized it can be applied to predict or control events. Since it's important for theories to organize knowledge, many discussion sections propose a theory to account for the results of the experiment and then mention other published experiments showing results that also are explained by the theory.

A second reason that theories are useful is that they direct research in fruitful directions. They make predictions, often called hypotheses, that are then tested in additional experiments. The results of Loftus and Zanni's experiment caused them to formulate some theories. One theory, in turn, suggested a hypothesis that led them to perform another experiment, the one briefly described as Loftus and Palmer. In many discussion sections the authors suggest some follow-up experiments based on hypotheses predicted by their theory.

A third reason why scientists use theories is to try to satisfy intellectual curiosity. It's interesting to know that people asked *the* questions tend to report that they saw objects that weren't really present. But it's even more interesting to know *why*. In a sense theories attempt to explain why events occur as they do. As it turns out, the "explanations" often arouse as much curiosity as they satisfy.

*Evaluating Theories*

How can you decide whether a proposed theory is a good explanation of the results in an experiment?

First, the theory should be logically consistent with the results. If some results disagree with the theory, then the theory is inaccurate or incomplete in some way. The theories suggested by Loftus and Zanni to explain their results do seem to me to be logical explanations of their results. However, I think there's an error in the presentation of the extra experiment, the one reported as Loftus and Palmer. The results of that experiment showed that the phrasing of questions affected people's answers to *other* questions even a week later. Yet Loftus and Zanni tell us that these results did "not indicate that questions asked subsequent to an event can cause reconstruction in one's memory of that event." That interpretation seems illogical to me. The results really appear to support the memory reconstruction theory. I suspect that the word "not" is a typographical error. However, it certainly illustrates the necessity to be critical of authors' interpretations of their own findings.[1]

Looking more closely at Loftus and Zanni's results, I think there's additional support for the memory reconstruction theory. There's an interesting relationship in the data: With *the* questions there was almost no difference in the number of "yes" answers about objects that were or weren't present. "Yes" answers by people asked *the* questions were unreliable indicators of whether an object was present. According to the response-bias theory, people asked *the* questions require less objective evidence to answer either "yes" or "no." Since there should be more objective evidence of presence about objects that were really present, this theory suggests to me that those objects should have more "yes" answers. That didn't happen, so that theory apparently isn't a good explanation of the results. This relationship appears to agree more closely with the memory reconstruction theory, that subjects asked *the* questions assumed that the object was present and sometimes even added it to their memory representation. When the subjects answered "yes" they really "saw" the object in memory even though it hadn't been present in the film.

A second criterion of a good theory is that it also accounts for results not in the experiment. Good theories show the commonality of many experimental results. Loftus and Zanni don't discuss any other results that could be accounted for by their theories. However, the reader is free to find such results.

A third criterion of a useful theory is that it predicts other, untested results. The fact that the memory reconstruction theory caused Loftus and Palmer to perform an additional experiment suggests that theory is fruitful. Are there other experiments implied by their theories?

---

[1] Another interpretation of Loftus and Zanni's statement was pointed out by an alert reviewer. Loftus and Zanni could have meant that reconstruction did not occur *anew* in the second week as a result of the second week's question, but occurred only at the time of the first questions, lasting through the second week. In that case there's no typographical error but only an ambiguous sentence. In either case the point is the same: Don't trust authors to do your thinking for you.

Loftus and Zanni suggested that questions phrased with *the* cause people to assume that such objects were present. That could be tested in an experiment in which people were asked either *a* or *the* questions and were then asked to *guess* whether the objects were present. If *the* questions do cause people to assume that objects were present, then these people would guess "yes" more often than people asked *a* questions.

Thinking along a different line, one could ask how subjects asked *the* questions decided to answer "yes" or "no." One possibility is that sometimes they added the object to their memory representation and sometimes not; if they added it, they answered "yes." A second possibility is that they actually added *all* the suggested objects to their memory representations but tried to answer honestly about whether they had seen the object in the film. These two possibilities might be compared in an experiment in which all the subjects were questioned and then questioned again perhaps a month later. If the first theory were true, people would answer the questions the same way the second time. But if the second theory were true, people might give more "yes" answers on the second questioning, because they might answer on the basis of their reconstructed memories.

You may be able to think of other experiments suggested by these theories. According to this criterion it appears as if these theories might be quite useful.

A fourth criterion of a good theory is that it's as simple as possible. If theories are unnecessarily complicated, they're inefficient. Can we think of a simpler theory that might account for Loftus and Zanni's results? Perhaps people asked questions are generally in the habit of answering "I don't know" because they're lazy. However, when they're asked *the* questions they try to answer "yes" or "no" because they've learned from past experience that when people ask *the* questions, they expect an answer. In the past, when they've said "I don't know" to *the* questions, the questioners have said things such as "What do you mean? You either saw it or not. Which is it?" Thus, this theory states that people make the response that requires the least total effort. It's simpler than the explanation offered by Loftus and Zanni because it doesn't require us to know the subject's assumptions, question rephrasings, and so on. However, it doesn't account for the increase in "yes" answers about objects that weren't present. Perhaps you can think of a simpler theory that does account for all the results.

A fifth criterion for a good theory is that it offers a satisfying explanation of the results. But what's a satisfying explanation? Loftus and Zanni explained the increase in "yes" responses by saying that the subjects reconstructed their memories. Even assuming that's correct, it's hardly an ultimate explanation. To understand what it means to reconstruct a memory we need a theory about how memory works. In turn we're likely to need theories about anatomy and physiology, which might lead into biochemistry and the reasons for the behavior of molecules, atoms, and even electrons. At that point we can still ask why the electrons or other subatomic particles act as they do. We have no ultimate explanation. Alternatively, we could take a less reductionist approach. Instead of delving into chemistry and physics we might propose a theory that memory is structured as it is because of evolutionary pressures. We might then try to piece together what those pressures might have been and what other consequences they might have. Again, however, even if we were successful, we wouldn't have an ultimate answer. Why are there evolutionary pressures?

Scientific theories don't really attempt to offer ultimate explanations of events. They can't. What they do instead is suggest that the observed events are really members of a somewhat larger, more general class of events. Different theories can account for the same event on many different levels. We make the choice about which level to use by deciding which level appears to organize information in the most comprehensive, understandable, and useful way. Trying to explain Loftus and Zanni's results in terms of the activity of electrons or even neurons is likely to be fruitless at this stage of our knowledge. Trying to fit them into theories about how memory works is more likely to be satisfactory.

*Developing Your Own Theories*

As you read experiments you may want to develop theories of your own. Some people seem to develop theories automatically without appearing to think about them, but others find the process somewhat intimidating. While it's true that a great deal of knowledge is required to create theories that revolutionize our ways of thinking about a scientific area, theories can be quite small and unrevolutionary and still be useful. It's also helpful to recognize that there may be many useful theories to account for the same results. Some people have the impression that there's only one correct theory to account for an event and that all other interpretations are simply incorrect. However, that's not the way scientific theories function. People with different interests and areas of knowledge are likely to devise different theories, and they all may be useful.

Suppose an experiment were to show that students who drink a lot of alcoholic beverages in college are likely to get lower grades. A theorist whose interests were physiological might develop a theory about the effects of alcohol on nerve functioning. A more socially oriented theorist might devise a theory about how the social activities related to drinking affect grades. Another theorist might conceive of both drinking and low grades as resulting from past learning and try to pin down the life experiences that led to those habits. Someone else might theorize about the personality traits likely to occur in a person who drinks and gets low grades. All these theoretical viewpoints could be useful in understanding the nature of the relationship and in organizing and guiding research in the related areas. No single viewpoint is necessarily "right," though some may lead in more fruitful directions than others.

The more you know about an area, the better equipped you are to formulate theories. As a novice in an area, you might have some trouble thinking of theories, but as you learn more it gets easier.

Let's think of some theories to account for the results of Yeatts and Brantley. They showed that their treatment improved the typing performance of a child with cerebral palsy. One theoretical direction of interest concerns what this shows about the nature of the deficits caused by cerebral palsy. Yeatts and Brantley mentioned that the hospital personnel had doubts about the effectiveness of the teaching technique. This suggests that the hospital personnel had informal theories about the abilities of children with cerebral palsy that were in some way incorrect. Apparently the deficits caused by this disease can be overcome to some degree with proper training. Another theoretical direction concerns why this treatment was effective. One theory might suggest that the raisins motivated the child to try harder than in the past. Another theory might suggest that the raisins helped the child to distinguish between correct and incorrect words more clearly than in the past. Since this experiment wasn't designed as a comparison of the theories it doesn't allow conclusions about their relative accuracy. Other experiments would be needed to do that.

*Making and Testing
Predictions from Theories*

Thus far we've discussed only theories that were made up or applied after the experiment was completed. We might call these post hoc theories. Their purpose is to suggest explanations and to direct future research. Since post hoc theories are made up or applied after the experiment is completed the experiment can't really be thought of as a "test" of that theory. However, as we discussed in Chapter 1, experiments are often designed specifically to test theories. To do that the researchers make predictions based on the theory and then perform an experiment to determine whether the results are consistent with the theory. Since the theory led to the design of the experiment this kind of experiment *is* a proper test of the theory.

What predictions could be made from the theories we've created to account for Yeatts and Brantley's results? Our first theory is that an increase in motivation accounted for the improvement in performance. Based on that theory we might predict that if the child were given, say, dried peas rather than raisins there would be little or no improvement (unless of course the child liked dried peas). Our second theory is that increased information accounted for the improvement in performance. According to that theory there should be just as much learning from dried peas as from raisins. Since the two theories make contradictory predictions an experiment could be designed that would compare them directly. If the results showed that children rewarded with dried peas learned little or nothing while children rewarded with raisins improved their performance, that would suggest that the first theory was true. If the results showed that the peas were much the same as the raisins, that would support the second theory. Of course it's possible that the results might be equivocal, with the dried peas leading to less learning than raisins but still causing an improvement in performance. Neither theory taken alone would account for those results.

While some experiments are designed to compare two or more theories directly a more common experimental design simply tests predictions made by one theory. For example, suppose an experiment has shown that babies cry more as the temperature increases. One theoretical explanation is that babies like cooler temperatures better. A specific prediction based on that theory might be that if babies were given a ring to pull, they would pull it if it caused cool air to blow on them but not if it caused warm air. An experiment could be performed to test that prediction. A different theory to account for the original relationship is that cool temperatures require babies to use up more of their energy in keeping warm, so they don't have as much energy left for crying. One specific prediction based on that theory would be that babies in cooler temperatures not only cry less, but also play less. Again, an experiment could test that prediction. You may be able to think of other theories or predictions.

Like any experiment, experiments performed to test theories must be well designed. If there are confounded variables, poor operational definitions, or other problems, the results can't be interpreted clearly regardless of how they turn out. If the study is well designed, there are three possible results you might read about. One possibility is that the results are in the direction predicted by the theory. Such results support the theory but don't prove that it's correct. There may be another theory that accounts for the results and is better in some way. The second possibility is that the results clearly contradict the theory. These results mean that the elements of the theory leading to the prediction are incorrect; changes or a whole new theory are needed. The third possibility is that the results are equivocal. They agree in part and disagree in other areas. That means either that the theory is completely wrong or is incomplete in some way.

Recall now the Schutz and Keislar social class/word class study. They were specifically testing two predictions or hypotheses derived from a theory: (1) Deficiency in word recall of lower-class children in the kindergarten-primary grades is greater for function words than for nouns and verbs in relation to the performance of middle-class children. (2) At the kindergarten-primary level, the difference between social classes for recall of function words, relative to recall of nouns and verbs, is larger for children in the higher grades than for children in the lower grades. Their first experiment showed results in agreement with the first hypothesis, but the interaction relating to the second hypothesis wasn't statistically significant. In Chapter 4 we mentioned that there was a confound present in the study in that all the children received all the types of words in the same order. Because of that confound Schutz and Keislar felt that their experiment couldn't be clearly interpreted so, as you'll read next, they performed a second experiment. As you read it notice how they changed the second study to remove the confounded variable. When you get to the discussion section notice how they briefly point out the theoretical implications of their results and then suggest a few reasons to doubt the generality of the study.

## Study II of the Schutz and Keislar Word-Class Study

A major weakness in the design of the first study was that all subjects were given exactly the same order of items. If a fatigue effect was present and if the lower class children were more susceptible to this effect, it is possible that the interaction could be partly accounted for on this spurious basis. Consequently, in the second study, an improved design was adopted: each child was given one of six different forms of the test, the order of the word classes being systematically rearranged from one form to the next. Since the differences seem to be greater with the younger ages, in the second study, pre-kindergarten children were used and the scoring standards slightly relaxed. For the second study, the only concern was with Hypothesis 1, namely, that there is an interaction between word class and socio-economic status.

*Subjects*

The lower class subjects consisted of 30 four-year-old children who were attending day-care centers in connection with schools located in neighborhoods designated as poverty areas by the federal government and thus eligible for special funds. The middle class children were 30 four-year-olds attending three private nursery schools in upper middle class neighborhoods of Los Angeles.

*Method*

Six different versions of the test instrument were prepared using the same list of words as in the original version. However, for each of these forms the order of the class lists was different. For every set of three consecutive items, consisting of word lists of exactly the same length, there were six possible permutations of the three lists for the different word classes. One form was first constructed using one of these permutations for the first three items, a second for the next three and so forth with the last permutation being used for the last three items. Other forms were systematically constructed by using a different permutation for each set of three items. In this way, therefore, any given item consisted of a list for a given word class on two of the six forms (from 1 to 18). In order to keep the test from being too difficult for these younger children, scoring standards were relaxed slightly: for any item if the child repeated the list without omissions or additions, it was counted as correct even though the order of the words was changed by the child.

*Results*

The means and standard deviations for these two groups of 30 children each are given in Table 2. It should be noted that the results cannot be meaningfully compared with the scores obtained in the first study: the test instrument and scoring method were modified in significant ways. A two-way analysis of variance (social class by word class) was carried out. While the main effect of word class was significant, $F(2, 116) = 18.07$, that for social class was not. The interaction between social class and word class was significant, $F(2, 116) = 8.03$, $p < .01$; the major hypothesis is again supported by the results of the second study.

**Discussion.** The findings of these two studies give some support to one aspect of Basil Bernstein's theory of social class differences in language development, that it is in the function words that lower class children are particularly deficient. It is true that the sample of nouns and verbs used in the test instrument was limited;

Table 2
Means and Standard Deviations on Revised
Immediate Memory Test by Word Class and Socio-economic Level, Study II[a,b]

| Socio-economic Level | N | Word Class | | | | | |
|---|---|---|---|---|---|---|---|
| | | Function Words | | Nouns | | Verbs | |
| | | Mean | SD | Mean | SD | Mean | SD |
| Low | 30 | 3.0 | .99 | 4.0 | 1.00 | 3.7 | .73 |
| Middle | 30 | 3.8 | .75 | 4.0 | .63 | 3.8 | .60 |

[a] Score is the number of words in the longest list correctly repeated without regard to order.
[b] Based on 60 pre-kindergarten children.

greater differences between the two socio-economic classes may have emerged if less common words had been taken from the Thorndike-Lorge list. Furthermore, it is plausible that the differences found in ability to repeat these less "concrete" words may parallel differences on other measures such as mental ability.

Nevertheless, the results suggest one kind of deficiency in language that such disadvantaged children possess as they begin their schooling. The findings also imply, for other studies of language assessment, the utility of such an immediate memory technique involving "word span."

Schutz and Keislar emphasized the support of one aspect of Bernstein's theory of social class differences in language development. There was indeed a specific deficit in function words by lower-class children. Thus those portions of the theory may be correct, although there may be another theory that could account for the same results. The authors don't mention their lack of support for the second hypothesis. It's difficult to know just how to interpret that result. For one thing, that hypothesis was tested only in the first experiment in which the confounded variable was present. Another fact making it difficult to interpret these results is that the means were actually in the direction predicted by the theory, although not statistically significant. It may be that this experiment was too insensitive to detect the effect even though it was really present. For these reasons it's not clear whether this aspect of the theory is correct or incorrect. To make it clearer another experiment should be performed in which the hypothesis is retested in a more sensitive way, with the confounded variable removed.

### Combined Results and Discussion Sections

Sometimes articles have combined results and discussion sections. Often this form is used because the authors decided to perform statistical analyses they hadn't planned after seeing the results of the experiment. The data suggested new ideas for comparisons that the authors then carried out. Obviously the unplanned comparisons involve post hoc statistical tests and thus should be viewed with some caution. Primarily they're useful for suggesting future experiments.

Sometimes the results and discussion sections are combined when authors want to introduce their data gradually, presenting theoretical arguments and interpretations for each result in turn.

The "results section" presented in Chapter 5 for the Sullivan and Skanes teacher evaluation study was really the first two paragraphs of a combined results and discussion section. Recall that in that study Sullivan and Skanes obtained both teacher evaluations and final exam grades from students in 130 sections of different classes and calculated means for each section. The results presented in Chapter 5 showed that these two variables correlated +.39 overall, with some differences between different subject areas.

It's not quite clear just why Sullivan and Skanes decided to combine their sections. There are several comparisons that might not have been planned until after they saw the data. However, by using the combined sections, Sullivan and Skanes were able to present a coherent and satisfying explanation for previously conflicting and confusing results.

As you read the remainder of this article notice the combination of the presentation of new results and discussion of the theoretical and practical relevance of these findings. Keep in mind that the results could have interpretations other than those of the authors. Try to distinguish among three kinds of information presented in this section: (1) clear descriptions of events observed in the experiment, (2) statements of supposed fact concerning events that were *not* observed in this experiment but were presented as general knowledge, and (3) conclusions based on the experimental observations or general knowledge facts. Then answer the following questions.

1. What are the major results presented by Sullivan and Skanes?
2. What conclusions can *you* draw from those results?

## The Remainder of the Results and Discussion Section of the Sullivan and Skanes Teacher Evaluation Study

The relatively small magnitude of the correlation coefficient in this study may be partially accounted for by the restricted range for both variables. Faculty members were, in general, evaluated highly, with little variability. The mean rating was 3.7 and the standard deviation of the ratings was .41. Similarly, the range of achievement among the sections was narrow. For most of the courses, then, the instructors formed a homogeneous group, yet although the overall significant positive correlation was low, it was a clear indication of the ability of students to identify those teachers from whom they learn most.

In subjects where the population of instructors was more heterogeneous (e.g., Science 115, in that three of the eight instructors were teaching in our university for the first time), the correlation was relatively high, indicating that in this situation students find it easier to make a valid assessment. Further support for this interpretation and additional information concerning the factors involved in the validity of student evaluations may be found in an examination of the results in Psychology 1000. This subject has the largest number of sections and the greatest heterogeneity among instructors. Of the 40 sections included in this report, 27 were taught by full-time instructors and the remaining 13 by graduate students who were hired as part-time teaching assistants. The course consists of 10 modules which are taught in a highly structured manner with clearly specified objectives and frequent evaluations (Sullivan, 1969), but each instructor has considerable responsibility for the specific instructional material which is used and the method of presentation.

The relationship between evaluation and student achievement was tabulated, and correlation coefficients were calculated for the entire group and separately for the subgroups of full-time instructors and graduate students. For the group of full-time instructors, the correlation is +.528, which is significant at the .01 level. For the graduate students, the correlation is +.007 which does not differ significantly from zero. The overall correlation between rating and final examination mark for all 40 sections is +.410, which is significant at the .01 level.

The difference between the validity of ratings for part-time and full-time instructors is important and interesting. The two factors which appear to be of most importance in explaining the difference are the nature of the commitment to teaching and the amount of experience.

Full-time instructors are heavily committed to teaching and must accept major responsibility for the outcome of their instruction. A high level of student achievement is, therefore, a primary goal for them, and since they have considerable autonomy in the actual process of instruction, they are able to accept a high degree of responsibility for the attainment of this goal. Part-time teaching assistants and graduate students are not heavily committed to teaching and do not usually have much autonomy in planning the process of instruction or in carrying out their specific responsibilities. Student achievement may not be a primary goal for them. They may be more interested in alternative goals, such as getting along well with the students they teach or arousing interest in stimulating topics which are not necessarily relevant to the major objectives of the course. Student evaluation, therefore, might be based on the success that they have in attaining these other goals rather than on achievement.

This point has also been noted by Gessner (1973), who commented on the minimal level of responsibility for student achievement which characterized the teaching assistants in the Rodin and Rodin (1972) study.

The other factor, amount of experience, is also of considerable importance. When we divided our full-time instructors in psychology into those who were in their first year of full-time teaching (inexperienced) and those with one or more years of full-time teaching (experienced), we found that the correlation between evaluation and achievement was significant ($p < .01$) for the experienced ($r = .685$) but not the inexperienced ($r = .132$) instructors. This finding suggests that valid ratings are easier to obtain in the case of experienced instructors who presumably have developed a more consistent style of teaching than inexperienced instructors.

These data may help to provide an explanation of the contradictory results which have been reported in previous studies. The Rodin and Rodin (1972) results are based on part-time teaching assistants, while those of Gessner (1973) and Frey (1973) are based on full-time experienced instructors. It is likely that most of the studies which have reported no correlations or a negative correlation between student evaluation and achievement have used results from part-time or inexperienced instructors, while those which have reported positive correlations have gathered data from full-time and experienced faculty members.

On the basis of the results of this and other studies referred to above, it appears that under certain circumstances students are able to provide an accurate estimate of the amount they learn from an instructor and that student evaluations may provide a valid indication of the amount which the students have learned. Valid ratings are much more common and are easily obtained in the case of experienced and full-time instructors than in the case of inexperienced or part-time instructors.

In an attempt to find the characteristics associated with effective and highly evaluated teaching, the instructors in those subjects which have the largest number of sections (biology, mathematics, and psychology) were divided into four groups at the medians of the achievement and evaluation dimensions. Of the resulting four groups, two groups, the high-achievement–high-evaluation group and the low-achievement–low-evaluation group may be described as "consonant." The other two, the high-achievement–low-evaluation group and the low-achievement–high-evaluation group may be described as "dissonant."

An examination of the mean scores for the instructors in each quadrant on the specific evaluation questions (i.e., interest in students, clarity of presentation, and ability to arouse interest) and the percentage of students answering yes to each of the descriptive items revealed the following:

**1.** Positive answers to items associated with attitude toward students, for example, "is friendly and understanding" and "is usually available for help," had surprisingly little relation to either favorable evaluation or a high level of student achievement.

**2.** Positive answers to items associated with clarity of presentation, for example, "is well prepared," "uses the blackboard effectively," and "explains things clearly," were related to a favorable evaluation but were not necessarily associated with a high level of student achievement.

**3.** Positive answers to items associated with "task orientation," for example, "expects too much from students" and "lectures present too much material" were not related to a favorable evaluation but were associated with a high level of student achievement.

Incidentally, the items associated with a high level of achievement differed somewhat from subject to subject. In psychology and biology, success was associated with direct and continuing pressure for the student to work hard, whereas in mathematics, success tended to be associated with a supportive orientation which helped to prevent the students from becoming discouraged.

When a comparison is made for all subjects between the "consonant" instructor groups (i.e., those who are high on achievement and evaluation versus those who are low in both), it is apparent that successful and highly evaluated instructors are able to combine "task orientation" with the qualities associated with high evaluation. Thus, successful instructors are able to combine task orientation with the following: clarity of presentation, the encouragement of independent thinking, the expression of different points of view, and the ability to convey enthusiasm for the subject. Unsuccessful and poorly evaluated instructors on the other hand are obviously not able to present the material clearly, to arouse enthusiasm for the subject, or to induce the students to work hard.

When a comparison is made between the "dissonant" instructor groups, that is, those high on one dimension but not on the other, interesting findings emerge. Some examples of these are as follows.

Positive answers to questions such as "conveys enthusiasm for his subject" and "encourages the expression of different points of view" tend to be associated with a low rather than a high level of achievement. These "dissonant" instructors, then, were not able to combine arousing enthusiasm for the subject with task orientation. Some, the high-achievement–low-evaluation group, placed most emphasis on student achievement while others, the low-achievement–high-evaluation group, placed greater emphasis on arousing interest and enthusiasm.

Students may at times, often in the case of a "dissonant" instructor, have an unfavorable reaction to a successful and achievement-oriented instructor. He may be seen as a hard and demanding task master who discourages questions and comments about interesting but irrelevant material in his determination to ensure that all of the required material of the course is covered and that high standards are maintained.

On the other hand, the highly evaluated but academically less successful instructor obviously impresses the students by the efforts which he makes to arouse interest and enthusiasm and to generate discussion, but he does so at the expense of the standard of work which he requires of the students. Although it is difficult to relate these findings to those obtained from other studies, since such data are rarely reported, these results are similar to and entirely consistent with those reported by Peck and Veldman (1973).

Many will argue that it is of more fundamental importance for the students to acquire an interest in the subject in a first course than for them to acquire information or knowledge. The argument states that students who acquire information but not interest are "turned off" from the subject while those who acquire an interest are "turned on" and, therefore, elect to do further courses in the subject and to attain a higher level of achievement in these subsequent courses.

It is, of course, difficult to provide evidence which is crucial to this argument, but we were able to gather some relevant data by examining the second-year academic performance of those students who had completed and passed Psychology 1000 in 1971–1972. Table 2 gives the percentage of students taught by each of the four groups of instructors who elected to take a second-year course in psychology and the average mark obtained in these second-year psychology courses.

Table 2
Performance of Students in Second-Year Psychology
Courses Associated with Various Types of Instructor Characteristics

| Characteristics of Instructors | | | Academic Performance in Second-Year Psychology Courses | |
|---|---|---|---|---|
| Achievement | Evaluation | $n$ | % Students Taking Courses | M Mark |
| High | High | 8 | 40.7 | 61.5 |
| High | Low | 4 | 43.3 | 66.1 |
| Low | High | 5 | 30.8 | 57.9 |
| Low | Low | 7 | 26.1 | 55.8 |

The proportion of students taught by a high-achievement instructor who elect to take further courses in psychology compared with that of those taught by a highly evaluated instructor is higher although not significantly so ($F = 4.10$, $df = 1/18$, $.05 < p < .10$). The mean marks in subsequent psychology courses are more closely related to previous achievement than to instructor evaluation ($F = 9.04$, $df = 1/99$, $p < .01$).

It is interesting to note that the one group which demonstrates both the highest percentage of students continuing and the highest marks in second-year courses is that of students taught by high-achievement–low-evaluation instructors.

When the specific courses for which students registered in the second year were examined, it was found that a smaller percentage of students taught by high-achievement–low-evaluation instructors tended to register for second-year major courses in psychology than of students taught by low-achievement–high-evaluation instructors (5.3% versus 8.4%). However, even in these major courses, the marks obtained by the students of high-achievement–low-evaluation instructors were higher (61.0% versus 56.5%). Also, a higher proportion of students taught by low-achievement–high-evaluation instructors dropped psychology major courses after registration.

It seems, then, that in introductory psychology, at any rate, if an instructor concentrates on producing a high level of achievement, his students are at least as likely to take subsequent courses and are more likely to do well in those courses than they would have if his emphasis had been on arousing interest and enthusiasm in the subject. On the other hand, the instructor who concentrates on arousing interest in the subject without at the same time taking steps to ensure a high level of achievement may be doing his students a disservice in that they may elect to major in the subject, but lacking the necessary background, they may do poorly in or fail the required courses at the second-year level.

This finding may be of considerable importance since most beginning instructors tend, and are in fact encouraged, to spend a great deal of time finding ways of making the subject interesting but do not spend much time in finding ways to encourage the students to work hard and are apprehensive about insisting on high standards of performance.

Our results certainly suggest that because students are likely to be initially interested in a subject such as psychology and because they have had no previous experiences with the subject which were discouraging, the instructor would be well advised to concentrate on student achievement and not to be overly concerned with arousing interest in the subject. It is of crucial importance, however, that this result should not be overgeneralized since in a subject for which initial interest is not likely to be high and in which the student may have had previous experiences that were discouraging, for example, mathematics, the interest-arousing role of the instructor may be of much more importance.

---

What were the major results presented by Sullivan and Skanes? One set of results indicates that experienced, full-time instructors show a strong, positive correlation between student achievement and teacher evaluation. However, inexperienced instructors show no such relationship. Although Sullivan and Skanes don't use these terms, they've demonstrated an interaction among their variables: For experienced instructors there's a positive relationship, but for inexperienced instructors there's no relationship.

The second set of results concerns the relationships between answers to specific questions on the evaluation questionnaire to the two major variables. These results are summarized in the three numbered paragraphs in the section.

The third set of results concerns the continuation of students into subsequent psychology courses from the various kinds of instructors. Sullivan and Skanes's Table 2 suggests that students coming from high-achievement classes are more likely to continue and to do well.

What conclusions do I draw from these results? One conclusion is that for experienced instructors, I'd expect student evaluations to be positively correlated with how much the student learned in the class. However, it's important to note that these data tell nothing about differences between students *within* any one class. As an instructor I shouldn't expect my high-achieving students to give me higher ratings than my low-achieving students. I would expect to get higher ratings, on the average, than an instructor whose students learned less than mine.

Another conclusion is that for inexperienced instructors, I wouldn't expect such a correlation. Presumably many inexperienced instructors are what Sullivan and Skanes call "dissonant" instructors, high on one variable and low on the other. To exaggerate, after looking at the answers on the questionnaire, I find myself thinking of the two types of dissonant instructors as unfriendly slave-drivers and likable marshmallows.

The third conclusion I would draw from these results is that students coming from high-achievement classes would be more likely to continue.

Your conclusions might be different from mine, because your interests are likely to be somewhat different. However, there are some conclusions that wouldn't be appropriate. For example, we could not conclude on the basis of these data that individual instructors could improve their evaluations by becoming more achievement oriented. The instructors are all individuals and have many personality traits possibly confounded with their teaching emphases. An individual might become more achievement-oriented and still receive low evaluations because of his other characteristics.

Similarly these data wouldn't allow us to conclude that stressing achievement would cause more students to continue in psychology classes. The students in this study might have continued because they learned a lot, but there could be other explanations as well. In this setting the students who learned more received higher grades, which could have influenced them to continue. Another explanation is that they chose to continue because of other characteristics of their instructors. You may be able to think of additional explanations.

Not all the relationships described in this combined section were equally reliable. The only comparison clearly planned by Sullivan and Skanes before seeing the results involved the correlation between the two major variables, evaluation and achievement. All the data about the amount of experience of the instructors, the "dissonant" and "consonant" teachers, the continuation of students to additional courses, and the different needs of different courses seem to have been analyzed post hoc. The theories and ideas have been created after the experimenters saw the results. As such they need replication before they can be thought of as trustworthy. Nonetheless, they offer a satisfying explanation for this previously confusing area of research.

Now it's time to read the discussion sections of your sample articles and to try to answer these questions:

**1.** What conclusions do the authors draw about the relationships between the variables?

2. Can the results be explained on the basis of confounded variables?

3. Are the results trustworthy or likely to be due to chance factors?

4. Do the conclusions follow logically from the results? Are the relevant comparisons made in the experiment? Are the levels of the variables adequate to support the conclusions? Are the operational definitions adequate? Are there any unwarranted hidden assumptions of causality?

5. Do the authors suggest any practical implications? If so, are the implied generalizations warranted?

6. Do the authors suggest any theoretical implications? If so, are these theoretical suggestions useful and satisfying?

7. Can you think of any additional practical or theoretical implications of the results?

**Section B
Interpreting
Your Results**

You've completed your experiment. If you haven't already done so, you must figure out what it means. Then, as a scientist, you should think about the theoretical implications of your results: where they fit into what's already known and what they show about existing theory. Based on that you should think of additional experiments that will show more about the area. If it's appropriate, you also may want to think of the practical implications of your findings.

All the care you took in planning, designing, and performing your experiment pays off when it's time to draw conclusions. However, even if you've been careful, there are several common problems that might make it difficult for you to draw conclusions easily. First we'll discuss drawing conclusions based on your results. Then we'll discuss some problems that might complicate your task.

### Drawing Conclusions

The conclusions you draw depend on the original purpose of your experiment.

If your experiment was designed to tell you the relationship between variables, your conclusion consists of a description of the demonstrated relationship. If it's causal, you should say so.

If your experiment was supposed to answer a question, the conclusion consists of an answer.

If your experiment was supposed to test a theory, your conclusion consists of a statement of the relationship of your results with the predictions of the theory. The results might agree with the predictions, might clearly disagree, or might be equivocal.

**Problems**

These problems will give you practice in drawing conclusions from experimental results. For each one draw the appropriate conclusion.

**Q.** An experiment is designed to discover the relationships among athletic ability, the average amount of sleep per night, and the average time of getting up in the morning. The results showed that people high in athletic ability tend to get more sleep and to get up earlier than other people.

**A.** Since this experiment was designed only to discover the relationships among the variables, the conclusion is simply a restatement of the results.

**Q.** An experiment is designed to find out how much of a decrease in returns to jail was caused by a new rehabilitation program involving work outside the prison. The results showed that there were as many repeat offenders after the new program as after the old programs.

**A.** This experiment was designed to answer a specific question. The answer to the question is that there was no decrease caused by the new program.

**Q.** An experiment is designed to test a theory that says women are more influenced by external events than men. The specific prediction is that women are more depressed by rainy weather than men. The results are shown in Figure 6.2.

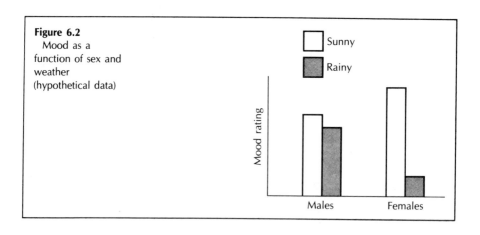

**Figure 6.2**
 Mood as a function of sex and weather (hypothetical data)

**A.** The results are in the direction predicted by the theory. The differences between sunny and rainy weather were much greater for the women than for the men, supporting the notion that women are more influenced by external events than men.

**Q.** An experiment is designed to test a theory that says people are more likely to lie on a test if they believe that the material being measured by the test is very serious. All the subjects took a test that was really designed to elicit reports of mild neurotic symptoms. The items included statements such as "I often feel anxious before an exam" and "I sometimes worry that people don't like me." Since most college students normally agree with a large number of such statements, the experimenters assumed that people reporting few such symptoms were probably lying. The experiment used four randomly selected groups and manipulated the instructions given before the test. The results are shown in Figure 6.3.

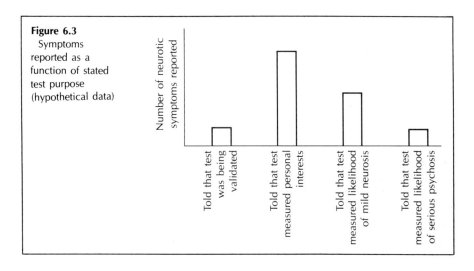

Figure 6.3 Symptoms reported as a function of stated test purpose (hypothetical data)

**A.** The results of three of the groups suggest that the prediction is accurate, but the group told that the test was being validated reported many fewer symptoms than some of the other groups. The experimenter had anticipated that this group would report more symptoms than any other group, since the people in it didn't know at all what was being measured. Apparently something in the experimenter's logic or theory was incorrect, because the prediction wasn't supported.

### Events That Can Make It Difficult to Draw Conclusions

This book has taught you to avoid many of the common problems that can make it difficult to interpret results. We'll assume that your experiment was well planned and carefully executed. Despite that you might have not planned quite carefully enough or might have just suffered some bad luck. As you'll see, in all these cases, the only complete cure is to repeat the experiment with more complete safeguards. You may draw conclusions from experiments with these problems, but they'll be somewhat unclear.

### *Unreliable Results*

Little confidence can be placed in results based on only a few measurements or on measurements that vary a great deal. Inferential statistics are supposed to help determine the reliability of your results. But what can you do if they tell you that your very interesting looking results aren't statistically significant? What can you do if you can't even test the reliability of your results statistically?

One possibility is to repeat the experiment. If you repeat it exactly and obtain the same results, it appears more likely that your results are trustworthy even if they again fail to be statistically significant. Instead of repeating the experiment exactly, you might make some changes to reduce random variability and thereby increase the likelihood that any real effects will be statistically detectable. If there was much variability in your experimental situation, it could be reduced by standardizing your procedures and eliminating distractions. Variability between subjects can be reduced by selecting subjects from a more limited group or can be controlled by adding one or more subject variables as independent variables. However, these steps reduce the degree to which you can generalize your results.

Another way to increase the reliability of your findings is to increase the number of subjects, the number of measurements on each subject, or both.

It's important to keep in mind that your results might appear to be accidental because they really are. Your variables in fact may be unrelated. In that case repeating the experiment with reduced variability may make the lack of relationship more clear. If this occurs, then the only conclusion available is that the variables are unrelated.

### *Damaging Confounded Variables*

You may have known all along that there was a possibly damaging confound present in your experiment. That was the problem encountered by Schutz and Keislar in their social-class/word-class experiments. On the other hand, you might have had bad luck such as your machine breaking down 10 times during one level of your independent variable and never breaking down during the other levels.

Regardless of the source of your confound you can't tell whether the relationships are between the variables of interest or between the confounded variable and the variables of interest. If the effects of your confounded variable are in the same direction as the effects of your independent variable, the differences will look bigger than they really are. If the effects of your confounded variable are in the opposite direction, the effects of your independent variable will look smaller than they really are or may disappear entirely. The only cure for this problem is to repeat the experiment while eliminating the confounded variable, either by holding it constant or by distributing its effects equally over all levels of the independent variables, as Schutz and Keislar did. If that's impossible, you can draw conclusions which take account of the possible confound. For example, Schutz and Keislar might have drawn a conclusion from their first study saying that there was an interaction involving some combination of the variables social class, word class, and specific word order.

*Initial Differences between Randomly Assigned Groups*

If you assigned subjects randomly to different levels of an independent variable, there's a chance that your groups differed significantly just by accident. Differences between the groups' performance on the dependent variable might occur just because of the characteristics of the subjects assigned to each group. Obviously, this is just one kind of confound that might occur, but it's common enough to warrant separate mention.

If there's no reason to believe that this happened in your experiment, then you needn't be concerned. However, you might have noticed as you performed the experiment that for some reason you got 20 math majors in one group and none in the others. Or it might happen that subjects disappeared from some levels. This is just bad luck, and there's nothing that can be done about it after your experiment is completed. Your conclusion will have to take account of the possible subject differences. To correct the problem, the experiment should be repeated, perhaps with matched groups.

*Confusing Order Effects*

If you used a within-subject independent variable, you probably used several orders to prevent practice and fatigue effects from being confounded with the independent variable. If the orders differ significantly, or if they interact with your independent variable, it may complicate your conclusions.

For example, suppose an experiment varied problem difficulty as a within-subject variable and order as a between-subject variable and found the results shown in Figure 6.4.

What may have happened is that there was a practice effect, with the problem that came second being a bit easier than if it came first. It's still possible to conclude that the hard problems caused more errors, but that conclusion should be qualified with respect to the order. The difference is much greater in one order than in the other.

If order has confusing effects in your experiment, you must do the best you can to make sense of them. If you can't make any sense of your results, the only cure is to repeat the experiment with between-subject variables.

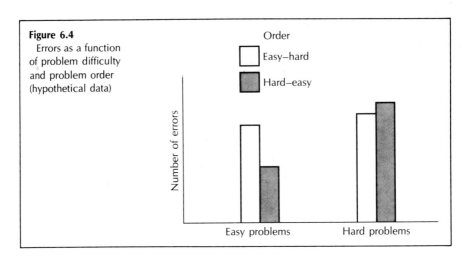

Figure 6.4
Errors as a function of problem difficulty and problem order (hypothetical data)

### Apparently Meaningless Results

Another problem that can arise is that your results make no sense at all to you. For example, an experiment might be designed to discover whether the familiarity (or frequency) of overheard words determines how much they interfere with another task. The results might show that of four familiarity levels, all are identical except for level 3, which interfered more. Such a relationship is surprising from a theoretical point of view; it just doesn't make much sense.

One possible explanation is that the results were caused by some unrecognized confounded variable. In this example there could be something special about the level 3 words, such as their spelling or the number of syllables. If there is a confound, it might have caused the otherwise surprising results. A second possible explanation is that the results are accidental. A replication of the experiment would test that possibility. A third explanation is that the surprising relationship is real; the relationship between the variables is more complicated than you anticipated. If you're sure there aren't any confounds and the same results can be repeated, then the relationship must be real, even if it doesn't appear to make much sense. Your next job is to understand it.

### Problems

The following problems should give you practice in recognizing problems that can make it difficult to draw conclusions from experimental results. In each case try to describe clearly the problem that has arisen.

**Q.** A public opinion analyst performed a study in which people were interviewed after seeing a presidential election debate on TV. The viewers were divided into two categories: those who watched only the debate and those who watched the debate and the subsequent press analysis. The results showed that the people who watched only the debate tended to favor Candidate R while those who also watched the subsequent discussion tended to favor Candidate D. The analyst wanted to conclude that the discussion tended to sway people in favor of Candidate D.

**A.** The analyst can't draw that conclusion because there are many variables confounded with the independent variable, which was nonmanipulated. To draw that conclusion the experiment must be repeated with all the confounded variables controlled. To do that the analyst would have to manipulate the independent variable.

**Q.** An astrology fan performed a study in which the material acquisitiveness of individuals born under various signs was compared. The experiment was performed by stopping people on the street, asking their birthday, and classifying them into signs according to the dates commonly given in newspapers. Material acquisitiveness was measured by asking them to rate themselves on a 10-point scale. The results showed that Leo people received higher ratings and Pisces people lower ratings than all other signs, which was in agreement with the expected results. However, the differences failed to reach statistical significance. The astrologist wants to conclude that the differences are real.

**A.** The astrologist can't draw that conclusion because the results might be accidental. To cure the problem the experiment should be repeated, but with several changes. One might be to move to a less distracting experimental situation. Another would be to refine the measurements of both variables to reduce random variability. Another change might be to use a larger number of people. If the repeated experiment failed to show a significant relationship, the astrology fan might consider changing her theory.

**Q.** The owner of a grocery store checked back over his records and found that over the preceding 6 months about 20 lb. of salt were sold every week, except for one week in September when 42 lb. of salt were sold. The difference was statistically significant, but the grocer can't see any logical reason for it. He wants to conclude that there's something special about that week and that he should stock up on salt for the following year.

**A.** The grocer shouldn't stock up on salt without considering two other possibilities. One is that the result was caused by a confounded variable he failed to recognize, such as a press report that salt containers would be in short supply. The other possibility is that the week was just an accident. If that's so, then weeks in September in other years wouldn't show the same pattern. However, if he gets the same result in subsequent years and can't discover any confounds, he can just conclude that it's a surprising finding.

★ Does your experiment have any problems that make it difficult to draw conclusions? What are they? What will you do about them?

★ What conclusions can you draw from your experiment?

### Deriving Theoretical Implications

In a sense the conclusions you've drawn from your study *are* a very small theory. The variables are related in a particular way; anyone repeating your experiment could predict that they would get the same results. But you may want to develop or use a more complete theory than that.

*Post Hoc Theories*

If your experiment wasn't originally designed to test a theory, you may want to try to think of some theoretical explanations for your results, that is, make up a post hoc (after-the-fact) theory. Why were your variables related in the way that you found? If you're new to the area, it's a good idea to read more about it. By reading you can learn what other people have found and what theories already exist. Do other experiments show the same patterns of results as yours? Do they show different patterns? Do any existing theories seem to explain your results?

If no theory exists, you may want to make up one of your own, as we discussed in Section A of this chapter. In making up a theoretical explanation try to think of general rules that account for your own findings and for as many other experimental results as seem practical. Remember that your theory should organize as many results as possible, should be simple, and should make clear predictions. You'll find that sometimes the different goals conflict. For example, a theory that says sometimes men are smarter than women and other times women are smarter encompasses all the possible results. However, it's quite useless because it doesn't allow you to make predictions. To do that it would have to specify the conditions under which the different relationships occur. But if there were a long, complicated list of different conditions, the theory might become too complex to be a satisfying explanation.

Theories created after the data are collected, post hoc theories, are used primarily to make predictions that can be tested in future experiments. Thus, after you've created one or more theories to account for your results, use the theories to make predictions. If you have more than one theory perhaps you can design an experiment that would help you to decide between them.

*Tested Theories*

If your experiment was designed to test a theory, you'll certainly want to consider the theoretical implications of your results.

Results agreeing with theoretical predictions tend to support the theory that led to the prediction, but there's more to say. Are there other theoretical interpretations of your results? Is your experiment really a good test of the theory?

Results disagreeing with theoretical predictions suggest that the theory leading to the prediction is incorrect. Is there another, better theoretical explanation?

Results partly agreeing and partly disagreeing with a theory are probably the most common finding. Either the entire theory is incorrect or it's incomplete and needs changes to encompass the results. Can you make the necessary changes in the theory? Does that make the theory too complicated to be satisfying? Are there better theories?

As with post hoc theories the next step is to make predictions to test whatever theories you think might account for your results.

- ★ What theoretical implications do your results have?
- ★ What predictions are suggested by your theories?

**Practical Implications**

Depending on the nature of your findings, it may be appropriate to consider their practical implications. In deriving these implications, remember to consider whether the generalizations are warranted.

- ★ Do your results have practical implications that you believe are appropriate?

Chapter Seven
Putting It
Together

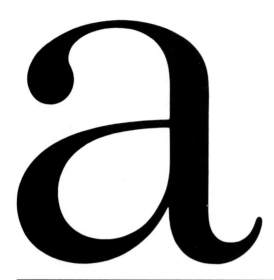

**Section A
Reading the
Whole Report**

Now you've learned how to read all the sections of an experimental report except the title and the abstract. In this section we'll discuss first how to read these sections and then how to read whole reports and finally how to interpret reports that don't have all four sections.

**Titles**

The title of an article is often what originally arouses your interest; it's almost always the first thing you read. It's worthwhile to spend a moment studying the title before going on; it's surprising how much you can learn. Titles normally name the most important variables in the experiment and may give other information as well. These are some of the titles of the articles you've read in this book:

Therapeutic abortion in a midwestern city
College students' adjustment and frequency of marijuana use
The effect of "imaging" on mirror image drawing
Improving a cerebral palsied child's typing with operant techniques
Issue relevance and source credibility as a determinant of retention
Young children's immediate memory of word classes in relation to social class

Notice how in each title the most important variables are named. Also something about the experimental design can be inferred in most cases. The abortion study was a simple survey that made no attempt to demonstrate relationships between particular variables. Therefore the title simply named the topic of primary interest. The marijuana study attempted to show a relationship between the two variables named. A careful reading of the title suggests that causal conclusions are probably not appropriate in that study. In contrast, the mirror image, typing, and source credibility studies all allowed causal conclusions to be drawn. That fact can be inferred from their titles from phrases such as "effect of ... on ... ," "improving ... with ... ," and "as a determinant of ...". The word-class/social-class study was primarily concerned with the interaction between these two variables. That's hinted at by the phrase "in relation to" in the title.

In addition to naming the variables and to hinting at the experimental design some of the titles also mentioned the subjects who participated in the experiment.

**Problems**

Q. Here's a list of titles to articles you've never seen. In each case pick out the variables and try to decide whether the experiment is likely to have shown a causal relationship between the variables.

1. Interpersonal attraction and attitude similarity
2. Sensory alterations and aggressive behavior in the rat
3. The stare as a stimulus to flight in human subjects
4. Status of frustrator as an inhibitor of horn-honking responses
5. Responses to voice of mother and stranger by babies in their first year
6. Perception as a function of retinal locus and attention
7. Eye contact in children as a function of age, sex, and social and intellective variables
8. Somatotropic organization of trigeminal-ganglion neurons

A. The variables should have been easy for you to pick out, although you might not know what all the words mean. From only the grammar used it would appear that causal conclusions might have been drawn in the articles Number 3 through 7. However, it's important to also look at the variables themselves. They sometimes tell a different story. In Number 2 the grammar would suggest that causal conclusions weren't drawn, but it seems likely that the actual experiment involved making the sensory alterations and seeing how that affected aggressive behavior. If the subjects were human, of course, that would be less likely. In Number 7 the phrase "as a function of" suggests that causal relationships are being discussed. However, the nature of the variables makes clear that that couldn't be so. In Number 8 it appears that no relationships are studied.

### Abstracts and Summaries

Abstracts appear at the beginning of articles, and summaries appear at the end; otherwise, they're similar. Older articles tend to have summaries, while more recent works have abstracts. Reading either can help you decide whether to read the whole article or can serve as a useful reminder of the main points after you've completed the article. In some cases you may have access only to the abstract and title. That occurs in the case of unpublished dissertations, in articles written in foreign languages you can't read, or in articles you can find in *Psychological Abstracts* but can't locate the original. Thus it's important for you to know how to read abstracts and to know their limitations.

A good abstract summarizes the most important aspects of the experiment in the article. It describes the subjects, the most important elements of the experimental design and procedure, and the most important results. If it's necessary for clarity, a description of the operational definitions of the variables may be included. Sometimes a brief theoretical discussion also is present.

You've been reading abstracts of hypothetical experiments throughout this book in the problems, so they're not new to you. Now look at the abstracts from some of the studies you've read. Within each abstract try to locate the answers to these questions:

1. Who were the subjects?
2. What was the experimental design?
3. What was the experimental procedure?
4. What were the results?
5. What other information is given?

### The Abstract from the Yeatts and Brantley Typing Study

A 7-yr.-old cerebral palsied girl was taught to correct responses on two typing tasks using reinforcement contingent on number of correctly typed words. Maintenance of the behavior remained contingent on the presence of tangible reinforcement.

### The Abstract from the Cross and Davis Marijuana Study

Since marijuana use by college students is widespread and controversial, data which bear on the adjustment of student marijuana users are needed. 178 students were surveyed about their drug use and administered the Rotter Incomplete Sentences Blank. Subjects were divided into five categories according to frequency of use. Maladjustment scores and frequency of use were unrelated; however, the very heaviest drug users were more maladjusted. Users also were more likely to have been in psychotherapy than nonusers. Very heavy marijuana users are, like alcoholics, in considerable psychological distress, whereas moderate users show no personality dysfunction.

## The Abstract from the McDaniel and Vestal Source Credibility Study

Forty-four college students were randomly assigned to four groups in which they read a single communication varying in levels (high-low) of source credibility and issue relevance. Immediate retention was assessed by six questions contained on one of two randomly distributed questionnaires. Long-term retention assessment consisted of combining items on the immediate retention questionnaires—thus obtaining two measures, one for identical items and another for new items. Results indicated that source credibility and issue relevance interact with the high source credibility/high issue relevance group retaining a statistically significant greater amount of issue content than the high source credibility/low issue relevance and low source credibility/high issue relevance groups.

## The Abstract from the Loftus and Zanni Question-Wording Study

Two experiments are reported in which subjects viewed a film of an automobile accident and then answered questions about events occurring in the film. Relative to questions containing an indefinite article (e.g., Did you see *a* broken headlight?), questions which contained a definite article (e.g., Did you see *the* broken headlight?) produced (1) fewer uncertain or "I don't know" responses, and (2) more "recognition" of events that never, in fact, occurred. The results, which are consistent with the view that questions asked subsequent to an event can cause a reconstruction in one's memory of that event, have important implications for courtroom practices and eyewitness investigations.

## The Abstract from the Schutz and Keislar Word-Class Study

To investigate immediate recall of nouns, verbs, and function words, using a word analogue to the digit-span memory test, two studies were carried out, one with 60 children in the kindergarten, first and second grades and the other with 60 prekindergarten children. Half of the children in each set lived in a lower class neighborhood, the other half in a middle class area. In each study, lower class children, in comparison with middle class children, showed a significantly greater recall of verbs and nouns than of function words. The results are consistent with the proposal of Basil Bernstein that children from poverty environments use relatively fewer function words when compared with their peers from more affluent homes.

Section A  Reading the Whole Report							271

You probably noticed that the abstracts were very concise and often difficult to read. That's because the authors are trying to compress so much information into so little space. If you remember to look for the subjects, design, procedure, and results, however, you should be able to understand most abstracts.

You also may have noticed that none of the shortcomings of the studies that we've discussed were mentioned in the abstracts. That's common practice in published articles, unless the shortcomings are themselves of interest. For that reason it's always desirable to read the whole article rather than depending on only the authors' abstract.

**Read the Whole Report**

The normal sequence to be followed in reading an experiment is to start at the beginning and read straight through. From the title and abstract you can decide whether to read the report. Then, if you decide to go on, you just read each section in turn.

Remember that the article is arranged in the form of an argument or demonstration. In the introduction section a question is raised in a particular context, and a means for answering the question is explained. In the method section the authors tell the procedures they followed to answer that question. The results section tells what they found. And the discussion section tells how their findings relate to the question raised in the introduction.

In a normal article with four sections the introduction and discussion sections form a unit, placing the experiment in a practical or theoretical context. The method and results sections form another unit, limited as closely as possible to descriptions of observable events. Sometimes you may be interested in just the experimental facts because you want to fit them into a theoretical framework different from the one of interest to the authors. In that case you'd concentrate on the method and results sections and might not even read the other sections. Of course, if the results and discussion sections are combined, you must be especially careful not to confuse the results with the interpretations of the results.

*Articles with No Formal*
*Divisions between Sections*

When an experiment is very short there may be no formal divisions at all. In fact, the Yeatts and Brantley study of the child with cerebral palsy was such an article, although I divided it into sections for this book. Normally these brief articles do have sections corresponding to those we've discussed; they just aren't labeled.

Read now one very short, complete article. It was written by Michelle Dugan and Charles Sheridan and was titled "Effects of instructed imagery on temperature of hands." It appeared in *Perceptual and Motor Skills* in 1976 in Volume 42 on page 14.

As you read the article note the parts corresponding to the introduction, method, results, and discussion sections. (*Hint:* In my opinion, the paragraphs don't correspond to the section divisions.)

## The Dugan and Sheridan Hand Temperature Experiment

There is, at present, controversy over whether voluntary control of manual temperature can be attained by way of biofeedback (Taub, 1975). Difficulty in obtaining control of manual temperature with biofeedback seems paradoxical in light of the extensive literature indicating that such control can be attained by Pavlovian conditioning (Gottschalk, 1946), induction of attitudes related to such diseases as urticaria and Raynaud's disease (Gottlieb, Gleser, & Gottschalk, 1967), and hypnosis (Maslach, Marshall, & Zimbardo, 1972). It is important to submit research using methods other than biofeedback to careful scrutiny.

The present study focussed on the effectiveness of instructed imagery in producing thermal control. Sixteen college-student volunteers served as subjects. Baseline thermal measures were taken for each of 15 min. Both hands warmed for 13 subjects during baseline, both cooled for one subject and the temperature of hands changed in opposite directions for the remaining two subjects (binomial $p < .03$). Mean change for right and left hands was identical at 5.6°F.

Next, these subjects were randomly assigned to two equal groups, in one of which they were instructed to imagine their hands in very warm water and in the other they were to imagine their hands in ice cold water. Manual temperatures were taken each minute from 0 to 15. The mean of the temperature of baseline and for Min. 0 was compared to the last 10 readings. If at least 9 of the 10 (binomial $p < .03$)[1] readings differed in the appropriate direction from that mean, a subject was considered to have achieved the desired change. All 10 subjects with the image related to cooling cooled reliably on at least one hand, and six of them did so for both hands. Mean peak change for this group was −1.91°F. Five of the subjects with the image related to warmth showed reliable warming of at least one hand, but only one did so for both hands. Indeed, four of them exhibited reliable cooling of at least one hand. Mean peak increase for this group was 0.66°F. These findings indicate the control of temperature of the hands can be achieved through merely giving instructions to produce relevant images, without hypnosis, conditioning, or feedback.

Sokolov (1963) has pointed out that drops in the temperature of the hands are a component of orientation responses. It is possible that the tendency for cooling to occur even with imagery related to warmth was due to competition between this component and a tendency for shifts in temperature to occur in correspondence with the quality of the image. Better control might be achieved if

---

[1] An alert reviewer pointed out that this use of the binomial test, an inferential statistic, is questionable because the 10 temperature readings aren't independent.

this orientational component could be minimized, perhaps through long practice on a given image. It is possible that some of the difficulties encountered in obtaining warming by way of biofeedback are due to the inherent requirement that attention be directed toward the stimulus providing feedback which creates orientational responses sometimes competitive with the target behavior.

---

I divided the sections as follows: The first paragraph places the experiment in context; it clearly belongs in the introduction section. The first sentence of the second paragraph states the purpose of this experiment, so it also belongs in the introduction. At that point the method section begins with a description of the subjects. That section continues through the description of the operational definition of "achieving the desired change." Then the results begin with the statement that all 10 subjects with cooling images cooled at least one hand. The results continue through a few more sentences, including the statement about the mean peak increase for the warm image group. The last sentence of that paragraph presents the authors' conclusions and thus belongs in the discussion section. The last paragraph presents a theoretical explanation for the results.

By making these divisions mentally it's easy to separate the experimental observations from the theoretical context just as you would in an article with formal divisions. Then you can analyze the article just as you would any experimental report. (As an aside, it seems likely to me that these authors had difficulty getting increases in temperature because almost all the subjects' temperatures had increased during baseline. They just couldn't go any higher.)

*Incomplete Descriptions*
*of Experiments*

In many cases, of course, you don't have access to complete experimental reports. When you read one article it briefly describes other experiments. In review articles, textbooks, and classroom lectures only brief descriptions are presented. You may also see brief descriptions of experiments in newspapers or magazines or hear descriptions on radio or on TV. How can you evaluate them?

Of course, the only way really to evaluate the experiment is to return to the original. However, if enough information is given so that you can understand the nature of the experiment, you have many critical tools that you can use to determine whether the conclusions that are drawn are warranted by the facts presented. In fact almost all the problems throughout this book have consisted of very brief descriptions of experiments, so you've already had a good deal of practice at this task.

In the public communications media it's common to present only the conclusions from experiments. In that case it's usually not possible to evaluate the conclusions critically. For example, if a TV newsreader reports only that a study showed 65 percent of all fatal auto accidents involve alcohol, it's difficult to evaluate that conclusion. There just isn't enough information. Is that the percentage of total accidents or of total fatalities? How much alcohol? Is there any evidence that the alcohol caused the accidents? There's no way to know. The best you can do is to figure out what questions should be asked and what alternative interpretations could be made depending on the various answers.

Some conclusions call attention to themselves even when they're presented alone because it's very unlikely that an experiment has been performed that would actually warrant these conclusions. For example, a report that being near-sighted causes children to do better in school should immediately arouse your suspicions. How could such a study be performed? The experimenters would have to randomly assign some children to be near-sighted! It seems far more likely that an association has been discovered between the variables and that someone has misinterpreted that finding.

**Figure 7.1**
Example cards

```
Cross, H. J., & Davis, G. L.  College students'
adjustment and frequency of marijuana use.  J.
of Counseling Psyc., 1972, 19, 65-67.  178
college students answered drug use questionnaires
and took an adjustment test.  No demonstrated
relationship.  Weak evidence for negative corre-
lation.  (More reports of psychotherapy from
users; heavy users of all drugs were
maladjusted.)
```

```
Schutz, S. R., & Keislar, E. R.  Young child-
ren's immediate memory of word classes in
relation to social class.  JVLVB, 1972, 11,
13-17.  60 K, 1st, & 2nd graders and 60 4-yr.-
olds all did immediate recall of short lists
of nouns, verbs, and function verbs.  Lower-
class children recalled proportionately fewer
function words.  Nonsignificant support of age
x social class x word class interaction in
Study 1, but confound with practice or fatigue
effects possible.  Relevant to Bernstein
hypotheses.
```

Section A   Reading the Whole Report 275

If you pay close attention to these very brief reports it's surprising (alarming?) how often the conclusions presented appear to be in error. The near-sightedness/school performance example just described was actually misinterpreted in a newspaper article. Students learning to interpret experiments always can find examples of errors to bring to class. It's been the goal of this book to prevent people from making such errors and to allow you to detect when such errors have been made.

### Keep Records of What You've Read

If you're learning about a specific topic or just reading for general interest, you might want to develop a system for reminding yourself of the general outline and results of each article you've read. You may already have a system of your own or be able to develop one more to your liking, but I'll suggest one here to get you started.

An approach used by many workers is to develop a card file. In a card file you make up one card for each experiment. Most workers label the top of the card with the reference to the article, giving the authors' names, the title, and the journal information: title, year, volume, and page number. Then write a brief abstract of the experiment. Sometimes it may be sufficient to copy the authors' abstract, but it's usually more useful to write your own abstract, emphasizing aspects of most interest to you. Be sure you write enough to remind yourself of the experimental procedure and results. Next write the conclusions you drew from the study, and add any shortcomings of the study that might argue against those conclusions.

Figure 7.1 shows some example cards for two of the studies you've read. Notice how these abstracts differ from the ones written by the authors.

Once the information is recorded on cards you have a permanent file of what you've read. The cards can be sorted in different ways to look at relationships among related studies. This can be a great help when you're writing a research paper or planning an experiment of your own.

★ Now return to your sample articles. Read the abstracts. Then, taking everything in all the preceding chapters into account, write your own abstracts.

**Section B
Publicizing Your
Findings**

Your next task as a responsible scientist is to report your experiment to the widest appropriate audience. As you'll recall from Chapter 1, one of the characteristics of science is that it progresses by combining the efforts of many workers. Even if you consider your experiment a "failure," you should let others know about it. At least it may prevent others from repeating your errors. At best someone else may be able to interpret your findings in a useful way, even if you couldn't.

The reports of experiments can take different final forms, but the first step is always to prepare a complete written report.

### Write a Formal Report

In this section you'll learn how to write a report in the style in which articles are submitted for publication in journals published by the American Psychological Association (APA). Complete instructions can be found in the *Publication Manual*, which you can obtain by writing to APA. (It costs a few dollars; check the current price with APA before sending in your order.)

If you plan to go on to more advanced work it's very important for you to learn this style. You'll use it over and over again for papers, theses, dissertations, and original experiments. On the other hand, if you plan never to perform another experiment, you may think that learning this style is unnecessary. However, the format is surprisingly useful in writing proposals, reports, and case histories and in many kinds of jobs. Writing these reports is one of the more valuable general skills you can acquire. Moreover, there's always a possibility that you might change your mind and end up doing more research in the future.

*General Style*

As you've learned from reading formal reports, the style emphasizes efficiency and accuracy, often at the expense of originality of expression and esthetic values. Your primary task is to present information about your experiment in a concise, comprehensible form. Clear communication is the key. Don't try to snow the reader with lots of unnecessary jargon and technical terms. Keep the overall structure of the paper in mind so that the transitions between your sections are logical and easy for the reader to follow.

You might wonder how long your paper should be, but there's no easy answer. An average report of a single experiment is perhaps 8 to 10 double-spaced typed pages long, but it may be considerably shorter or longer. Don't be so concise that it's not clear what you did or why. You should say everything that's needed to make your experiment clear and relevant. On the other hand, don't repeat yourself unnecessarily or waste words on unsupported statements such as "this is an interesting area" or "further research is needed."

In the introduction and discussion sections a good rule to follow is "when in doubt, leave it out." You may have many references that don't clarify your study in any way, or you may have lots of ideas that are related only marginally to your experiment. If that happens, you should grit your teeth and edit out all the unnecessary material. In the method and results sections, where you're presenting original observations, you should be more inclined to include material unless you're sure it's irrelevant. If you fail to report events that occurred during your experiment that might have influenced your results, people may not be able to replicate your experiment. If you fail to report all your results, others may waste time duplicating work you've already completed.

Your final paper may include these sections: cover page, abstract, main body of text, reference notes, references, footnotes, tables, figure captions, figures, and appendices. We'll discuss first the main body of text, which includes the introduction, method, results, and discussion sections. Then we'll discuss the other sections and how to type the final paper.

*Introduction Section*

This section gives the reader a context in which to consider your experiment. It normally starts with very broad subjects and gradually narrows into the very specific area covered in your experiment. You gathered information appropriate for the introduction section in Section B of Chapter 2. You may want to return to the library now to do more research into the background of your experiment.

A good plan for the introduction section is to start with the general area of study, which is usually related to the dependent variable. Explain to the reader what the topic is, defining difficult words. If it seems useful, you can place the topic in historical perspective by mentioning early demonstrations.

Example: "Naming colors is quite difficult if the ink colors form words that name other colors (for example, the word "red" printed in yellow ink). This interference is called the Stroop effect, after an early worker in the field (Stroop, 1935)."

Next you may want to convince the reader that this is an important topic and worth knowing about. If so, you should refer to work indicating recent interest. Review articles are appropriate here along with some information about the general nature of the studies. Only refer to these articles, telling the reader where to go for more information.

Example: "Recent research on aggression has been reviewed by Packworm (1973). Studies have focused on the roles of parents (Smith & Smith, 1963), socioeconomic status (Rich, 1956), siblings (Grogan, Hogan, & Logan, 1972), and the weather (Rainmaker, 1974). Phylogenetic perspective has been provided for primates (Monk, 1942), cats (Felix, 1955), and ants (Fred, 1977)."

After the general background describe references directly relevant to the reasons behind your experiment. Describe these in more detail, mentioning the method and results, so the reader can understand your point and conclusions without looking up the original article. Describe only those aspects of the study relevant to your argument, but don't describe only the conclusions. Give the reader enough information to partly evaluate the conclusions. Articles here can be of three types: (1) those presenting theories to be tested in your experiment, (2) those making the same point as yours in a somewhat different way, and (3) those that you believe to be in error and that you're showing to be false or misinterpreted.

Example: "Posy (1931) originally proposed that circular seating arrangements are more conducive to a free interchange of ideas. Testing that theory, Ring and Round (1957) compared the number of different ideas expressed by group therapy clients seated in a circle to those expressed by clients seated in a row and found the number to be significantly higher for the circle clients. More recently, however, Downfall (1975) proposed that Ring and Round's results were caused not by circle effects in themselves but by facing effects. She showed that not only circle seating but also across-the-table seating resulted in more ideas than row seating. The present experiment proposed to compare the circle-seating to the facing-seating hypothesis by including a condition in which the clients are seated in a circle but facing outward rather than inward."

As in the preceding example, you finally describe your experiment and its purpose. State the theory, question, problem, or other reason for your experiment and describe how your experiment is related. Make sure that you give the reader a general feel for the whole experiment and that the logical relation between it and its purpose is clear. You might want to tell how various patterns of results could be interpreted.

*Method Section*

This section tells the reader what you've done. You should include everything necessary to repeat your experiment in *essential* detail. You shouldn't tell how you did something if it could be changed without affecting the results. If you're in doubt, it's safer to assume that things did affect the results and should be reported, but don't go overboard. Your data sheets which were never seen by the subjects, the brand of your stopwatch, and the way you wore your hair are probably irrelevant to the results and should be omitted. Most of the information you'll need to write your method section has already been collected in the second sections of Chapters 3 and 4.

Your main goal in the method section is to be clear. Avoid unnecessary technical terms and abbreviations. If you use labels, make them meaningful. For example, "Group 1" is meaningless while "Recall Instructions Group" gives the reader more help.

You should mention any methodologically questionable aspects of your experiment. You could have performed your experiment in many ways. Why did you do it the way you did? References can be used to support your decisions. These are some topics that might need to be mentioned: (1) You have reasonable operational definitions of your variables. Have others used the same or similar operational definitions? (2) Your methods are sensitive enough to show a result if it's there. Have others used similar procedures and had successful results? (3) You've recognized any confounds in your experiment and have reason to believe they're not damaging. Have others shown those confounds to be irrelevant?

You can usually improve the organization of your method section by using subsections. The following subsections are the most common, but you should use them only if they're useful; you also may use subsections other than these. If your experiment is very simple, don't use any subsections at all. The subsections can appear in any order. Try to organize your information in such a way that the reader can go straight through the method section, understanding each section as it's presented. Don't repeat yourself unless it's absolutely unavoidable.

*Subjects:* Describe the nature and number of subjects. Include enough information so that someone else can obtain similar subjects and so that readers can decide to what other subjects to generalize. Be sure to tell about any subjects who dropped out and from what levels they dropped and for what reasons. If you used a between-subject variable, you may describe in this subsection how the subjects were determined to be in their level of the independent variable.

*Apparatus:* Describe the important physical objects used in your experiment, if any. Automatic stimulus displays, recording devices, complex observation setups, and the like may be described verbally or diagramed in a figure, whichever is more efficient. Don't describe a very common apparatus such as a stopwatch, memory drum, reaction timer, tape recorder, or Skinner box unless there were special details that might have influenced your results. You can just mention common apparatus in your procedure subsection.

*Materials:* Describe the lists of words, questionnaires, pictures, or other "soft" materials used in your experiment, if any. Give enough detail so that essential aspects can be repeated by another experimenter. Examples or complete tables may be used if they give more adequate information in less space. Descriptions of how the materials were ordered and physically displayed are also appropriate. Don't include descriptions of data sheets or verbatim instructions to the subject unless they're important to the outcome of your experiment.

*Design:* Describe the independent and dependent variables, giving operational definitions as necessary. Tell how the variables were combined. Explain the orders of conditions and sessions, how subjects were assigned to various treatments, and so on. If your experimental design is simple, don't use a design subsection but include this information in other places.

*Procedure:* Tell what happened during the experiment. Often a chronological description of a session is the most clear organization for this subsection. Verbatim instructions are usually omitted.

### Results Section

This section describes your observations. If you didn't describe the dependent variable completely in your method section, your results section should start with that information. How were the raw data categorized, scored, averaged, or otherwise manipulated to obtain the reported results?

*Example:* "Each subject's recall sheet was scored for number correct and number of errors. Misspelled items were scored as correct if two judges agreed on what was intended; synonyms were scored as errors, as were intrusions and misspellings on which the judges didn't agree. Then the error-to-correct ratio was determined for each subject by dividing that subject's errors by the number correct."

Next present your summarized results. They may be in text, tables, or figures, as described in Chapter 5. Raw data aren't usually included unless you're using the minimal-statistics approach. Thus the results are usually frequencies (transformed into proportions or percentages), means, or correlations.

Also describe the results in words, telling which independent variables were related to which dependent variables. If useful, these verbal descriptions may be in terms of main effects and interactions.

If you used inferential statistics, they may be used to show how likely it is that the relationships are accidental.

*Example:* The mean score for females was 18.2, but for males it was 12.6, showing that the females made more errors. However, that difference failed to reach significance, $t(44) = 1.02$, $p > .05$."

### Discussion Section

This section tells the reader how you interpreted your results. Much of this information was collected in Section B of Chapter 6.

First, for clarity's sake, repeat your interesting results and the problem under investigation in your experiment, as explained in the introduction section. Then tell how your results are related to that problem. Be sure you explain the logical relation between your results and your conclusions.

At this point, if you have nothing further to say, stop writing. However, there may be more to say. It's often instructive to compare and contrast your findings with other studies in the literature. If others have shown the same results as you, it's proper to point that out. If your results suggest a new interpretation for others' results, it's also proper to point that out. If there are theoretical or practical implications of your study, you should discuss them. If some aspect of your study suggests other fruitful areas of investigation, indicate them briefly. If your study has serious shortcomings, you should also point them out. Self-criticism is rare in published papers, primarily because in order to get published most studies must be free of major flaws. However, for in-class purposes, it's appropriate to add some description of the limitations of your study.

*References*

You should list all the references you mentioned in any of your sections, but no others. There are two types of references: nonretrievable references such as personal communications, lecture notes, and papers read at meetings, and retrievable references from journals and books. The former should be listed as reference notes; the latter, as references.

The purpose of listing the references is to make possible their use by the reader; this purpose can be accomplished only if the reference citations are accurate and complete. Check the list of references carefully against the original publications. If you weren't able to read the original article, the secondary source from which you obtained the reference should be indicated.

For each reference be sure you have the authors' last names and initials, the title of the article or book, and the date of publication. For journal articles also include the journal title, volume number, and page number. For books the publisher and city of publication should be included; if only certain chapters or pages were referenced, they should be indicated. For reference notes there should be as much information as seems useful about each reference, including names, dates, and places as appropriate.

*Abstract*

Limit your abstract to between 100 and 150 words. Be sure you tell who the subjects were, what the experimental design was, what the experimental procedure was, and what the results were. If necessary, you should describe the operational definitions of your variables. You may also include a brief theoretical or practical statement to place your experiment in context.

Because they're so short, abstracts are difficult, but keep at it. Remember that many people see only your abstract, so it should be as clear and complete as you can make it.

*Title*

Consider your title carefully; it's often the most difficult part of the report to write. It should be a concise statement of the exact topic of the article, usually mentioning the main variables. If a causal relationship has been demonstrated, phrases such as "The effect of . . . on . . . " or " . . . as a function of . . . " can suggest that to the reader. If no causal relationship has been studied, phrases such as "The relation between . . . and . . . " or a simple " . . . and . . . " can help the reader understand your study. Avoid words that serve no useful purpose such as "An experimental investigation of . . . " or "A study of . . . " In some instances as many as 15 words may be needed to define the topic, but shorter titles are preferred.

*Other Sections*

Tables and figures should be prepared as discussed in Chapter 5 and numbered consecutively. For publication the figure legends should appear on a separate page, but for in-class use, the legends can appear on the same page as the figures.

Footnotes should generally be avoided. If they're absolutely necessary, they should be indicated with superscript numbers ([1,2]) and listed together on a separate footnote page.

Your instructor may require appendices, including your raw data, instructions, and statistical calculations. They're not usually included in published reports.

**Type a Formal Report**

Your report should be typed. There are many details to be learned about typing a paper in this format. They make more sense when you realize that they're conventions developed for the printer's convenience. However, even if you don't plan to run out and publish your experiment immediately, the format is useful. It aids people who read your report in following the information you present. Moreover, it's commonly used for master's theses and doctoral dissertations as well as published reports in the social sciences.

In Appendix D of this book is a copy of a typed paper with some explanatory notes. Follow that model rather slavishly in preparing your paper, including spacing and punctuation.

Of course, misspellings, poor grammar, and unclear writing should be avoided. Good sentence and paragraph structure are particularly important when complicated ideas are being presented. Don't submit the report without proofreading it carefully. Better still, have someone else proofread it critically.

You should type the report in black ink on ordinary white typing paper. It should be double-spaced throughout, including references, tables, and quotations. There should be generous margins as well. That allows room for editors (and teachers) to write notes.

Begin each of the following parts of the manuscript on a new page and arrange the sections you have in this sequence: cover page, abstract, main body of text, reference notes, references, footnotes, tables, figure legends, figures, and appendices. All the pages should be numbered consecutively except the cover page and figure pages. They should be paper-clipped together.

*Spacing around Punctuation*

Use normal spacing around punctuation: one space after commas and semicolons; two spaces after colons, except in ratios (3:1); two spaces after periods ending sentences, headings, or parts of a reference; one space after periods in initials in personal names (J. R. Jones or Jones, J. R.); no space after internal periods in abbreviations (a.m., U.S.); no space around hyphens (trial-by-trial); no space around dashes (studies—published and unpublished—are now . . . ); and one space around minus signs $(x - y)$.

*Common Spelling Errors*

Avoid these common errors. Use stimul*us* when there's only one and stimul*i* when there's more than one. Usually use effect as a noun and affect as a verb (an effect, to affect). *Data* are plural, so you say "the data *were*" rather than "the data *was*." *Procedure* has only one *e* in the middle.

*Cover Page*

The cover page should present the title of the article, the authors' names, their affiliations, and their writing address. The authors' names are given in the normal order, without titles such as Mr. or PhD. The affiliations are the school or organization where the authors worked while performing the study. That's the writing address which is given.

Those elements should be centered separately, one under the other, and typed in capital and lowercase letters (not all capitals). At the bottom of the page you should type a shortened title of one or two words. That's called the running head. All the subsequent pages in the published article are identified by the running head.

*Page Numbers*

All pages coming after the cover page should have the running head typed in the upper-right corner and be numbered directly beneath the running head.

*Abstract Page*

Center and underline the word *Abstract.* Then type the abstract itself.

*Headings*

On the first page of the text retype the complete title at the top of the page. Then start the introduction; don't use a heading for the introduction.

Main headings are typed centered with important words capitalized. If there are several experiments, headings of Experiment 1 and Experiment 2 are main headings. If there's only one experiment, then the method, results, and discussion sections are main headings.

Side headings are typed flush to the left margin with important words capitalized. If there are several experiments, then the method, results, and discussion sections of each are side headings. If there's only one experiment, then the subjects, procedure, and other subsections are side headings.

Paragraph headings are typed with a paragraph indentation with only the first word capitalized. They're followed by a period, two spaces, and then the text. They're used for the subjects, procedure, and other subsections when there is more than one experiment.

Main, side, and paragraph headings are underlined.

*References in the Text*

Show all references in the text by citing in parentheses the author's surname and the year of publication as entered in the list of references at the end of the article. *Example:* "A recent study (Jones, 1968) has shown . . . " If the name of the author occurs in the text, the reference citation in parentheses includes only the year of publication. *Example:* "Jones (1958) has shown . . . "

If a reference has two authors, the citation includes the surnames of both authors each time it appears in the text. If a reference has more than two authors, the original citation includes all the names but later ones include only the name of the first author and the abbreviation et al. *Example:* (Jones et al., 1958).

If there's more than one publication by the same author in the same year, distinguish them in your paper by adding a, b, and so on to the year of publication. *Example:* (Jones, 1958a).

Several references cited at the same point in the text are separated by semicolons and enclosed in a pair of parentheses. The order is alphabetical by authors' surnames or, if there are several works by the same author, in sequence by year of publication. *Example:* (Brown, 1972; Jones, 1956, 1961a, 1961b; Smith, 1969).

To refer to a particular page, pages, or chapter of a reference, give the information following the reference citation. *Examples:* (Jones, 1958, p. 510), (Jones, 1958, pp. 510–512), (Jones, 1958, Ch. 2).

### *Abbreviations*

Avoid them unless they're widely known. Never start a sentence with an abbreviation except in an abstract.

Long technical names may be abbreviated if they occur frequently. They should be spelled out when they first occur and followed by their abbreviations in parentheses. Thereafter only the abbreviation is used. The abbreviations don't have internal periods. *Example:* "The Thematic Apperception Test (TAT) was used. The TAT . . .".

### *Numbers*

For numbers larger than nine, use numerals rather than spelling words out. The only exception is if you *must* start a sentence with a number, for example, "Eighteen men and 10 women replied."

You should spell out the numbers zero through nine if they occur naturally in the text and are *not* any of the following (that is, use numerals in these cases): data, a unit of measurement or time, a point on a scale, an age, a time, a date, a percentage, a sum of money, a page number, or part of a series. *Example:* "There were three experiments performed in which six 5-year-olds rated their liking of receiving $1 or $3 on a 7-point scale."

### *Tables and Figures*

Mention each one by number in the text, and describe the information verbally. After the reference to the table include a message to the printer such as this:

```
-------------------------
Insert Table 1 about here
-------------------------
```

Put the actual table or figure at the back of the paper. For publication purposes figures must be prepared to rather exacting standards. See the APA *Publication Manual* for complete instructions.

*Statistics*

To present an inferential statistic in text give its name, degrees of freedom, value, and probability level. *Examples:* "The girls showed significantly greater liking for school than did the boys, $t(22) = 2.62$, $p < .01$. The analysis of variance indicated a significant retention interval effect, $F(1, 34) = 123.07$, $p < .001$."

Letters used as statistical symbols always are underlined. Greek letters should be handprinted and labeled.

Either the word percent or the % sign may be used after a number; be consistent within a paper. Use the word percentage when no number is given.

*Reference Notes*

The page is titled Reference Note(s) as a main heading. Each note begins with the contributor's surname, followed by initials. Each initial is followed by a period and a space. Each note contains information about a nonretrievable reference cited in the text. They're arranged in alphabetical order by author's surname. *Example:*

Jones, J. R. Lectures on developmental psychology. California State University, Hayward, 1969–1970.

Smith, P. P. Personal communication, 1976.

*References*

The page is titled Reference(s) as a main heading. Each entry includes the authors' surnames and initials, followed by the title and all information necessary to retrieve the reference. They're arranged in alphabetical order by authors' surname. See the reference section of this book and Appendix D for examples.

**Report Your Findings to Your Local Colleagues**

You should tell your local colleagues (in most cases your instructor and classmates) about your results. In the most common case you'll turn in your typed report to your instructor. Most teachers have additional requirements as well. A common requirement is to attach a copy of your raw data to the report as an appendix. This is a good practice in any case, because it ensures that you have a clean, well-organized copy of your data where you can find it easily in the future.

You may also want to include copies of your instructions, stimulus materials, and data sheets as additional appendices. Statistical calculation sheets can be helpful to the instructor.

Be sure you have a copy of your report for yourself for future reference.

**Report Your Experiment Orally**

If you want to describe your experiment orally, you'll find that it just doesn't work to read your written report out loud. An oral report is usually less complete and less formal than a written one, but somehow you must get most of the same information across. Why was the experiment performed? Exactly what did you do? What did you find? What conclusions can be drawn?

You should plan your talk in detail. Don't plan to read from notes, and don't memorize anything verbatim unless it's absolutely necessary. Know which information you're going to present and in what order. Find out beforehand how long you'll be speaking, and choose an appropriate amount of detail to fill that interval. If you have only 5 min., you can present only a bare outline of your procedure and results. If you have 20 min., you can spend more time on the reasons for the experiment and your interpretations of the results.

Keep your use of technical jargon to an absolute minimum. It's usually clearer to say "There were four groups of subjects. One group did thus-and-so . . . " than to say "The independent variable was manipulated between subjects, and was operationally defined as . . . " You should probably omit descriptions of statistical details, and of theoretically uninteresting variables such as order of treatments.

You must be extremely careful about the sequence in which you present information orally. People should be able to understand each sentence as you say it; they can't look forward or back for clarification as they can with a written report.

It's often helpful to bring copies of your stimulus materials for illustrative purposes. Results can be presented in a table or figure more easily than they can be described verbally. Make sure you have charts big enough to be seen, or plan to use the chalkboard while you speak.

**Decide Whether to Report
Your Results to a Wider Audience**

Reporting your results in writing or orally to your local colleagues is the absolute minimum. In some cases it may be the end of the road. In others you may want to perform follow-up experiments and eventually go on to report your findings to the larger scientific community or to people in a position to apply the results in practical situations. We'll discuss some of the issues that arise in making that decision.

*Clear Interpretation of Results*

If you have damaging confounded variables or it's not clear that your results aren't accidental, then the interpretation of your results isn't clear.

★ Can you interpret your results clearly?

If you said "no," there's little sense in publicizing your results widely. Either the experiment should be redesigned to remove the problems, or you should just move on to a new area. If you said "yes," then continue.

*Well-known Results*

If you've replicated a well-known, well-publicized result, then you've shown that the result is reliable. You can feel confident in making practical application of that result or in building theories based on it. However, there's little point to informing the larger scientific community of something it already knows.

★ Are your demonstrated results well known?

If you answered "yes," then you shouldn't publicize these results. Move on to more original questions. If you said "no," then continue.

*Interesting Results*

Results can be interesting practically, theoretically, or both. For example, a finding that people prefer one brand of peanut butter over another is primarily of practical interest. A finding that people make fewer errors under imagery instructions is primarily of theoretical interest. A finding that people answer differently when asked *the* questions rather than *a* questions has both practical and theoretical interest. It's difficult to come up with examples of completely uninteresting results; looked at with an open mind, many facts are suggestive and interesting. However, there are degrees of interest level, and some findings are of little interest to anyone.

★ Are your results of practical interest?

If you said "yes," and if your results are clear and not well known, you should try to report them to people who are in a position to make use of the information. In the peanut butter example, the manufacturers, merchants, and peanut butter eaters might be interested in the results. Unfortunately, there are no good ways to reach those audiences. You can write letters and try to attract the attention of the public communications media; that's about it.

★ Are your results of theoretical interest?

If you answered "no" to both questions, then you've been working on an uninteresting question. Drop it, and switch to a more interesting area. If you said "yes," and if your results are clear and not well known, you should consider reporting them to a larger segment of the scientific community. However, they probably need more work.

**Perform Follow-up Studies**

There is one case in which it's particularly important that you perform follow-up studies: when the interesting result is a lack of difference. The finding that two levels of an independent variable are the same is often quite interesting. For example, suppose you performed an experiment in which you injected different amounts of a drug into hamsters to see how it influenced their learning. Because that drug had previously facilitated rat learning you would be surprised and interested to find that it had no effect on your hamsters. Similarly, if you performed an experiment in which you expected to replicate a well-known result, you would be surprised if you failed.

Such demonstrations are very troublesome because of the statistical limitations discussed in Chapters 1 and 5: Inferential statistics can tell us only when levels are reliably different, not that they're the same. When levels appear to be the same there's no way to determine how likely it is that the result is accidental. A lack of difference might mean that the levels are the same, but it might mean instead that the experiment was too insensitive to detect real differences that were there. Your hamsters might not have differed because you ran the experiment badly or because you had some unnoticed confounded variables that counteracted the effect.

★ Is your interesting result a lack of difference?

If you answered "yes," you must demonstrate that your experiment was sensitive enough to detect any differences that might exist.

One way to do that is to repeat your experiment several times, with a great deal of control over the extraneous variables and a large number of subjects. If you consistently find no difference, then your claim appears more reasonable. However, that's a lot of work.

Another way to demonstrate that your experiment is sensitive enough to detect any differences that exist is to enlarge your experiment so that you both replicate your original finding of no difference *and* compare it to a demonstrated difference in the same experiment. In the hamster experiment you could repeat the experiment using both hamsters and rats. If everything goes as expected, you'll find a species-by-drug interaction, with the drug influencing the rats and not the hamsters. That makes it clear that your experimental technique can detect differences if they're there to detect and that your lack of difference is reliable.

Even when your result involves a clearly demonstrated difference it's important to perform one or more replications of your experiment. If you repeat your experiment with no changes in method, it makes it quite clear that your results are reliable. If you make changes in your method and continue to obtain the same result, the generality of your findings is increased. Obviously the more general and reliable the finding, the more valuable it is to the larger scientific community.

★ What follow-up studies will you perform?

These new experiments should be reported with the original one in a paper with several experiments.

### Report Your Experiments to the Larger Scientific Community

You can submit your finished paper to a journal for publication, or you can submit it for oral presentation at a professional meeting. In either case you must submit your findings in the particular written form required by that journal or professional association.

If you'd like to have your paper accepted by a journal, you should select one that publishes experimental articles of your length in your subject area. Look in the journal itself for "Information for Authors" or "Submission of Manuscripts." It will tell you the kind of articles to submit, how to prepare your manuscript, and where to send it. Most journals will give your article to one or more reviewers who will decide whether it's valuable enough to be published. They may reject the article, accept it, or accept it with the provision that you make some changes or additions.

Journals vary a good deal in prestige value. The high prestige journals are more likely to reject your paper, because they have a much higher submission rate in general. The least prestigious journals tend to have the least stringent review policies and thus are more likely to accept your article. Ask knowledgeable people in your area about the prestige of different journals so that you can send your article to a journal at an appropriate level.

If you'd like to present your paper orally at a professional meeting, contact a member of the appropriate professional organization and find out the policies about who can submit such papers. Many organizations require that the paper be sponsored by a member. After you locate a member find out the details for submitting your study. Usually a long abstract is requested, giving more details than a usual abstract but fewer than a complete paper.

Your submission will be reviewed and accepted or rejected. If it's accepted, you'll tell other members of the organization about your experiment in something between 5 and 15 min. It's not uncommon for such presentations to be read verbatim in the interest of efficiency. If you elect to read, practice your presentation several times to improve your intonation and to make the presentation less boring. Aside from the possibility that you'll be reading aloud, this presentation is much like the oral presentation to your local colleagues. Be especially careful about the sequence in which you present information. Minimize your use of technical jargon. Omit irrelevant methodological details and uninteresting results. Use visual aids to make it easier for people to understand your method and results.

### What Next?

Figure 7.2 is a flow chart outlining all the decisions you made while deciding whether to report your experiment to a wider audience. At the bottom of the chart the

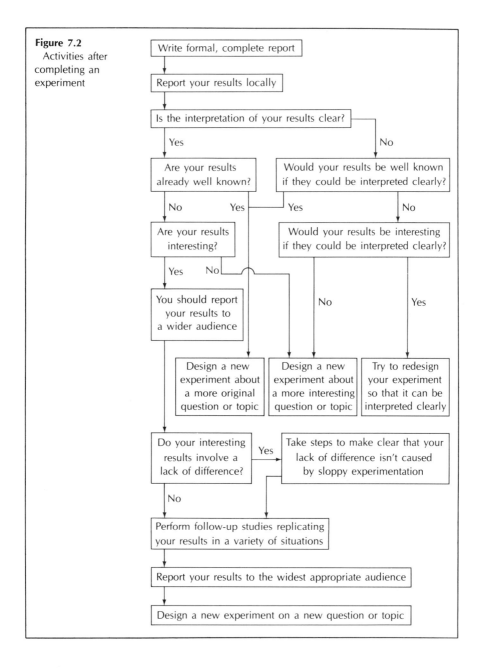

**Figure 7.2** Activities after completing an experiment

final step in each chain consists of designing a new experiment. These suggestions are for the type of experiment you should design to move in a fruitful direction.

If you enjoyed the process of designing, performing, analyzing, and understanding an experiment, you'll probably need little urging to design and perform additional experiments when you have the opportunity. If you didn't find the process rewarding—well, not everyone gets bitten by the science bug. At least you've learned how experiments work or sometimes don't work.

# Appendix A
# Some Information about Journals, with Library of Congress Call Numbers

## Sources of Information about Journals Relevant to Psychology and Allied Fields

**BF1 P65**   *Psychological Abstracts* presents abstracts from about 1000 journals, most of them presenting at least some original research. The "Coverage List" presents the names of the journals in alphabetical order; most of the names are quite informative about the contents of the journal.

**BF6 M34**   Markle, A., and Rinn, R. C. *Author's guide to journals in psychology, psychiatry, and social work.* Haworth Press, 1977. Gives information primarily of interest to prospective authors of articles but does provide a general description of the types of articles included in about 500 journals. The main listing is alphabetical, but there's a useful keyword index that classifies the journals by subject matter.

**BF6 T659**   Tompkins, M., and Shirley, N. *A checklist of serials in psychology and allied fields.* Whitson Publishing Co., 1969. Lists the objectives and fields of interest covered by about 700 journals.

### Some Journals with Articles Summarizing Original Research in Selected Areas

BF1 A15 — *Canadian Psychological Review.* Organizes and summarizes selected research areas.

BF1 A55 — *American Psychologist.* Also presents much information primarily of professional interest to active psychologists and some original research.

BF1 F7 — *Psychological Review.* The purpose of these papers is primarily to present theoretical discussions leading to original research.

BF1 P75 — *Psychological Bulletin.* Organizes and summarizes selected research areas.

BF30 A56 — *Annual Review of Psychology.* Summarizes recent literature annually in particularly active areas.

In addition to those general sources there are many more specialized summary journals, limited to such topics as experimental social psychology, child development, biofeedback, mental retardation, and psychobiology and physiological psychology. A good librarian can help you locate them.

### Some Journals Presenting at Least Some Original Empirical Research in English

BF1 A12 — *Acta Psychologica*

BF1 A158 — *International Journal of Psychology*

BF1 A5 — *American Journal of Psychology*

BF1 B7 — *British Journal of Psychology*

BF1 B723 — *British Journal of Social and Clinical Psychology*

BF1 B74 — *Australian Journal of Psychology*

BF1 C27 — *Canadian Journal of Behavioural Science*

BF1 C3 — *Canadian Journal of Psychology*

# Appendix A  Some Information about Journals

| | |
|---|---|
| **BF1** I39 | *Indian Journal of Psychology* |
| **BF1** J55 | *Journal of Applied Psychology.* Nonclinical applications. |
| **BF1** J57 | *Journal of Comparative and Physiological Psychology* |
| **BF1** J575 | *Journal of Consulting and Clinical Psychology* |
| **BF1** J6 | *Journal of Experimental Psychology, General* |
| **BF1** J61 | *Journal of Experimental Psychology, Human Learning and Memory* |
| **BF1** J612 | *Journal of Experimental Psychology, Human Perception and Performance* |
| **BF1** J613 | *Journal of Experimental Psychology, Animal Behavior Processes* |
| **BF1** J62 | *Journal of Research in Personality* |
| **BF1** J64 | *Journal of General Psychology.* Authors pay for publication; some articles are of lower quality than in other journals. |
| **BF1** J66 | *Journal of Personality* |
| **BF1** J67 | *Journal of Psychology.* Authors pay for publication; some articles are of lower quality than in other journals. |
| **BF1** J69 | *Journal of the Experimental Analysis of Behavior* |
| **BF1** L4 | *Learning and Motivation* |
| **BF1** P68 | *Psychological Record.* Authors pay for publication; some articles are of lower quality than in other journals. |
| **BF1** P883 | *Psychonomic Society Bulletin.* Many short articles. |
| **BF1** S35 | *Scandinavian Journal of Psychology* |
| **BF21** P843 | *Psychological Reports.* Many short articles. Authors are required to purchase a certain number of reprints, but articles *are* reviewed. |
| **BF39** M85 | *Multivariate Behavioral Research.* These articles require knowledge of multivariate statistics. |

| | |
|---|---|
| BF233 P4 | *Perception and Psychophysics.* Some short articles. |
| BF309 C62 | *Cognitive Psychology* |
| BF311 P348 | *Perception* |
| BF311 P36 | *Perceptual and Motor Skills.* Many short articles. Authors are required to purchase a certain number of reprints, but articles *are* reviewed. |
| BF371 M45 | *Memory and Cognition* |
| BF455 A1J6 | *Journal of Verbal Learning and Verbal Behavior* |
| BF636 A1J6 | *Journal of Applied Behavior Analysis* |
| BF636 A107 | *Organizational Behavior and Human Performance* |
| BF637 C6J6 | *Journal of Counseling Psychology* |
| BF698 J65 | *Journal of Personality Assessment* |
| BF699 D46 | *Developmental Psychology* |
| BF701 P4 | *Journal of Genetic Psychology.* Authors pay for publication; some articles are of lower quality than in other journals. |
| BF271 J64 | *Journal of Experimental Child Psychology* |
| BF728 J65 | *Journal of Cross-Cultural Psychology* |
| BF789 D404 | *Omega.* Death and dying. |
| BF1001 J66 | *Journal of Parapsychology.* Some assumptions are different than in other scientific journals. |
| CV201 R48 | *Research Quarterly.* Health, physical education, and recreation. |
| HF5415 2J66 | *Journal of Marketing Research* |
| HF5415 3J68 | *Journal of Consumer Research* |

# Appendix A Some Information about Journals

**HF5549** *Personnel Psychology*
**A2P53**

**HM251** *Journal of Social Psychology.* Authors pay for publication; some articles are of
**A1J6** lower quality than in other journals.

**HM251** *European Journal of Social Psychology*
**E8**

**HM251** *Journal of Applied Social Psychology*
**J52**

**HM251** *Journal of Experimental Social Psychology*
**J53**

**HQ1** *Journal of Marriage and the Family*
**J48**

**HQ5** *Journal of Sex Research*
**J6**

**HQ750** *Child Development*
**A1**

**HQ750** *Social Biology*
**A1E84**

**HQ793** *Adolescence*
**A44**

**HQ793** *Youth and Society*
**Y6**

**HQ796** *Journal of Youth and Adolescence*
**J625**

**HQ1060** *International Journal of Aging and Human Development*
**A33**

**HG1060** *Journal of Gerontology*
**J6**

**L11** American Educational Research Journal
**A66**

**L11** Education
**E2**

**L11** Elementary School Journal
**E6**

**L11** Journal of Experimental Education
**J77**

**LB1028** *Educational Research*
**E3**

**LB1050** *Journal of Reading.* Primarily essays rather than empirical studies.
**J6**

**LB1050** *Reading Research Quarterly.* Primarily essays rather than empirical studies.
**R42**

**LB1051** *British Journal of Educational Psychology*
**A2B7**

**LB1051** *Journal of Education Psychology.* Has many articles that are relatively easy to
**A2J6** read.

**LB1101** *Child Study Journal*
**C43**

**LB1101** *Psychology in the Schools*
**P75**

**LB1134** *Journal of Learning Disabilities*
**J68**

**LB1139** *Journal of Child Language.* Primarily essays rather than empirical studies.
**L3J6**

**LB1573** *Reading Teacher.* Primarily essays rather than empirical studies.
**R28**

**LB3013** *Journal of School Psychology*
**6J6**

**ML1** *Journal of Music Therapy*
**J57**

**ML1** *Journal of Research in Music Education*
**J6**

**QL785** *Animal Learning and Behavior*
**A725**

**QP303** *Journal of Motor Behavior*
**J64**

**QP376** *Brain Research*
**B72**

**QP461** *Journal of Auditory Research*
**A1J6**

**QP474** *Vision Research*
**V5**

**QP351** *Quarterly Journal of Experimental Psychology*
**E95234**

**QP351** *Journal of Neurobiology*
**J55**

# Appendix A  Some Information about Journals

**QP351** *Journal of Neurophysiology*
**J6**

**QP351** *Physiology and Behavior*
**P55**

**QP351** *Psychophysiology*
**P79**

**QP360** *Physiological Psychology*
**P47**

**RC321** *Behaviour Research and Therapy*
**B4**

**RC321** *Biological Psychiatry*
**B55**

**RC321** *Brain*
**B79**

**RC321** *Cortex*
**C6**

**RC321** *Electroencephalography and Clinical Neurophysiology*
**E43**

**RC321** *Experimental Neurology*
**E94**

**RC321** *Journal of Abnormal Psychology*
**J7**

**RC321** *Journal of Clinical Psychology*
**J74**

**RC321** *Journal of Mental Deficiency Research*
**J78**

**RC321** *Journal of Nervous and Mental Diseases*
**J83**

**RC321** *Journal of Psychiatric Research*
**J838**

**RC321** *Social Psychiatry*
**S6**

**RC326** *American Journal of Mental Deficiency*
**A415**

**RC331** *Psychopharmacologia*
**P7**

**RC475** *Psychotherapy.* Many case studies.
**P73**

| | |
|---|---|
| **RC488 A53** | *International Journal of Group Psychotherapy.* Many case studies. |
| **RC488 G7** | *Group Psychotherapy and Psychodrama.* Many case studies. |
| **RC488 5A1F3** | *Family Process.* Many case studies. |
| **RC489 B4B435** | *Behavior Therapy* |
| **RC489 B4J65** | *Journal of Behavior Therapy and Experimental Psychiatry* |
| **RC52 J6** | *Journal of Psychosomatic Research* |
| **RC570 A1M4** | *MR, Mental Retardation* |
| **RJ499 A1J6** | *Journal of Child Psychology and Psychiatry* |
| **RT1 N8** | *Nursing Research* |
| **RT71 A1J6** | *Journal of Nursing Education* |
| **T58 A2H8** | *Human Factors* |
| **T58 A2N35** | *Occupational Psychology* |
| **TA166 E7** | *Ergonomics* |

Appendix B
Summary of
Ethical
Principles in the
Conduct of
Research with
Human
Participants[1]

### The Ethical Principles

The decision to undertake research should rest upon a considered judgment by the individual psychologist about how best to contribute to psychological science and to human welfare. The responsible psychologist weighs alternative directions in which personal energies and resources might be invested. Having made the decision to conduct research, psychologists must carry out their investigations with respect for the people who participate and with concern for their dignity and welfare. The Principles that follow make explicit the investigator's ethical responsibilities toward participants over the course of research from the initial decision to pursue a study to the steps necessary to protect the confidentiality of research data. These Principles should be interpreted in terms of the context provided in the complete document offered as a supplement to these Principles.

**1.** In planning a study the investigator has the personal responsibility to make a careful evaluation of its ethical acceptability, taking into account these Princi-

---

[1] Reprinted from *Ethical Principles in the Conduct of Research with Human Participants* prepared by the Ad hoc Committee on Ethical Standards in Psychological Research and published by the American Psychological Association in 1973. Copyright 1973 by the American Psychological Association. Reprinted by permission.

ples for research with human beings. To the extent that this appraisal, weighing scientific and humane values, suggests a deviation from any Principle, the investigator incurs an increasingly serious obligation to seek ethical advice and to observe more stringent safeguards to protect the rights of the human research participant.

2. Responsibility for the establishment and maintenance of acceptable ethical practice in research always remains with the individual investigator. The investigator is also responsible for the ethical treatment of research participants by collaborators, assistants, students, and employees, all of whom, however, incur parallel obligations.

3. Ethical practice requires the investigator to inform the participant of all features of the research that reasonably might be expected to influence willingness to participate and to explain all other aspects of the research about which the participant inquires. Failure to make full disclosure gives added emphasis to the investigator's responsibility to protect the welfare and dignity of the research participant.

4. Openness and honesty are essential characteristics of the relationship between investigator and research participant. When the methodological requirements of a study necessitate concealment or deception, the investigator is required to ensure the participant's understanding of the reasons for this action and to restore the quality of the relationship with the investigator.

5. Ethical research practice requires the investigator to respect the individual's freedom to decline to participate in research or to discontinue participation at any time. The obligation to protect this freedom requires special vigilance when the investigator is in a position of power over the participant. The decision to limit this freedom increases the investigator's responsibility to protect the participant's dignity and welfare.

6. Ethically acceptable research begins with the establishment of a clear and fair agreement between the investigator and the research participant that clarifies the responsibilities of each. The investigator has the obligation to honor all promises and commitments included in that agreement.

7. The ethical investigator protects participants from physical and mental discomfort, harm, and danger. If the risk of such consequences exists, the investigator is required to inform the participant of that fact, secure consent before proceeding, and take all possible measures to minimize distress. A research procedure may not be used if it is likely to cause serious and lasting harm to participants.

8. After the data are collected, ethical practice requires the investigator to provide the participant with a full clarification of the nature of the study and to remove any misconceptions that may have arisen. Where scientific or humane values justify delaying or withholding information, the investigator acquires a special responsibility to assure that there are no damaging consequences for the participant.

9. Where research procedures may result in undesirable consequences for the participant, the investigator has the responsibility to detect and remove or correct these consequences, including, where relevant, long-term aftereffects.

10. Information obtained about the research participants during the course of

an investigation is confidential. When the possibility exists that others may obtain access to such information, ethical research practice requires that this possibility, together with the plans for protecting confidentiality, be explained to the participants as a part of the procedure for obtaining informed consent.

Appendix C
Calculational
Procedures for a
Few Very
Common
Statistics

**General Remarks**

This appendix is not a substitute for learning about statistics; it's only a calculational guide to some very commonly used statistics. For many experiments, statistics not included here would be appropriate.

By using this appendix, individuals having little or no statistical background can calculate some common statistics. However, keep in mind that statistics are meaningless unless they're appropriate for the purpose and fully understood by the user. Information on selecting appropriate statistics is provided in Chapter 5 of this book, along with some information on interpretation. However, many people require much more information than provided in this text to apply statistics to their own experimental data in a meaningful way. If you're not *sure* that you understand the meaning of the statistics you want to calculate with respect to your experiment, discuss them with a knowledgeable person.

To carry out most of these procedures you should have a calculator that can collect a sum of squared numbers, for example, $2^2 + 3^2 + 3^2 = 4 + 9 + 9 = 22$, without requiring you to write down all the squares. Otherwise, you're very likely to make errors. You should always check any statistical calculation, usually by repeating each step before proceeding. If your calculator has an independently accessible memory, you can often use it to provide other kinds of checks.

## Appendix C  Calculational Procedures

All the data presented in this appendix are hypothetical and not realistic. There are relatively few data, and the effects are generally stronger than in real experimental findings.

### Descriptive Statistics

*Differences-Between-Frequencies Approach: Mode, Median, Range*

*Mode*

The mode tells which level of a variable occurs most frequently.

*Example.* In this hypothetical study the variable is type of car driven. Twenty subjects were observed.

| S | Type of Car | S | Type of Car |
|---|---|---|---|
| 1 | Sedan | 11 | Sedan |
| 2 | Station wagon | 12 | Sedan |
| 3 | Station wagon | 13 | Sedan |
| 4 | Truck | 14 | Truck |
| 5 | Sedan | 15 | Sports car |
| 6 | Sports car | 16 | Truck |
| 7 | Sedan | 17 | Truck |
| 8 | Sports car | 18 | Sedan |
| 9 | Sedan | 19 | Station wagon |
| 10 | Sedan | 20 | Truck |

**Step 1.** Count the number of scores at each level.

| Type of Car | Number of Scores |
|---|---|
| Sedan | 9 |
| Station wagon | 3 |
| Truck | 5 |
| Sports car | 3 |

**Step 2.** Determine the mode. The mode is the level with the highest frequency of occurrence.

$$\text{Mode} = \text{Sedan}$$

# Appendix C  Calculational Procedures

*Median*

The median is the middle score.
*Example.* In this hypothetical study the variable is number of published papers. Twelve subjects were observed.

| S | Number of Papers | S | Number of Papers |
|---|---|---|---|
| 1 | 7 | 7 | 15 |
| 2 | 0 | 8 | 1 |
| 3 | 5 | 9 | 1 |
| 4 | 1 | 10 | 1 |
| 5 | 0 | 11 | 0 |
| 6 | 30 | 12 | 2 |

**Step 1.** Arrange the scores in order from lowest to highest.

$$0, 0, 0, 1, 1, 1, 1, 2, 5, 7, 15, 30$$

**Step 2.** Determine the median. The median is the value of the score having the same number of scores above and below; the median is the value of the middle score. (If you have an even number of scores, there will be two middle scores; in that case, take the mean of the two scores.)

Median = 0, 0, 0, 1, 1, <u>1</u>, <u>1</u>, 2, 5, 7, 15, 30

5 scores         5 scores

Median = 1

*Range*

The range is equal to the difference between the largest and smallest scores in the distribution. In the "Published Papers" example, the range is $30 - 0 = 30$.

*Differences-Between-Means Approach: Mean, Variance, Standard Deviation*

*Mean*

These instructions are suitable for determining the mean of a group of scores. In published results sections the most common symbol for the mean is *M*, but in statistical terminology other symbols are usually used. If the scores have been called X scores, the mean is often symbolized by $\bar{X}$.

*Example.* These hypothetical data are error scores by 12 subjects on a task. The same data are used in the calculation of the variance and standard deviation, which follow.

**Step 1.** Table your data as shown. No particular order is required.

| S | Number of Errors |
|---|---|
| 1 | 4 |
| 2 | 9 |
| 3 | 10 |
| 4 | 7 |
| 5 | 5 |
| 6 | 5 |
| 7 | 8 |
| 8 | 6 |
| 9 | 7 |
| 10 | 4 |
| 11 | 10 |
| 12 | 9 |

**Step 2.** Determine N, the number of scores.

$$N = 12$$

**Step 3.** Calculate $\Sigma X$, the sum of the scores. Add up all the scores.

$$\Sigma X = 4 + 9 + 10 + \text{etc.} = 84$$

**Step 4.** Calculate $\bar{X}$, the mean. Divide Step 3 by Step 2.

$$\bar{X} = \frac{\Sigma X}{N} = \frac{84}{12} = 7.00$$

*Variance and Standard Deviation*

These instructions are suitable for estimating the variance and standard deviation of a population of scores from a sample of scores. Some common symbols for the variance are *Var.*, $S^2$, and $\sigma^2$. Some common symbols for the standard deviation are

# Appendix C  Calculational Procedures

*SD*, *S*, and $\sigma$. The example data are the same as those used for calculating the mean; the steps continue in sequence. Since the standard deviation is the square root of the variance, the variance is always calculated first.

*Variance.* This is the formula for the variance.

$$\text{Var.} = \frac{\Sigma X^2 - (\Sigma X)^2/N}{N - 1}$$

**Step 5.** Calculate $\Sigma X^2$, the sum of the squared scores. Square each score in the group and add them together. (You'll square N numbers.)

$$\Sigma X^2 = 4^2 + 9^2 + 10^2 + \text{etc.} = 642$$

**Step 6.** Calculate $(\Sigma X)^2$. Square Step 3.

$$(\Sigma X)^2 = 84^2 = 7056$$

**Step 7.** Divide Step 6 by Step 2, N.

$$\frac{(\Sigma X)^2}{N} = \frac{7056}{12} = 588.00$$

**Step 8.** Subtract Step 7 from Step 5.

$$\Sigma X^2 - \frac{(\Sigma X)^2}{N} = 642 - 588.00 = 54.00$$

**Step 9.** Calculate $N - 1$. Subtract 1 from Step 2.

$$N - 1 = 12 - 1 = 11$$

**Step 10.** Divide Step 8 by Step 9. This yields Var.

$$\text{Var.} = \frac{54.00}{11} = 4.91$$

*Standard Deviation.* The formula for the standard deviation is the square root of the formula for the variance.

$$SD = \sqrt{\text{Var.}} = \sqrt{\frac{\Sigma X^2 - (\Sigma X)^2/N}{N - 1}}$$

**Step 11.** Take the square root of Step 10. This yields *SD*.

$$SD = \sqrt{4.91} = 2.22$$

*Correlational Approach:*
*Pearson Product-Moment*
*Correlation Coefficient (r)*

*Pearson Product-Moment Correlation Coefficient (r)*

These instructions are suitable for determining the correlation between two linearly related groups of scores. There must be a group of subjects, with scores on two variables for each subject.

*Example.* In this hypothetical study the two variables are the amount of sleep reported by each subject and the number of trials required by each subject to learn a paper and pencil maze the next day.

*Table Your Data*

**Step 1.** Table your data as shown, keeping the scores for each subject together in rows.

| S | Amount of Sleep (hours) | Number of Trials |
|---|---|---|
| 1 | 5.3 | 8 |
| 2 | 6.0 | 7 |
| 3 | 9.5 | 2 |
| 4 | 8.1 | 3 |
| 5 | 7.6 | 4 |
| 6 | 7.5 | 5 |
| 7 | 7.4 | 4 |
| 8 | 4.9 | 6 |
| 9 | 8.5 | 4 |
| 10 | 7.2 | 4 |
| 11 | 6.5 | 6 |
| 12 | 6.1 | 7 |
| 13 | 10.0 | 4 |
| 14 | 8.0 | 3 |
| 15 | 8.6 | 4 |

*Preliminary Steps*

**Step 2.** Determine $N$, the number of subjects in your experiment, that is, the number of *pairs* of scores.

$$N = 15$$

## Appendix C  Calculational Procedures

**Step 3.** Arbitrarily designate one of your measures as X and the other as Y.

$$X = \text{Amount of sleep} \qquad Y = \text{Number of trials}$$

**Step 4.** Calculate $\Sigma X$, the sum of the scores on measure X.

$$\Sigma X = 5.3 + 6.0 + 9.5 + \text{etc.} = 111.2$$

**Step 5.** Calculate $\Sigma Y$, the sum of the scores on measure Y.

$$\Sigma Y = 8 + 7 + 2 + \text{etc.} = 71$$

**Step 6.** Calculate $\Sigma X^2$, the sum of the squared X scores. Square each X score, and add the squares together. (You'll square N numbers.)

$$\begin{aligned}\Sigma X^2 &= 5.3^2 + 6.0^2 + 9.5^2 + \text{etc.} \\ &= 28.09 + 36.00 + 90.25 + \text{etc.} \\ &= 854.24\end{aligned}$$

**Step 7.** Calculate $\Sigma Y^2$, the sum of the squared Y scores. Square each Y score and add the squares together. (You'll square N numbers.)

$$\Sigma Y^2 = 8^2 + 7^2 + 2^2 + \text{etc.} = 64 + 49 + 4 + \text{etc.} = 377$$

**Step 8.** Calculate $\Sigma XY$, the sum of the cross-products of the two variables. Multiply each subject's X score times that subject's Y score; then add all the cross-products together.

$$\begin{aligned}\Sigma XY &= (5.3 \times 8) + (6.0 \times 7) + (9.5 \times 2) + \text{etc.} \\ &= 42.4 + 42.0 + 19.0 + \text{etc.} = 497.5\end{aligned}$$

*Calculate the Correlation Coefficient, r*

This is the formula for the Pearson r.

$$r = \frac{N\Sigma XY - (\Sigma X)(\Sigma Y)}{\sqrt{[N\Sigma X^2 - (\Sigma X)^2][N\Sigma Y^2 - (\Sigma Y)^2]}}$$

**Step 9.** Multiply Step 2 times Step 6.

$$N\Sigma X^2 = 15 \times 854.24 = 12{,}813.6$$

**Step 10.** Square Step 4. Subtract the answer from Step 9.

$$N\Sigma X^2 - (\Sigma X)^2 = 12{,}813.6 - 111.2^2 = 12{,}813.6 - 12{,}365.44 = 448.16$$

**Step 11.** Multiply Step 2 times Step 7.

$$N\Sigma Y^2 = 15 \times 377 = 5655$$

**Step 12.** Square Step 5. Subtract the answer from Step 11.

$$N\Sigma Y^2 - (\Sigma Y)^2 = 5655 - 71^2 = 5655 - 5041 = 614$$

**Step 13.** Multiply Step 10 times Step 12. Take the square root of the answer.

$$\sqrt{[N\Sigma X^2 - (\Sigma X)^2][N\Sigma Y^2 - (\Sigma Y)^2]} = \sqrt{448.16 \times 614} = \sqrt{275{,}170.24} = 524.57$$

**Step 14.** Multiply Step 2 times Step 8.

$$N\Sigma XY = 15 \times 497.5 = 7462.5$$

**Step 15.** Multiply Step 4 times Step 5. Subtract the answer from Step 14.

$$N\Sigma XY - (\Sigma X)(\Sigma Y) = 7462.5 - (111.2)(71) = 7462.5 - 7895.2 = -432.7$$

**Step 16.** Divide Step 15 by Step 13. This yields $r$.

$$r = \frac{-432.7}{524.57} = -.82$$

Thus the correlation between these sets of numbers is $-.82$. To evaluate the statistical significance of a correlation coefficient, look up its value in Table C.3. The degrees of freedom are $N - 2$. For this example the correlation is statistically significant at the .001 level.

### Inferential Statistics

*Chi-Square, or $\chi^2$*

These instructions are suitable for determining whether there is a significant relation between two sets of scores. The data should consist of frequency values telling the joint occurrence of two variables. In the most common case there are many subjects; the level occupied by each subject on both variables must be known.

*Example.* In this hypothetical study 200 subjects were interviewed. For each subject two things were determined: their self-reported political affiliation (Republican,

## Appendix C  Calculational Procedures

Democrat, or neither) and their monthly income (less than $500, $500–999, $1000–1500, or more than $1500).

*Table the Data*

**Step 1.** Organize the data into a contingency table showing the number of scores falling into each of the combined categories.

|            | Less Than $500 | $500–999 | $1000–1500 | More Than $1500 |
|------------|---------------|----------|------------|-----------------|
| Republican | 5             | 6        | 18         | 21              |
| Democrat   | 15            | 45       | 26         | 14              |
| Neither    | 10            | 13       | 22         | 5               |

*Preliminary Steps*

**Step 2.** Add the numbers in each row to obtain row sums.

Sum for Republicans = 5 + 6 + 18 + 21 = 50
Democrats = 100
Neither = 50

**Step 3.** Add the numbers in each column to obtain column sums.

Sum for less than $500 = 5 + 15 + 10 = 30
$500–999 = 64
$1000–1500 = 66
More than $1500 = 40

**Step 4.** Obtain the grand sum. Add up all the row sums (Step 2). This yields the grand sum. Then add up all the column sums (Step 3). This also yields the grand sum. Your answers should match.

Grand sum = 50 + 100 + 50 = 200 (addition of row sums)
Grand sum = 30 + 64 + 56 + 40 = 200 (addition of column sums)

**Step 5.** Make a new table of *expected values*. It should have the same structure as the contingency table in Step 1 but have new numbers called expected values. To calculate expected values for each cell in the table multiply the row sum (Step 2) times the column sum (Step 3) for the cell in which you're interested. Then divide that product by the grand sum (Step 4).

# Appendix C  Calculational Procedures

Expected value for Republican/Less than $500 $= \dfrac{50 \times 30}{200}$

$$= \dfrac{1500}{200} = 7.50$$

Democrat/$500–999 $= \dfrac{100 \times 64}{200}$

$$= \dfrac{6400}{200} = 8.00$$

Etc.

### Table of Expected Values

|            | Less Than $500 | $500–999 | $1000–1500 | More Than $1500 |
|------------|----------------|----------|------------|-----------------|
| Republican | 7.50           | 16.00    | 16.50      | 10.00           |
| Democrat   | 15.00          | 32.00    | 33.00      | 20.00           |
| Neither    | 7.50           | 16.00    | 16.50      | 10.00           |

*Calculate Chi-Square and Evaluate It*

The formula for calculating chi-square is this:

$$\chi^2 = \Sigma \dfrac{(O - E)^2}{E}$$

**Step 6.** For each cell in the table, take the observed or $O$ value from your original table (Step 1) and subtract the corresponding expected or $E$ value from your expected value table (Step 5). Square the answer. Then divide that answer by that expected value. Repeat this for each cell.

$\dfrac{(O - E)^2}{E}$ for Republican/Less than $500 $= \dfrac{(5 - 7.50)^2}{7.50} = \dfrac{(-2.50)^2}{7.50}$

$$= \dfrac{6.25}{7.50} = .8333$$

|            | Less Than $500 | $500–999 | $1000–1500 | More Than $1500 |
|------------|----------------|----------|------------|-----------------|
| Republican | .83            | 6.25     | .14        | 12.10           |
| Democrat   | .00            | 5.28     | 1.48       | 1.80            |
| Neither    | .83            | .56      | 1.83       | 2.50            |

# Appendix C  Calculational Procedures

**Step 7.** Add up all the values computed in Step 6. This yields $\chi^2$.

$$\chi^2 = .83 + 6.25 + .14 + \text{etc.} = 33.60$$

**Step 8.** Evaluate the $\chi^2$ by looking it up in Table C.4, and following the instructions there. Your degrees of freedom are equal to the number of rows minus 1 times the number of columns minus 1.

In the example, the degrees of freedom are $(3 - 1)(4 - 1) = 2 \times 3 = 6$. Thus, the $\chi^2$ is significant at the .001 level; the two variables of political affiliation and income are significantly related.

*Note:* Chi-square must always be positive in value. A negative value always means that a mistake has been made.

### Independent-Measures t Test

These instructions are suitable for comparing two means gathered on two independent groups of subjects, that is, designs having one independent variable with two levels that is varied between subjects.

*Example.* In this hypothetical experiment the dependent variable was the score obtained by each subject on a standardized test of creativity. One group received special instructions about how to appear more creative; the other group received no such special instructions.

### Table Your Data

**Step 1.** Table your data as shown, keeping the scores for each group of subjects together.

| | Score on Creativity Test | | |
|---|---|---|---|
| S | Special Instructions | S | No Special Instructions |
| 1 | 80 | 11 | 95 |
| 2 | 103 | 12 | 66 |
| 3 | 76 | 13 | 80 |
| 4 | 91 | 14 | 72 |
| 5 | 110 | 15 | 60 |
| 6 | 89 | 16 | 91 |
| 7 | 87 | 17 | 82 |
| 8 | 95 | 18 | 84 |
| 9 | 72 | 19 | 77 |
| 10 | 99 | 20 | 77 |
| | | 21 | 90 |
| | | 22 | 87 |

# Appendix C  Calculational Procedures

*Preliminary Steps*

**Step 2.** Arbitrarily designate one level of your independent variable as Level 1 and the other as Level 2.

Level 1 = Special instructions    Level 2 = No special instructions

**Step 3.** Determine $n_1$, the number of scores at level 1.

$$n_1 = 10$$

**Step 4.** Determine $n_2$, the number of scores at level 2.

$$n_2 = 12$$

**Step 5.** Calculate $\Sigma X_1$, the sum of the subjects' scores at level 1.

$$\Sigma X_1 = 80 + 103 + 76 + \text{etc.} = 902$$

**Step 6.** Calculate $\Sigma X_2$, the sum of the subjects' scores at level 2.

$$\Sigma X_2 = 95 + 66 + 80 + \text{etc.} = 961$$

**Step 7.** Calculate $\Sigma X_1^2$, the sum of the squared scores of the subjects at level $X_2$. Square *each* score and add them together. (You'll square $n_1$ numbers.)

$$\Sigma X_1^2 = 80^2 + 103^2 + 76^2 + \text{etc.} = 82{,}666$$

**Step 8.** Calculate $\Sigma X_2^2$, the sum of the squared scores of the subjects at level $X_2$. Square *each* score and add them together. (You'll square $n_2$ numbers.)

$$\Sigma X_2^2 = 95^2 + 66^2 + 80^2 + \text{etc.} = 78{,}153$$

**Step 9.** Calculate $\overline{X}_1$, the mean of the scores at level 1. Divide $\Sigma X_1$ (Step 5) by $n_1$ (Step 3). This yields $\overline{X}_1$.

$$\overline{X}_1 = \frac{\Sigma X_1}{n_1} = \frac{902}{10} = 90.2$$

**Step 10.** Calculate $\overline{X}_2$, the mean of the scores at level 2. Divide $\Sigma X_2$ (Step 6) by $n_2$ (Step 4). This yields $\overline{X}_2$.

$$\overline{X}_2 = \frac{\Sigma X_2}{n_2} = \frac{961}{12} = 80.08$$

# Appendix C  Calculational Procedures

*Calculate the t Score and Evaluate It*

This is the formula for the independent-measures $t$ test.

$$t = \frac{\overline{X}_1 - \overline{X}_2}{\sqrt{\dfrac{\Sigma X_1^2 - \dfrac{(\Sigma X_1)^2}{n_1} + \Sigma X_2^2 - \dfrac{(\Sigma X_2)^2}{n_2}}{n_1 + n_2 - 2} \times \left(\dfrac{1}{n_1} + \dfrac{1}{n_2}\right)}}$$

**Step 11.** Square Step 5. Divide the answer by Step 3.

$$\frac{(\Sigma X_1)^2}{n_1} = \frac{90^2}{10} = \frac{813{,}604}{10} = 81{,}360.4$$

**Step 12.** Square Step 6. Divide the answer by Step 4.

$$\frac{(\Sigma X_2)^2}{n_2} = \frac{961^2}{12} = \frac{923{,}521}{12} = 76{,}960.08$$

**Step 13.** Add together Steps 7 and 8; then subtract Steps 11 and 12.

$$\Sigma X_1^2 - \frac{(\Sigma X_1)^2}{n_1} + \Sigma X_2^2 - \frac{(\Sigma X_2)^2}{n_2}$$

$$= 82{,}666 - 81{,}360.4 + 78{,}153 - 76{,}960.08 = 2498.52$$

**Step 14.** Add together Steps 3 and 4; then subtract the number "2."

$$n_1 + n_2 - 2 = 10 + 12 - 2 = 20$$

**Step 15.** Divide Step 13 by Step 14.

$$\frac{2498.52}{20} = 124.926$$

**Step 16.** Calculate $1/n_1 + 1/n_2$. Divide 1 by $n_1$ (Step 3); divide 1 by $n_2$ (Step 4); add those numbers together.

$$\frac{1}{n_1} + \frac{1}{n_2} = \frac{1}{10} + \frac{1}{12} = .100 + .083 = .183$$

**Step 17.** Multiply Step 15 times Step 16. Take the square root of the answer. This yields the denominator of the $t$ test.

$$\sqrt{124.926 \times .183} = \sqrt{22.8614} = 4.7814$$

**Step 18.** Subtract Step 10 from Step 9. This yields the numerator.

$$\bar{X}_1 - \bar{X}_2 = 90.20 - 80.08 = 10.12$$

**Step 19.** Divide Step 18 by Step 17. This yields the $t$ value. (*Note:* The value of $t$ may be negative.)

$$t = \frac{10.12}{4.7814} = 2.12$$

**Step 20.** Evaluate the $t$ score by looking it up in Table C.5 and following the instructions there. Your degrees of freedom are $n_1 + n_2 - 2$, which you calculated in Step 14. In the example the degrees of freedom are 20.

For this example, the difference between the means was significant at the .05 level.

*Related-Measures t Test*

These instructions are suitable for determining whether the difference between the two means gathered on one group of subjects is significantly different from zero. That is, it is suitable for designs having one independent variable with two levels that is varied within subjects.

*Example.* In this hypothetical experiment the dependent variable was the reading rate of subjects. The subjects were tested under two conditions of lighting, good and poor.

*Table Your Data*

**Step 1.** Table your data as shown, keeping the scores for each subject together in rows.

|   | Reading Rate (w.p.m.) | |
|---|---|---|
| S | Good Light | Poor Light |
| 1 | 320 | 310 |
| 2 | 240 | 260 |
| 3 | 610 | 550 |
| 4 | 590 | 500 |
| 5 | 450 | 410 |
| 6 | 350 | 380 |
| 7 | 380 | 380 |
| 8 | 220 | 200 |
| 9 | 400 | 410 |
| 10 | 510 | 480 |

# Appendix C   Calculational Procedures

*Preliminary Steps*

**Step 2.** Determine N, the number of subjects in your experiment, that is, the number of *pairs* of scores.

$$N = 10$$

**Step 3.** Calculate D scores, or difference scores for each subject. To calculate D scores, subtract the subject's score on one level of the independent variable from that same subject's score on the other level. Be sure you keep track of any minus scores.

| S  | Good Light | Poor Light | D Score |
|----|------------|------------|---------|
| 1  | 320        | 310        | 10      |
| 2  | 240        | 260        | −20     |
| 3  | 610        | 550        | 60      |
| 4  | 590        | 500        | 90      |
| 5  | 450        | 410        | 40      |
| 6  | 350        | 380        | −30     |
| 7  | 380        | 380        | 0       |
| 8  | 220        | 200        | 20      |
| 9  | 400        | 410        | −10     |
| 10 | 510        | 480        | 30      |

**Step 4.** Calculate $\Sigma D$, the sum of the D scores. Add all the D scores together. Be sure you keep careful track of the minus scores.

$$\Sigma D = 10 + (-20) + 60 + \text{etc.} = 190$$

**Step 5.** Calculate $\Sigma D^2$, the sum of the squared D scores. Square each D score and add the squares. (You'll square N numbers.) Remember that a squared minus score is positive.

$$\Sigma D^2 = 10^2 + (-20)^2 + \text{etc.} = 100 + 400 + \text{etc.} = 16{,}100$$

**Step 6.** Calculate $\overline{D}$, the mean difference score. Divide $\Sigma D$ (Step 4) by N (Step 2). This yields $\overline{D}$.

$$\overline{D} = \frac{\Sigma D}{N} = \frac{190}{10} = 19.00$$

*Calculate the t Score and Evaluate It*

This is the formula for the related-measures $t$ test.

$$t = \frac{\overline{D}}{\sqrt{\frac{\Sigma D^2 - (\Sigma D)^2/N}{N(N-1)}}}$$

**Step 7.** Square Step 4. Divide the answer by Step 2.

$$\frac{(\Sigma D)^2}{N} = \frac{190^2}{10} = \frac{36{,}100}{10} = 3610.00$$

**Step 8.** Subtract Step 7 from Step 5.

$$\Sigma D^2 - \frac{(\Sigma D)^2}{N} = 16{,}100 - 3610.00 = 12{,}490.00$$

**Step 9.** Calculate $N(N-1)$.

$$N(N-1) = 10(10-1) = 10 \times 9 = 90$$

**Step 10.** Divide Step 8 by Step 9. Take the square root of the answer.

$$\sqrt{\frac{\Sigma D^2 - (\Sigma D)^2/N}{N(N-1)}} = \sqrt{\frac{12{,}490.00}{90}} = \sqrt{138.7778} = 11.78$$

**Step 11.** Divide Step 6 by Step 10. This yields the $t$ score. (*Note:* $t$ may be negative.)

$$t = \frac{19.00}{11.78} = 1.61$$

**Step 12.** Evaluate the $t$ score by looking it up in Table C.5 and following the instructions given there. Your degrees of freedom are $N - 1$. In the example the degrees of freedom are 9.

For this example, the difference between the means is *not* significantly different from zero; $p > .10$.

*Analyses of Variance*

For any analysis of variance the final step in the calculation is to find one or more $F$ scores. If the $F$ score is found to be significant it means that the relationship

# Appendix C  Calculational Procedures

being tested is significant. If there are more than two means being compared in that term, the F score does *not* tell specifically which means differ from each other. After finding a significant *F* score, additional comparisons can be made by using the Newman-Keuls test or other individual comparisons tests.

Note that *F* must always be positive. If you find negative scores at any point in your calculations it always means that an error has been made.

The analyses in this book assume that there are an equal number of subjects at each level of any between-subject variable. It is possible to calculate analyses of variance for unequal numbers of subjects, but those instructions are not included in this book. If you have unequal numbers of subjects you may either consult a more advanced text for instructions or eliminate subjects randomly from your groups to attain equal numbers.

*One-Way Between-Groups Analysis of Variance*

These instructions are suitable for designs having one independent variable that is varied between subjects. Each group of subjects should have an equal number.

*Example.* In this hypothetical experiment the dependent variable was the number of words recalled by each subject, and the independent variable was the type of study instructions: make up a story, alphabetize the words, and repeat each word. Thus there were three groups of subjects. Each group had seven subjects, making a total of 21 subjects and 21 measures in all.

*Table Your Data*

**Step 1.** Table your data as shown, keeping the scores for each group of subjects together.

| | Number of Words Recalled | | | | |
|---|---|---|---|---|---|
| | Type of Instructions | | | | |
| S | Make Up a Story | S | Alphabetize Each Word | S | Repeat Each Word |
| 1 | 15 | 8 | 12 | 15 | 8 |
| 2 | 12 | 9 | 10 | 16 | 9 |
| 3 | 8 | 10 | 9 | 17 | 7 |
| 4 | 10 | 11 | 11 | 18 | 6 |
| 5 | 13 | 12 | 8 | 19 | 9 |
| 6 | 11 | 13 | 9 | 20 | 10 |
| 7 | 12 | 14 | 10 | 21 | 6 |

## Appendix C   Calculational Procedures

*Preliminary Steps*

**Step 2.** Determine $N$, the total number of measures in your table. This is equal to the number of subjects in the experiment.

$$N = 21$$

**Step 3.** Determine $G$, the number of groups of subjects. This is equal to the number of levels of your independent variable.

$G = 3$, since there were three types of instructions

**Step 4.** Determine $n$, the number of subjects in each group. To check, $G$ (Step 3) $\times n$ should equal $N$ (Step 2).

$$n = 7 \quad G \times n = 3 \times 7 = 21 = N$$

**Step 5.** Calculate *group sums*. Add up all the scores in each group. You'll add $n$ scores in each group. Check these carefully.

Group sum for story = $15 + 12 + 8 + 10 + 13 + 11 + 12 = 81$
Alphabetize = 69
Repeat = 55

**Step 6.** Calculate the *grand sum*. Add up all the group sums (Step 5). This yields the grand sum. Check it carefully.

$$\text{Grand sum} = 81 + 69 + 55 = 205$$

**Step 7.** Calculate $C$, the correction term. Square the grand sum (Step 6), and divide the answer by $N$ (Step 2). This yields $C$.

$$\text{Correction term} = C = \frac{205^2}{21} = 2001.1905$$

*Make an Analysis of Variance Summary Table and Fill in the Degrees of Freedom*

**Step 8.** Make an analysis of variance summary table in the form shown here. Fill in the names of each of your variables as appropriate. Leave all the columns except the first column blank for now. You'll fill them in with the steps shown in the table.

## Appendix C  Calculational Procedures

| Analysis of Variance of Dependent Variable | | | | | |
|---|---|---|---|---|---|
| Source of Variance | df | Sum of Squares | Mean Square | F | p |
| Between groups (your independent variable) | Step 9 | Step 11 | Step 13 | 14 | 15 |
| Within-groups error term | Step 9 | Step 12 | Step 13 | | |
| Total | Step 9 | Step 10 | | | |

**Step 9.** Calculate degrees of freedom for each of the terms in your analysis of variance summary table by following the formulas given here. Record the answers in your table.

| Source of Variance | Formula for Degrees of Freedom | Step Reference |
|---|---|---|
| Between groups | $G - 1$ | Step 3 |
| Within-groups error term | $(N - 1) - (G - 1)$ | Steps 2 and 3 |
| Total | $N - 1$ | Step 2 |

| Analysis of Variance of Words Recalled | |
|---|---|
| Source of Variance | df |
| Instructions | $3 - 1 = 2$ |
| Within-groups error term | $(21 - 1) - (3 - 1) = 20 - 2 = 18$ |
| Total | $21 - 1 = 20$ |

*Calculate Sums of Squares*

**Step 10.** Calculate SS tot, the total sum of squares. Square each of the scores in your original table (Step 1). [You'll square $N$ (Step 2) numbers.] Add the squares. Finally, subtract $C$ (Step 7) from the answer. This yields SS tot. Record it in your summary table.

$$SS\ tot = (15^2 + 12^2 + etc.) - C$$
$$= 2105.0000 - 2001.1905 = 103.8095$$

**Step 11.** Calculate SS bg, the sum of squares between-groups, which corresponds to your independent variable. Square each of your group sums (Step 5). [You'll square $G$ (Step 3) numbers.] Add the squares. Divide that sum by $n$ (Step 4). Then subtract $C$ (Step 7). This yields SS bg. Record it in your summary table.

$$\text{SS instructions} = \frac{81^2 + 69^2 + 55^2}{7} - C$$

$$= \frac{14{,}347}{7} - C$$

$$= 2049.5714 - 2001.1905 = 48.3809$$

**Step 12.** Calculate SS wg, the sum of squares for the within-groups error term. Simply subtract SS bg (Step 11) from SS tot (Step 10). This yields SS wg. Record it in your summary table.

$$\text{SS wg} = \text{SS tot} - \text{SS bg}$$
$$= 103.8095 - 48.3809 = 55.4285$$

*Calculate Mean Squares and F Scores, and Evaluate the F Score*

**Step 13.** Calculate mean squares for the between-groups and within-groups terms in your summary table. To calculate mean squares divide the sum of squares by the degrees of freedom (Step 9) for that term.

$$\text{Mean square instructions} = \frac{\text{SS instructions}}{\text{df instructions}} = \frac{48.3809}{2} = 24.1905$$

$$\text{Mean square wg} = \frac{\text{SS wg}}{\text{df wg}} = \frac{55.4285}{18} = 3.0794$$

**Step 14.** Calculate an $F$ score corresponding to your independent variable by dividing the MS bg by the MS wg.

$$\text{Instructions } F \text{ score} = \frac{\text{MS instructions}}{\text{MS wg}} = \frac{24.1905}{3.0794} = 7.86$$

**Step 15.** Look up the $F$ score in Table C.6, following the instructions there.

| Analysis of Variance of Words Recalled | | | | | |
|---|---|---|---|---|---|
| Source of Variance | df | Sum of Squares | Mean Square | F | p |
| Instructions | 2 | 48.3809 | 24.1905 | 7.86 | <.001 |
| Within-groups error term | 18 | 55.4285 | 3.0794 | | |
| Total | 20 | 103.8095 | | | |

Thus the effect of instructions was significant at the .001 level.

## Appendix C  Calculational Procedures

*One-Way Within-Subjects Analysis of Variance*

These instructions are suitable for designs having one independent variable which is varied within subjects.

*Example.* In this hypothetical experiment the dependent variable was the score obtained on a reading comprehension test. The independent variable was the amount of noise present while the subjects were reading and taking the test. There were four levels of noise: none, low, medium, and high. Actually each subject was measured a number of times at each noise level in varying orders, but these measurements were combined to yield a single "mean comprehension score" for each subject at each noise level. There were 12 subjects. Since there were four scores for each subject, there were a total of 48 scores.

*Table Your Data*

**Step 1.** Table your data as shown, keeping the scores for each subject together in rows. Arrange the different treatments so that the scores can be combined in columns.

| | Mean Score on Reading Comprehension Test | | | |
|---|---|---|---|---|
| | | Level of Noise | | |
| S | None | Low | Medium | High |
| 1 | 8.0 | 6.0 | 6.2 | 5.5 |
| 2 | 8.8 | 7.3 | 5.0 | 5.5 |
| 3 | 7.0 | 5.1 | 4.0 | 5.0 |
| 4 | 6.8 | 5.1 | 6.3 | 6.0 |
| 5 | 7.2 | 6.1 | 4.5 | 4.0 |
| 6 | 8.4 | 7.6 | 5.0 | 4.5 |
| 7 | 7.9 | 6.4 | 6.8 | 4.1 |
| 8 | 6.5 | 6.0 | 5.8 | 5.0 |
| 9 | 8.0 | 6.8 | 7.0 | 7.2 |
| 10 | 7.2 | 6.5 | 6.4 | 6.0 |
| 11 | 8.5 | 7.5 | 7.2 | 7.4 |
| 12 | 8.1 | 6.0 | 5.5 | 4.0 |

*Preliminary Steps*

**Step 2.** Determine $S$, the number of subjects.

$$S = 12$$

**Step 3.** Determine $T$, the number of times each subject was measured. This is equal to the number of levels of your independent variable.

$T = 4$, since there were four levels of noise; each subject was measured four times

**Step 4.** Determine $N$, the total number of measures in your table. To check, $S$ (Step 2) $\times$ $T$ (Step 3) should equal $N$.

$$N = 48 \qquad S \times T = 12 \times 4 = 48$$

**Step 5.** Calculate *treatment sums*. Add up all the scores at each level of your independent variable, or treatment. You'll add $S$ scores in each treatment; you'll have $T$ sums.

Treatment sum for No noise = $8.0 + 8.8 +$ etc. = $92.4$
Low noise = $6.0 + 7.3 +$ etc. = $76.4$
Medium noise = $6.2 + 5.0 +$ etc. = $69.7$
High noise = $5.5 + 5.5 +$ etc. = $64.2$

**Step 6.** Calculate *subject sums*. Add up all $T$ scores for each subject. You'll have $S$ subject sums.

Subject sum for subject 1 = $8.0 + 6.0 + 6.2 + 5.5 = 25.7$
Subject sum for subject 2 = $26.6$
Etc.

**Step 7.** Calculate the *grand sum*. Add up all the treatment sums; this yields the grand sum. Then add up all the subject sums; this also yields the grand sum. Your answers should match.

Grand sum = $92.4 + 76.2 + 69.7 + 64.2$
= $302.7$ (addition of treatment sums)
Grand sum = $25.7 + 26.6 +$ etc.
= $302.7$ (addition of subject sums)

**Step 8.** Calculate $C$, the correction term. Square the grand sum (Step 7) and divide the answer by $N$ (Step 4). This yields $C$.

$$\text{Correction term} = C = \frac{302.7^2}{48} = 1908.9019$$

*Make an Analysis of Variance Summary Table and Fill in the Degrees of Freedom*

**Step 9.** Make an analysis of variance summary table in the form shown here. Fill in the names of your variables as appropriate. Leave all the columns except the first blank for now. You'll fill them in with the steps shown in the table.

# Appendix C  Calculational Procedures

## Analysis of Variance of Dependent Variable

| Source of Variance | df | Sum of Squares | Mean Square | F | p |
|---|---|---|---|---|---|
| Subjects | Step 10 | Step 12 | Step 15 | | |
| Treatment (your independent variable) | Step 10 | Step 13 | Step 15 | 16 | 17 |
| Error term | Step 10 | Step 14 | Step 15 | | |
| Total | Step 10 | Step 11 | | | |

**Step 10.** Calculate degrees of freedom for each of the terms in your analysis of variance summary table by following the formulas given here. Record the answers in your summary table.

| Source of Variance | Formula for Degrees of Freedom | Step Reference |
|---|---|---|
| Subjects | $S - 1$ | Step 2 |
| Treatment | $T - 1$ | Step 3 |
| Error term | $(N - 1) - (S - 1) - (T - 1)$ | Steps 2, 3, and 4 |
| Total | $N - 1$ | Step 4 |

## Analysis of Variance of Test Scores

| Source of Variance | df |
|---|---|
| Subjects | $12 - 1 = 11$ |
| Noise level | $4 - 1 = 3$ |
| Error term | $(48 - 1) - (12 - 1) - (4 - 1) = 47 - 11 - 3 = 33$ |
| Total | $48 - 1 = 47$ |

*Calculate Sums of Squares*

**Step 11.** Calculate SS tot, the total sum of squares. Square *each* of the scores in your original table (Step 1). [You'll square $N$ (Step 4) numbers.] Add the squares. Finally, subtract $C$ (Step 8). This yields SS tot. Record it in your summary table.

$$\text{SS tot} = (8.0^2 + 8.8^2 + \text{etc.}) - C$$
$$= 1985.8900 - 1908.9019 = 76.9881$$

**Step 12.** Calculate SS Ss, the subject sum of squares. Square each of your subject sums (Step 6). [You'll square S (Step 2) numbers.] Add the squares. Divide that sum by T (Step 3). Finally, subtract C (Step 8). This yields SS Ss. Record it in your summary table.

$$\text{SS Ss} = \frac{26.6^2 + 21.2^2 + \text{etc.}}{4} - C$$

$$= \frac{7717.85}{4} - C = 1929.4625 - 1908.9019 = 20.5605$$

**Step 13.** Calculate SS treat, the sum of squares for the treatment, your independent variable. Square each of your treatment sums (Step 5). [You'll square T (Step 3) numbers.] Add the squares. Divide that sum by S (Step 2). Then subtract C (Step 8). This yields SS treat. Record it in your summary table.

$$\text{SS noise} = \frac{92.4^2 + 76.4^2 + 69.7^2 + 64.2^2}{12} - C$$

$$= \frac{23354.45}{12} - C = 1946.2042 - 1908.9019 = 37.3023$$

**Step 14.** Calculate SS error term, the sum of squares for the error term in this analysis. Simply subtract SS Ss (Step 12) and SS treat (Step 13) from SS tot (Step 11). This yields SS error term. Record it in your summary table.

$$\text{SS error term} = \text{SS tot} - \text{SS Ss} - \text{SS noise}$$

$$= 76.9881 - 20.5606 - 37.3023 = 19.1252$$

*Calculate Mean Squares and an F Score, and Evaluate the F Score*

**Step 15.** Calculate mean squares for all the terms in your analysis of variance summary table except the "total." To calculate mean squares, divide each sum of squares by the degrees of freedom (Step 10) for that term. Record the mean squares in your summary table.

$$\text{Mean square subjects} = \frac{\text{SS Ss}}{\text{df Ss}} = \frac{20.5606}{11} = 1.8691$$

$$\text{Mean square instructions} = \frac{\text{SS noise}}{\text{df noise}} = \frac{37.3023}{3} = 12.4341$$

$$\text{Mean square error} = \frac{\text{SS error}}{\text{df error}} = \frac{19.1252}{33} = .5796$$

# Appendix C  Calculational Procedures

**Step 16.** Calculate an $F$ score corresponding to your independent variable, or treatment, by dividing MS treat by MS error.

$$\text{Instructions } F \text{ score} = \frac{\text{MS instructions}}{\text{MS error}} = \frac{12.4341}{.5796} = 21.45$$

**Step 17.** Look up the $F$ score in Table C.6, following the instructions there.

| | | Analysis of Variance of Test Scores | | | |
|---|---|---|---|---|---|
| Source of Variance | df | Sum of Squares | Mean Square | F | p |
| Subjects | 11 | 20.5606 | 1.8691 | | |
| Noise level | 3 | 37.3023 | 12.4341 | 21.45 | < .001 |
| Error term | 33 | 19.1252 | .5796 | | |
| Total | 47 | 76.9881 | | | |

Thus the effect of noise was significant at the .001 level. (Notice that the effect of "subjects" can't be evaluated.)

*Two-Way Between-Groups Analysis of Variance*

These instructions are suitable for designs having two independent variables, both varied between subjects. Each group of subjects representing the combination of the two variables should have an equal number of subjects.

*Example.* In this hypothetical study the "independent variables" were sex and age, and the dependent variable was the number of addition problems solved correctly by each subject. There were two sexes and three ages—5 to 6 years, 7 to 8 years, and 9 to 10 years—making a total of six groups. There were seven subjects in each sex/age group, making a total of 42 subjects.

*Table Your Data*

**Step 1.** Table your data as shown, keeping the scores for each group of subjects together.

| | | Number of Problems Solved | | | |
|---|---|---|---|---|---|
| 5–6 Years | | 7–8 Years | | 9–10 Years | |
| Male | Female | Male | Female | Male | Female |
| 7 | 7 | 7 | 8 | 10 | 11 |
| 7 | 8 | 8 | 7 | 12 | 9 |
| 6 | 6 | 9 | 7 | 11 | 12 |
| 8 | 5 | 10 | 9 | 9 | 10 |
| 5 | 8 | 7 | 7 | 12 | 9 |
| 6 | 5 | 10 | 8 | 13 | 10 |
| 6 | 6 | 8 | 9 | 11 | 10 |

*Preliminary Steps*

**Step 2.** Determine $N$, the total number of measures in your table. This is equal to the number of subjects in the experiment.

$$N = 42$$

**Step 3.** Arbitrarily designate one of your variables as Variable A and the other as Variable B.

$$A = \text{Age}, B = \text{Sex}$$

**Step 4.** Determine $a$, the number of levels of Variable A.

$a = 3$, since there were three age levels

**Step 5.** Determine $b$, the number of levels of Variable B.

$b = 2$, since there were two sexes, male and female

**Step 6.** Determine $G$, the total number of groups of subjects. To check, $a$ (Step 4) $\times$ $b$ (Step 5) should equal $G$.

$$G = 6 \quad 3 \times 2 = 6$$

**Step 7.** Determine $n$, the number of subjects in each group. To check, $G$ (Step 6) $\times$ $n$ should equal $N$ (Step 2).

$$n = 7 \quad 6 \times 7 = 42 = N$$

**Step 8.** Calculate *group sums*. Add up all the scores in each group. You'll add $n$ scores in each group. Check these carefully.

Group sum for 5- to 6-year-old males
$= 7 + 7 + 6 + 8 + 5 + 6 + 6 = 45$

Group sums for all groups are these:

| 5–6 | | 7–8 | | 9–10 | |
|---|---|---|---|---|---|
| M | F | M | F | M | F |
| 45 | 45 | 59 | 55 | 78 | 71 |

# Appendix C  Calculational Procedures

**Step 9.** Calculate the *grand sum*. Add up all the group sums (Step 8). This yields the grand sum. Check it carefully.

$$\text{Grand sum} = 45 + 45 + 59 + 55 + 78 + 71 = 353$$

**Step 10.** Calculate C, the correction term. Square the grand sum (Step 9), and divide the answer by N (Step 2). This yields C.

$$\text{Correction term} = C = \frac{353^2}{42} = 2966.8810$$

*Make an Analysis of Variance Summary Table and Fill in the Degrees of Freedom*

**Step 11.** Make an analysis of variance summary table in the form shown here. Fill in the names of each of your variables as appropriate. Leave all the columns except the first blank for now. You'll fill them in with the steps shown in the table.

| | Analysis of Variance of Dependent Variable | | | | |
|---|---|---|---|---|---|
| Source of Variance | df | Sum of Squares | Mean Square | F | p |
| Variable A | Step 12 | Step 16 | Step 21 | 22 | 23 |
| Variable B | Step 12 | Step 18 | Step 21 | 22 | 23 |
| A × B | Step 12 | Step 19 | Step 21 | 22 | 23 |
| Within-groups error term | Step 12 | Step 20 | Step 21 | | |
| Total | Step 12 | Step 13 | | | |

**Step 12.** Calculate degrees of freedom for each of the terms in your analysis of variance summary table by following the formulas given here. Record the answers in your table.

| Source of Variance | Formula for Degrees of Freedom | Step Reference |
|---|---|---|
| Variable A | a − 1 | Step 4 |
| Variable B | b − 1 | Step 5 |
| A × B | (a − 1) × (b − 1) | Steps 4 and 5 |
| Within-groups error term | (N − 1) − (G − 1) | Steps 2 and 6 |
| Total | N − 1 | Step 2 |

| Analysis of Variance of Problems Solved | |
|---|---|
| Source of Variance | df |
| Age | $3 - 1 = 2$ |
| Sex | $2 - 1 = 1$ |
| Age × sex | $(3 - 1) \times (2 - 1) = 2 \times 1 = 2$ |
| Within-groups error term | $(42 - 1) - (6 - 1) = 41 - 5 = 36$ |
| Total | $42 - 1 = 41$ |

*Calculate Sums of Squares*

**Step 13.** Calculate SS tot, the total sum of squares. Square each of the scores in your original table (Step 1). [You will square $N$ (Step 2) numbers.] Then add the squares. Finally, subtract $C$ (Step 10) from the answer. This yields SS tot. Record it in your summary table.

$$\text{SS tot} = (7^2 + 7^2 + \text{etc.}) - C = 3145 - 2966.8810$$

$$= 178.1190$$

**Step 14.** Calculate SS bg, the sum of squares between the groups. Square each of your group sums (Step 8). [You will square $G$ (Step 6) numbers.] Add the squares. Divide that sum by $n$ (Step 7). Then subtract $C$ (Step 10). This yields SS bg. Save it for future use.

$$\text{SS bg} = \frac{45^2 + 45^2 + \text{etc.}}{7} - C = \frac{21{,}681}{7} - C$$

$$= 3097.2857 - 2966.8810 = 130.4047$$

**Step 15.** Calculate sums for each level of Variable A. You can do that by adding the group sums (Step 8) at each level of Variable A across Variable B. You will have $a$ (Step 4) sums.

Sum for 5 to 6 years old = 45 + 45 = 90
Sum for 7 to 8 years old = 59 + 55 = 114
Sum for 9 to 10 years old = 78 + 71 = 149

**Step 16.** Calculate SS A, the sum of squares for Variable A. Square each of your A sums (Step 15). [You will square $a$ (Step 4) numbers.] Add the squares. Divide that sum by $b$ (Step 5) × $n$ (Step 7). Then subtract $C$ (Step 10). This yields SS A. Record it in your summary table.

# Appendix C Calculational Procedures

$$\text{SS age} = \frac{90^2 + 114^2 + 149^2}{2 \times 7} - C = \frac{43{,}297}{14} - C$$

$$= 3092.6429 - 2966.8810 = 125.7619$$

**Step 17.** Calculate sums for each level of Variable B. You can do that by adding the group sums (Step 8) for each level of Variable B across Variable A. You will have *b* (Step 5) sums.

$$\text{Sum for males} = 45 + 59 + 78 = 182$$
$$\text{Sum for females} = 45 + 55 + 71 = 171$$

**Step 18.** Calculate SS B, the sum of squares for Variable B. Square each of your B sums (Step 17). [You will square *b* (Step 5) numbers.] Add the squares. Divide that sum by *a* (Step 4) × *n* (Step 7). Then subtract C (Step 10). This yields SS A. Record it in your summary table.

$$\text{SS sex} = \frac{182^2 + 171^2}{3 \times 7} - C = \frac{62{,}365}{21} - C$$

$$= 2969.7619 - 2966.8810 = 2.8809$$

**Step 19.** Calculate SS A × B, the sum of squares for the interaction of Variables A and B. Simply subtract SS A (Step 16) and SS B (Step 18) from SS bg (Step 14). This yields SS A × B. Record it in your summary table.

$$\text{SS age} \times \text{sex} = \text{SS bg} - \text{SS age} - \text{SS sex}$$
$$= 130.4047 - 125.7619 - 2.8809$$
$$= 1.7619$$

**Step 20.** Calculate SS wg, the sum of squares within groups, which is the error term in this analysis. Subtract SS bg (Step 14) from SS tot (Step 13). This yields SS wg error term. Record it in your summary table.

$$\text{SS wg} = \text{SS tot} - \text{SS bg}$$
$$= 178.1190 - 130.4047 = 47.7143$$

*Calculate Mean Squares and F Scores, and Evaluate the F Scores*

**Step 21.** Calculate mean squares for all the terms in your analysis of variance summary table except the "total." To calculate mean squares, divide each sum of squares by the degrees of freedom (Step 12) for that term. Record the mean squares in your summary table.

$$\text{Mean square age} = \frac{SS\ age}{df\ age} = \frac{125.7619}{2} = 62.8810$$

$$\text{Mean square wg} = \frac{SS\ wg}{df\ wg} = \frac{47.7143}{36} = 1.3254$$

Etc.

**Step 22.** Calculate $F$ scores corresponding to A, B, and A × B by dividing their mean squares by the mean square for the error term, MS wg.

$$\text{Age } F \text{ score} = \frac{62.8810}{1.3254} = 47.44$$

Etc.

**Step 23.** Look up each $F$ score in Table C.6, following the instructions there. Evaluate each effect separately.

| Analysis of Variance of Problems Solved | | | | | |
|---|---|---|---|---|---|
| Source of Variance | df | Sum of Squares | Mean Square | F | p |
| Age | 2 | 125.7619 | 62.8810 | 47.44 | < .001 |
| Sex | 1 | 2.8809 | 2.8809 | 2.17 | > .10  n.s. |
| Age × sex | 2 | 1.7619 | .8810 | .66 | > .10  n.s. |
| Within-groups error term | 36 | 47.7143 | 1.3254 | | |
| Total | 41 | 178.1190 | | | |

Thus in the example given, the effect of age is significant at the .001 level and the effect of sex and the sex × age interaction did not approach significance.

*Two-Way Mixed Analysis of Variance*

These instructions are suitable for designs having two independent variables, one varied between subjects and the other varied within subjects. Each group of subjects should have an equal number of subjects.

*Example.* In this hypothetical experiment the dependent variable was the mood rating measured by a standardized questionnaire. The within-subject variable was the day of the week on which the questionnaire was filled out; the questionnaire was administered on Monday, Wednesday, and Friday to all subjects. The between-subject variable was the order of the three tests. One third of the subjects were tested first on Monday, then on Wednesday, and then on Friday; another third was tested first on Wednesday, then on Monday, and then on Friday; the remaining third was tested

# Appendix C  Calculational Procedures

first on Friday, then on Monday, and then on Wednesday. There were five subjects in each order group, making a total of 15 subjects. Since there were three measures on each subject there were a total of 45 measures.

(These data also could have been analyzed using a two-way within-groups analysis of variance. The two independent variables would have been day of the week and practice. Practice would have three levels: first administration, second administration, and third administration.)

*Table Your Data*

**Step 1.** Table your data as shown, keeping the scores for each subject together in rows and the scores for each group of subjects together. Arrange the levels of the within-subject variable in the same order in each group.

|   |   |   |   |   |   | Rated Mood |   |   |   |   |   |
|---|---|---|---|---|---|---|---|---|---|---|---|
|   | Order MWF |   |   |   | Order WFM |   |   |   | Order FMW |   |   |
| S | M | W | F | S | M | W | F | S | M | W | F |
| 1 | 54 | 51 | 50 | 6 | 40 | 51 | 53 | 11 | 44 | 49 | 58 |
| 2 | 49 | 53 | 57 | 7 | 43 | 53 | 56 | 12 | 48 | 46 | 52 |
| 3 | 51 | 50 | 50 | 8 | 46 | 51 | 51 | 13 | 46 | 44 | 51 |
| 4 | 46 | 50 | 52 | 9 | 43 | 54 | 56 | 14 | 44 | 47 | 57 |
| 5 | 44 | 49 | 52 | 10 | 45 | 51 | 50 | 15 | 41 | 44 | 55 |

*Preliminary Steps*

**Step 2.** Determine $S$, the number of subjects in the experiment.

$$S = 15$$

**Step 3.** Determine $T$, the number of times each subject was measured. This is equal to the number of levels of your within-subject variable.

$T = 3$, since each subject was measured three times, once on
each of the three days of the week.

**Step 4.** Determine $N$, the total number of measures in your table. To check, $S$ (Step 2) $\times$ $T$ (Step 3) should equal $N$.

$$N = 45 \quad S \times T = 15 \times 3 = 45$$

**Step 5.** Determine $G$, the number of groups of subjects. This is equal to the number of levels of your between-subject variable.

$G = 3$, since there were three groups of subjects, one for each order of testing

**Step 6.** Determine $n$, the number of subjects in each group. To check, $G$ (Step 5) × $n$ should equal $S$ (Step 2).

$$n = 5 \quad G \times n = 3 \times 5 = 15 = S$$

**Step 7.** Recognize what is meant by a *cell* in your table. Each collection of $n$ subject scores is a cell. Your table has $T$ (Step 3) × $G$ (Step 5) cells.

The upper left cell in this table has the scores 54, 49, 51, 46, and 44 in it. There are nine cells in the table, that is, 3 × 3.

**Step 8.** Calculate cell sums. Add up all the scores in each cell. You'll add $n$ scores in each cell.

Cell sum for left-most cell
$= 54 + 49 + 51 + 46 + 44 = 244$

Cell sums for all cells are these:

| Order MWF | | | Order WFM | | | Order FMW | | |
|---|---|---|---|---|---|---|---|---|
| M | W | F | M | W | F | M | W | F |
| 244 | 253 | 261 | 217 | 260 | 266 | 223 | 230 | 273 |

**Step 9.** Calculate subject sums. Add up all $T$ scores for each subject.

Subject sum for subject 1 $= 54 + 51 + 50 = 155$
2 $= 49 + 53 + 57 = 159$
Etc.

**Step 10.** Calculate the *grand sum*. Add up all the cell sums; this yields the grand sum. Then add up all the subject sums; that also yields the grand sum. Your answers should match.

Grand sum $= 244 + 253 +$ etc. $= 2227$
(addition of cell sums)

Grand sum $= 155 + 159 +$ etc. $= 2227$
(addition of subject sums)

## Appendix C  Calculational Procedures

**Step 11.** Calculate C, the correction term. Square the grand sum (Step 10), and divide the answer by N (Step 4). This yields C.

$$\text{Correction term} = C = \frac{(2227)^2}{45} = 110{,}211.7556$$

*Make an Analysis of Variance Summary Table and Fill in the Degrees of Freedom*

**Step 12.** Make an analysis of variance summary table in the form shown here. Fill in the names of each of your variables as appropriate. Leave all the columns except the first column blank for now. You'll fill them in with the steps shown in the table.

| | Analysis of Variance of Dependent Variable | | | | |
|---|---|---|---|---|---|
| Source of Variance | df | Sum of Squares | Mean Square | F | p |
| Between groups (your between-subjects variable) | Step 13 | Step 17 | Step 23 | 24 | 25 |
| Within-groups error term | Step 13 | Step 18 | Step 23 | 24 | 25 |
| Treatment (your within-subject variable) | Step 13 | Step 20 | Step 23 | 24 | 25 |
| Treatment × between-groups (the interaction) | Step 13 | Step 21 | Step 23 | 24 | 25 |
| Residual error term | Step 13 | Step 22 | Step 23 | | |
| Total | Step 13 | Step 14 | | | |

**Step 13.** Calculate degrees of freedom for each of the terms in your analysis of variance summary table by following the formulas given here. Record the answers in your summary table.

| Source of Variance | Formula for Degrees of Freedom | Step Reference |
|---|---|---|
| Between-groups variable | $G - 1$ | Step 5 |
| Within-groups error term | $(S - 1) - (G - 1)$ | Steps 2 and 6 |
| Treatment variable | $T - 1$ | Step 3 |
| Treatment × between groups | $(T - 1) \times (G - 1)$ | Steps 3 and 5 |
| Residual error term | $(N - 1) - (S - 1) - (T - 1) -$ $[(T - 1) \times (G - 1)]$ | Steps 2, 3, 4, and 5 |
| Total | $N - 1$ | Step 4 |

| Analysis of Variance of Mood Ratings | |
|---|---|
| Source of Variance | df |
| Order of testing | $3 - 1 = 2$ |
| Within-groups error term | $(15 - 1) - (3 - 1) = 14 - 2 = 2$ |
| Day of the week | $3 - 1 = 2$ |
| Day of the week $\times$ order | $(3 - 1) \times (3 - 1) = 2 \times 2 = 4$ |
| Residual error term | $(45 - 1) - (15 - 1) - (3 - 1) - [(3 - 1) \times (3 - 1)]$ |
|  | $= 44 - 14 - 2 - 4 = 24$ |
| Total | $45 - 1 = 44$ |

*Calculate Sums of Squares*

**Step 14.** Calculate SS tot, the total sum of squares. Square *each* of the scores in your original table (Step 1). [You will square $N$ (Step 4) numbers.] Add the squares together. Then subtract $C$ (Step 11) from the answer. This yields SS tot. Place this in your summary table.

$$SS\ tot = (54^2 + 51^2 + \text{etc.}) - C$$
$$= 111{,}101 - 110{,}211.7556 = 889.2444$$

**Step 15.** Calculate SS $Ss$, the subject sum of squares. Square each of your subject sums (Step 9). [You will square $S$ (Step 2) numbers.] Add the squares together. Divide that sum by $T$ (Step 3). Finally, subtract $C$ (Step 11). This yields SS $Ss$. Record it for future use.

$$SS\ Ss = \frac{155^2 + 159^2 + \text{etc.}}{3} - C$$
$$= \frac{331{,}007}{3} - C$$
$$= 110{,}335.6667 - 110{,}211.7556 = 123.9111$$

**Step 16.** Calculate group sums. You can do that by adding all the subject sums (Step 9) in each group. You should have $G$ (Step 5) sums.

Sum for MWF order group = $155 + 159 + 151 + 148 + 145 = 758$
WFM order group = 743
FMW order group = 726

*Note:* These sums also can be calculated by adding up all the cell sums (Step 8) in each group.

## Appendix C  Calculational Procedures

**Step 17.** Calculate SS bg, the sum of squares for the between-subject variable. Square each of your group sums (Step 16). [You will square G (Step 5) numbers.] Add the squares. Divide that sum by T (Step 3) × n (Step 6), that is, the number of scores in each group. Then subtract C (Step 11). This yields SS bg. Record it in your analysis of variance summary table.

$$\text{SS order} = \frac{758^2 + 743^2 + 726^2}{3 \times 5} - C$$

$$= \frac{1{,}653{,}689}{15} - C = 110{,}245.9333 - 110{,}211.7556$$

$$= 34.1777$$

**Step 18.** Calculate SS wg, the sum of squares for the within-groups error term. Simply subtract SS bg (Step 17) from SS Ss (Step 15). This yields SS wg error term. Record it in your analysis of variance summary table.

$$\text{SS wg error term} = \text{SS Ss} - \text{SS order groups}$$
$$= 123.9111 - 34.1777 = 89.7334$$

**Step 19.** Calculate treatment sums, sums for each level of your within-subject variable. You can do that by adding all the cell sums (Step 8) across the groups, that is, across all the subjects. You will have T (Step 3) sums.

$$\text{Sum for Mondays} = 244 + 217 + 223 = 684$$
$$\text{Sum for Wednesdays} = 253 + 260 + 230 = 743$$
$$\text{Sum for Fridays} = 261 + 266 + 273 = 800$$

**Step 20.** Calculate SS treat, the sum of squares for your within-subject variable, or treatment. Square each of your treatment sums (Step 19). [You will square T (Step 3) numbers.] Add the squares. Divide that sum by S (Step 2). Then subtract C. This yields SS treat, the sum of squares for your within-subject subject variable. Record it in your analysis of variance summary table.

$$\text{SS day of the week} = \frac{684^2 + 743^2 + 800^2}{15} - C$$

$$= \frac{1{,}659{,}905}{15} - C = 110{,}660.3333 - 110{,}211.7556$$

$$= 448.5777$$

**Step 21.** Calculate SS treat × bg, the sum of squares for the interaction between your two variables. Square each of your cell sums (Step 8). [You will square T (Step 3) × G (Step 5) numbers.] Add the squares. Divide that sum by n (Step 6). Then subtract C (Step 11) and subtract SS treat (Step 20) and subtract SS bg (Step 17). This yields SS treat × bg. Record it in your analysis of variance summary table.

$$\text{SS day} \times \text{order} = \frac{244^2 + 253^2 + \text{etc.}}{5} - C - \text{SS day} - \text{SS order}$$

$$= \frac{554{,}268}{5} - C - \text{SS day} - \text{SS order}$$

$$= 110{,}853.8 - 110{,}211.7556$$
$$- 448.5777 - 34.1777$$

$$= 159.2890$$

**Step 22.** Calculate SS error term, the sum of squares for the residual error term. Simply subtract all the following from SS tot (Step 14): SS $Ss$ (Step 15), SS treat (Step 20), and SS treat × bg (Step 21). This yields SS error term. Record it in your summary table.

$$\text{SS error term} = \text{SS tot} - \text{SS } Ss - \text{SS day} - \text{SS day} \times \text{order}$$

$$= 889.2444 - 123.9111 - 448.5777$$
$$- 159.2890$$

$$= 157.4666$$

*Calculate Mean Squares and F Scores, and Evaluate the F Scores*

**Step 23.** Calculate mean squares for all the terms in your analysis of variance summary table except for "total." To calculate mean squares, divide each sum of squares by the degrees of freedom (Step 13) for that term. Record the mean squares in your table.

$$\text{Mean square order} = \frac{\text{SS order}}{df \text{ order}} = \frac{34.1777}{2} = 17.0888$$

$$\text{Mean square wg} = \frac{\text{SS wg}}{df \text{ wg}} = \frac{89.7334}{12} = 7.4778$$

Etc.

**Step 24.** Calculate $F$ scores corresponding to your between-subject variable, your within-subject variable, and the interaction. To calculate the $F$ score for any term, divide the mean square for that term by the mean square of the appropriate error term. To calculate the $F$ score for your between-subject variable, divide MS bg by MS wg. To calculate the $F$ scores for your within-subject variable and the interaction divide MS treat by MS error term and divide MS treat × bg by MS error term.

# Appendix C  Calculational Procedures

**Step 25.** Evaluate each F score by looking it up in Table C.6, following the instructions there. Evaluate each F separately.

| Analysis of Variance of Mood Ratings | | | | | |
|---|---|---|---|---|---|
| Source of Variance | df | Sum of Squares | Mean Square | F | p |
| Order of testing | 2 | 34.1777 | 17.0888 | 2.28 | > .10  n.s. |
| Within-groups error term | 12 | 89.7334 | 7.4778 | | |
| Day of the week | 2 | 448.5777 | 224.2888 | 34.18 | < .001 |
| Day of the week × order | 4 | 159.2890 | 39.8222 | 6.06 | < .01 |
| Residual error term | 24 | 157.4666 | 6.5611 | | |
| Total | 44 | 889.2444 | | | |

Thus the main effect of order is not significant, the main effect of day of the week is significant at the .001 level, and the interaction is significant at the .01 level. Inspection of the raw data suggests that the reason for the interaction is that there was a significant practice effect. Another analysis could be performed to see if that were true.

*Newman-Keuls Test*

These instructions are suitable for comparing totals of groups, treatments, or cells that have been shown by analysis of variance to differ significantly. The Newman-Keuls test allows you to determine which of the totals differ from each other. The totals must be based on an equal number of scores.

*Example.* The hypothetical data to be used as an example are those from the two-way mixed analysis of variance example, involving the independent variables of day of the week and order of testing, and the dependent variable of mood. The analysis of variance showed a significant effect of day of the week and of the interaction of day with order. We'll use the Newman-Keuls test to look more closely at the pattern of significant differences.

The cell totals to be compared are these:

| Order MWF | | | Order WFM | | | Order FMW | | |
|---|---|---|---|---|---|---|---|---|
| M | W | F | M | W | F | M | W | F |
| 244 | 253 | 261 | 217 | 260 | 266 | 223 | 230 | 273 |

Each cell total was based on 5 scores, that is, 5 scores were added to give each total.

In the analysis of variance there were two error terms, one a within-groups error term for use with the between-subjects variable and the other a residual error term, for use with any terms involving a within-subject variable. Since these totals involve a within-subjects variable (day of the week) the appropriate error term would be the residual error term. The mean square of the residual error term was 6.5611; it had 24 degrees of freedom.

**Step 1.** Perform an appropriate analysis of variance. If it shows a significant difference between your groups, treatments, or cells, you may perform a Newman-Keuls test or another post hoc comparison. If not, you should not perform additional tests.

**Step 2.** Select the totals to be compared. Arrange them in order from smallest to largest. Number them in that order.

| 1 | 2 | 3 | 4 | 5 | 6 | 7 | 8 | 9 |
|---|---|---|---|---|---|---|---|---|
| 217 | 223 | 230 | 244 | 253 | 260 | 261 | 266 | 273 |

**Step 3.** Determine $n$, the number of scores that had to be added together to get each total. (This may not be equal to $n$ in your analysis of variance.)

$n = 5$, since five scores were added together to get each total

**Step 4.** Determine the appropriate mean square error term and its degrees of freedom. The appropriate error term is the one which would be used to calculate an F score comparing the totals. Call this MS error and $df$.

$$\text{MS error} = 6.5611 \qquad df = 24$$

**Step 5.** Multiply $n$ (Step 3) times MS error (Step 4). Take the square root of the product.

$$\sqrt{n \times \text{MS error}} = \sqrt{5 \times 6.5611} = \sqrt{32.8055} = 5.7276$$

**Step 6.** This is a table of $q$ values. Find the row of $q$ values corresponding to the degrees of freedom of your MS error (Step 4). If your degrees of freedom are not listed, then choose the next lowest number. For example, if your MS error term had 50 $df$, you would use the 40 $df$ row.

In this case the correct row is the 24 $df$ row.

# Appendix C  Calculational Procedures

Values of $q$ for the Newman-Keuls Test[a]
$p = .05$

| df for MS Error | Number of Steps | | | | | | | | | | |
|---|---|---|---|---|---|---|---|---|---|---|---|
| | 2 | 3 | 4 | 5 | 6 | 7 | 8 | 9 | 10 | 11 | 12 |
| 6 | 3.46 | 4.34 | 4.90 | 5.31 | 5.63 | 5.89 | 6.12 | 6.32 | 6.49 | 6.65 | 6.79 |
| 8 | 3.26 | 4.04 | 4.53 | 4.89 | 5.17 | 5.40 | 5.60 | 5.77 | 5.92 | 6.05 | 6.18 |
| 10 | 3.15 | 3.88 | 4.33 | 4.65 | 4.91 | 5.12 | 5.30 | 5.46 | 5.60 | 5.72 | 5.83 |
| 12 | 3.08 | 3.77 | 4.20 | 4.51 | 4.75 | 4.95 | 5.12 | 5.27 | 5.40 | 5.51 | 5.62 |
| 14 | 3.03 | 3.70 | 4.15 | 4.45 | 4.69 | 4.88 | 5.05 | 5.19 | 5.32 | 5.43 | 5.53 |
| 16 | 3.00 | 3.65 | 4.05 | 4.33 | 4.56 | 4.74 | 4.90 | 5.03 | 5.15 | 5.26 | 5.35 |
| 18 | 2.97 | 3.61 | 4.00 | 4.28 | 4.49 | 4.67 | 4.82 | 4.96 | 5.07 | 5.17 | 5.27 |
| 20 | 2.95 | 3.58 | 3.96 | 4.23 | 4.45 | 4.62 | 4.77 | 4.90 | 5.01 | 5.11 | 5.20 |
| 24 | 2.92 | 3.53 | 3.90 | 4.17 | 4.37 | 4.54 | 4.68 | 4.81 | 4.92 | 5.01 | 5.10 |
| 30 | 2.89 | 3.49 | 3.84 | 4.10 | 4.30 | 4.46 | 4.60 | 4.72 | 4.83 | 4.92 | 5.00 |
| 40 | 2.86 | 3.44 | 3.79 | 4.04 | 4.23 | 4.39 | 4.52 | 4.63 | 4.74 | 4.82 | 4.91 |
| 60 | 2.83 | 3.40 | 3.74 | 3.98 | 4.16 | 4.31 | 4.44 | 4.55 | 4.65 | 4.73 | 4.81 |
| 120 | 2.80 | 3.36 | 3.69 | 3.92 | 4.10 | 4.24 | 4.36 | 4.48 | 4.56 | 4.64 | 4.72 |
| ∞ | 2.77 | 3.31 | 3.63 | 3.86 | 4.03 | 4.17 | 4.29 | 4.39 | 4.47 | 4.55 | 4.62 |

[a] This is more generally known as the distribution of the studentized range statistic. This table is adapted from E. S. Pearson and H. O. Hartley, *Biometrika tables for statisticians*. Used by permission of Cambridge University Press.

**Step 7.** Multiply each of the $q$ values in your row times the result of Step 5 until you reach the column in the table corresponding to the number of totals you're comparing. These answers are your *critical values*.

$2.92 \times 5.7276 = 16.72 =$ critical value for 2 totals
$3.53 \times 5.7276 = 20.21 =$ critical value for 3 totals
$3.90 \times 5.7276 = 22.34 =$ critical value for 4 totals
$4.17 \times 5.7276 = 23.88 =$ critical value for 5 totals
$4.37 \times 5.7276 = 25.03 =$ critical value for 6 totals
$4.54 \times 5.7276 = 26.00 =$ critical value for 7 totals
$4.68 \times 5.7276 = 26.80 =$ critical value for 8 totals
$4.81 \times 5.7276 = 27.55 =$ critical value for 9 totals

*Compare Actual Differences to Critical Values*

Now you'll compare the actual differences between your totals to the critical values. If your difference is bigger than the critical value, then that difference is significant; if your difference is smaller, it's not significant. You must compare each difference to the critical value for the appropriate number of totals being compared. If you want to compare 273 (the largest total) to 217 (the smallest), there are nine totals involved; the correct critical value is 27.55. If you want to compare 273 to 266 (the next largest total), there are only two totals involved; the correct critical value is 16.72.

There is a prescribed sequence in which tests on the differences between totals must be made. If any early test shows two totals not to differ significantly, no further comparisons between those totals should be made.

**Step 8.** First compare the largest total to the smallest.

$273 - 217 = 56$. That's larger than 27.55, the appropriate critical value, so that difference is significant. If it weren't, you'd stop here.

**Step 9.** Next compare the largest total to the next-to-the-smallest.

$273 - 223 = 50$, which is larger than 26.80; therefore, this difference also is significant.

Continue comparing each successively larger total to the largest total until you reach one that does not differ significantly from it or until you run out.

$273 - 230 = 43$, which is larger than 26.00
$273 - 244 = 29$, which is larger than 25.03
$273 - 253 = 20$, which is smaller than 23.88. Therefore, this difference isn't significant. Moreover, the differences between 273 and 260, 261, and 266 are also shown not to be significant. A good way to keep track is to draw a line under the nonsignificantly different totals.

217   223   230   244   <u>253   260   261   266   273</u>

**Step 10.** Next compare the next-to-the-largest total to the smallest, then to the next-to-the-smallest total, and so on until you reach a nonsignificant difference or run out.

$266 - 217 = 49$, which is larger than 26.80 (note this value)
$266 - 223 = 43$, which is larger than 26.00
$266 - 230 = 33$, which is larger than 25.03
$266 - 244 = 22$, which is smaller than 23.88

217   223   230   <u>244   253   260   261   266</u>   273

**Step 11.** Continue making comparisons until all possible tests have been made.

$261 - 217 = 44$, which is larger than 26.00
$261 - 223 = 38$, which is larger than 25.03
$261 - 230 = 31$, which is larger than 23.88
$260 - 217 = 43$, which is larger than 25.03
$260 - 223 = 37$, which is larger than 23.88

260 − 230 = 30, which is larger than 22.34
253 − 217 = 36, which is larger than 23.88
253 − 223 = 30, which is larger than 22.34
244 − 217 = 27, which is larger than 22.34
244 − 223 = 21, which is larger than 20.21
244 − 230 = 14, which is smaller than 16.72
230 − 217 = 13, which is smaller than 20.21

<u>217   223   230   244   253   260</u>   261   266   273

**Step 12.** Reorganize your totals into a meaningful organization, and try to understand your pattern of results.

| Order MWF | | | Order WFM | | | Order FMW | | |
|---|---|---|---|---|---|---|---|---|
| M | W | F | M | W | F | M | W | F |
| 244 | 253 | 261 | 217 | 260 | 266 | 223 | 230 | 273 |

Thus for the MWF order there were no significant differences between the different days. For the WFM order the Monday mood scores were significantly lower. For the FMW order the Friday scores were significantly higher. There are probably other ways to view this pattern of results as well.

### Evaluating the Statistical Significance of $r$

To use Table C.3 to evaluate the statistical significance of $r$, the Pearson product-moment correlation coefficient, you must have a value of $r$ and know the degrees of freedom. The degrees of freedom are equal to $N - 2$, where $N$ equals the number of pairs of scores used in calculating $r$.

Look down the $df$ column until you reach a number equal to your degrees of freedom; if your number isn't there, use the next lowest number. Then compare your calculated $r$ with the numbers in that row. If your $r$ is bigger than a number in that row (regardless of sign), then your correlation is significant at the level shown at the top of that column.

*Example.* Suppose we had an $r = .43$ which had been calculated on 30 pairs of scores. The degrees of freedom would be equal to 28, so we would look at the row where $df = 25$. In that row we see that .43 is bigger than .381 but not bigger than .487. Therefore, we know that the correlation is significant at the .05 level but not at the .01 level. It would be correct to write that $r = .43, p < .05$.

If we had an $r = .27$ calculated on 38 pairs of scores, it would not be significant, $p > .05$. It also would be correct to write $.10 > p > .05$.

## Appendix C  Calculational Procedures

**Table C.1.** Table of Random Numbers[a]

| Col. | (1) | (2) | (3) | (4) | (5) | (6) | (7) | (8) | (9) | (10) | (11) | (12) | (13) | (14) |
|---|---|---|---|---|---|---|---|---|---|---|---|---|---|---|
| Line | | | | | | | | | | | | | | |
| 1 | 10480 | 15011 | 01536 | 02011 | 81647 | 91646 | 69179 | 14194 | 62590 | 36207 | 20969 | 99570 | 91291 | 90700 |
| 2 | 22368 | 46573 | 25595 | 85393 | 30995 | 89198 | 27982 | 53402 | 93965 | 34095 | 52666 | 19174 | 39615 | 99505 |
| 3 | 24130 | 48360 | 22527 | 97265 | 76393 | 64809 | 15179 | 24830 | 49340 | 32081 | 30680 | 19655 | 63348 | 58629 |
| 4 | 42167 | 93093 | 06243 | 61680 | 07856 | 16376 | 39440 | 53537 | 71341 | 57004 | 00849 | 74917 | 97758 | 16379 |
| 5 | 37570 | 39975 | 81837 | 16656 | 06121 | 91782 | 60468 | 81305 | 49684 | 60672 | 14110 | 06927 | 01263 | 54613 |
| 6 | 77921 | 06907 | 11008 | 42751 | 27756 | 53498 | 18602 | 70659 | 90655 | 15053 | 21916 | 81825 | 44394 | 42880 |
| 7 | 99562 | 72905 | 56420 | 69994 | 98872 | 31016 | 71194 | 18738 | 44013 | 48840 | 63213 | 21069 | 10634 | 12952 |
| 8 | 96301 | 91977 | 65463 | 07972 | 18876 | 20922 | 94595 | 56869 | 69014 | 60045 | 18425 | 84903 | 42508 | 32307 |
| 9 | 89579 | 14342 | 63661 | 10281 | 17453 | 18103 | 57740 | 84378 | 25331 | 12566 | 58678 | 44947 | 05585 | 56941 |
| 10 | 85475 | 36857 | 53342 | 53988 | 53060 | 59533 | 38867 | 62300 | 08158 | 17983 | 16439 | 11458 | 18593 | 64952 |
| 11 | 28918 | 69578 | 88231 | 33276 | 70997 | 79936 | 56865 | 05859 | 90106 | 31595 | 01547 | 85590 | 91610 | 78188 |
| 12 | 63553 | 40961 | 48235 | 03427 | 49626 | 69445 | 18663 | 72695 | 52180 | 20847 | 12234 | 90511 | 33703 | 90322 |
| 13 | 09429 | 93969 | 52636 | 92737 | 88974 | 33488 | 36320 | 17617 | 30015 | 08272 | 84115 | 27156 | 30613 | 74952 |
| 14 | 10365 | 61129 | 87529 | 85689 | 48237 | 52267 | 67689 | 93394 | 01511 | 26358 | 85104 | 20285 | 29975 | 89868 |
| 15 | 07119 | 97336 | 71048 | 08178 | 77233 | 13916 | 47564 | 81506 | 97735 | 85977 | 29372 | 74461 | 28551 | 90707 |
| 16 | 51085 | 12765 | 51821 | 51259 | 77452 | 16308 | 60756 | 92144 | 49442 | 53900 | 70960 | 63990 | 75601 | 40719 |
| 17 | 02368 | 21382 | 52404 | 60268 | 89368 | 19885 | 55322 | 44819 | 01188 | 65255 | 64835 | 44919 | 05944 | 55157 |
| 18 | 01011 | 54092 | 33362 | 94904 | 31273 | 04146 | 18594 | 29852 | 71585 | 85030 | 51132 | 01915 | 92747 | 64951 |
| 19 | 52162 | 53916 | 46369 | 58586 | 23216 | 14513 | 83149 | 98736 | 23495 | 64350 | 94738 | 17752 | 35156 | 35749 |
| 20 | 07056 | 97628 | 33787 | 09998 | 42698 | 06691 | 76988 | 13602 | 51851 | 46104 | 88916 | 19509 | 25625 | 58104 |
| 21 | 48663 | 91245 | 85828 | 14346 | 09172 | 30168 | 90229 | 04734 | 59193 | 22178 | 30421 | 61666 | 99904 | 32812 |
| 22 | 54164 | 58492 | 22421 | 74103 | 47070 | 25306 | 76468 | 26384 | 58151 | 06646 | 21524 | 15227 | 96909 | 44592 |
| 23 | 32639 | 32363 | 05597 | 24200 | 13363 | 38005 | 94342 | 28728 | 35806 | 06912 | 17012 | 64161 | 18296 | 22851 |
| 24 | 29334 | 27001 | 87637 | 87308 | 58731 | 00256 | 45834 | 15398 | 46557 | 41135 | 10367 | 07684 | 36188 | 18510 |
| 25 | 02488 | 33062 | 28834 | 07351 | 19731 | 92420 | 60952 | 61280 | 50001 | 67658 | 32586 | 86679 | 50720 | 94953 |

# Appendix C  Calculational Procedures

| | | | | | | | | | |
|---|---|---|---|---|---|---|---|---|---|
| 26 | 81525 | 72295 | 04839 | 96423 | 24878 | 82651 | 66566 | 14778 | 76797 | 14780 | 13300 | 87074 | 79666 | 95725 |
| 27 | 29676 | 20591 | 68086 | 26432 | 46901 | 20849 | 89768 | 81536 | 86645 | 12659 | 92259 | 57102 | 80428 | 25280 |
| 28 | 00742 | 57392 | 39064 | 66432 | 84673 | 40027 | 32832 | 61362 | 98947 | 96067 | 64760 | 64584 | 96096 | 98253 |
| 29 | 05366 | 04213 | 25669 | 26422 | 44407 | 44048 | 37937 | 63904 | 45766 | 66134 | 75470 | 66520 | 34693 | 90449 |
| 30 | 91921 | 26418 | 64117 | 94305 | 26766 | 25940 | 39972 | 22209 | 71500 | 64568 | 91402 | 42416 | 07844 | 69618 |
| 31 | 00582 | 04711 | 87917 | 77341 | 42206 | 35126 | 74087 | 99547 | 81817 | 42607 | 43808 | 76655 | 62028 | 76630 |
| 32 | 00725 | 69884 | 62797 | 56170 | 86324 | 88072 | 76222 | 36086 | 84637 | 93161 | 76038 | 65855 | 77919 | 88006 |
| 33 | 69011 | 65795 | 95876 | 55293 | 18988 | 27354 | 26575 | 08625 | 40801 | 59920 | 29841 | 80150 | 12777 | 48501 |
| 34 | 25976 | 57948 | 29888 | 88604 | 67917 | 48708 | 18912 | 82271 | 65424 | 69774 | 33611 | 54262 | 85963 | 03547 |
| 35 | 09763 | 83473 | 73577 | 12908 | 30883 | 18317 | 28290 | 35797 | 05998 | 41688 | 34952 | 37888 | 38917 | 88050 |
| 36 | 91567 | 42595 | 27958 | 30134 | 04024 | 86385 | 29880 | 99730 | 55536 | 84855 | 29080 | 09250 | 79656 | 73211 |
| 37 | 17955 | 56349 | 90999 | 49127 | 20044 | 59931 | 06115 | 20542 | 18059 | 02008 | 73708 | 83517 | 36103 | 42791 |
| 38 | 46503 | 18584 | 18845 | 49618 | 02304 | 51038 | 20655 | 58727 | 28168 | 15475 | 56942 | 53389 | 20562 | 87338 |
| 39 | 92157 | 89634 | 94824 | 78171 | 84610 | 82834 | 09922 | 25417 | 44137 | 48413 | 25555 | 21246 | 35509 | 20468 |
| 40 | 14577 | 62765 | 35605 | 81263 | 39667 | 47358 | 56873 | 56307 | 61607 | 49518 | 89696 | 20103 | 77490 | 18062 |
| 41 | 98427 | 07523 | 33362 | 64270 | 01638 | 92477 | 66969 | 98420 | 04880 | 45585 | 46565 | 04102 | 46880 | 45709 |
| 42 | 34914 | 63976 | 88720 | 82765 | 34476 | 17032 | 87589 | 40836 | 32427 | 70002 | 70663 | 88863 | 77775 | 69348 |
| 43 | 70060 | 28277 | 39475 | 46473 | 23219 | 53416 | 94970 | 25832 | 69975 | 94884 | 19661 | 72828 | 00102 | 66794 |
| 44 | 53976 | 54914 | 06990 | 67245 | 68350 | 82948 | 11398 | 42878 | 80287 | 88267 | 47363 | 46634 | 06541 | 97809 |
| 45 | 76072 | 29515 | 40980 | 07391 | 58745 | 25774 | 22987 | 80059 | 39911 | 96189 | 41151 | 14222 | 60697 | 59583 |
| 46 | 90725 | 52210 | 83974 | 29992 | 65831 | 38857 | 50490 | 83765 | 55657 | 14361 | 31720 | 57375 | 56228 | 41546 |
| 47 | 64364 | 67412 | 33339 | 31926 | 14883 | 24413 | 59744 | 92351 | 97473 | 89286 | 35931 | 04110 | 23726 | 51900 |
| 48 | 08962 | 00358 | 31662 | 25388 | 61642 | 34072 | 81249 | 35648 | 56891 | 69352 | 48373 | 45578 | 78547 | 81788 |
| 49 | 95012 | 68379 | 93526 | 70765 | 10592 | 04542 | 76463 | 54328 | 02349 | 17247 | 28865 | 14777 | 62730 | 92277 |
| 50 | 15664 | 10493 | 20492 | 38391 | 91132 | 21999 | 59516 | 81652 | 27195 | 48223 | 46751 | 22923 | 32261 | 85653 |

[a]Source: *Table of 105,000 Random Decimal Digits,* Statement No. 4914, File no. 261-A-1, Interstate Commerce Commission, Washington, D.C., May 1949.

**Table C.2.** One Hundred Permutations of the Numbers 1–16

| | | | | | | | | | | | | | | | | |
|---|---|---|---|---|---|---|---|---|---|---|---|---|---|---|---|---|
| 1 | 2 | 14 | 3 | 8 | 4 | 12 | 10 | 5 | 6 | 7 | 15 | 9 | 11 | 13 | 16 | 1 |
| 2 | 10 | 3 | 4 | 1 | 13 | 5 | 16 | 6 | 12 | 14 | 11 | 8 | 9 | 15 | 7 | 2 |
| 3 | 16 | 5 | 10 | 15 | 3 | 4 | 13 | 11 | 2 | 14 | 1 | 12 | 7 | 9 | 8 | 6 |
| 4 | 14 | 11 | 13 | 8 | 5 | 3 | 7 | 4 | 10 | 15 | 6 | 1 | 9 | 12 | 16 | 2 |
| 5 | 15 | 7 | 2 | 5 | 8 | 12 | 10 | 16 | 1 | 4 | 13 | 9 | 3 | 14 | 11 | 6 |
| 6 | 1 | 7 | 8 | 5 | 16 | 15 | 2 | 12 | 4 | 3 | 14 | 10 | 9 | 11 | 6 | 13 |
| 7 | 4 | 9 | 3 | 7 | 1 | 11 | 10 | 5 | 2 | 16 | 6 | 12 | 8 | 13 | 14 | 15 |
| 8 | 5 | 8 | 3 | 16 | 12 | 11 | 2 | 14 | 1 | 9 | 4 | 10 | 15 | 6 | 7 | 13 |
| 9 | 7 | 16 | 12 | 5 | 6 | 3 | 10 | 13 | 14 | 9 | 4 | 11 | 8 | 1 | 2 | 15 |
| 10 | 13 | 1 | 2 | 11 | 14 | 9 | 7 | 16 | 8 | 3 | 4 | 10 | 6 | 15 | 12 | 5 |
| 11 | 3 | 4 | 8 | 9 | 11 | 6 | 5 | 1 | 12 | 16 | 10 | 15 | 7 | 13 | 2 | 14 |
| 12 | 12 | 7 | 16 | 3 | 8 | 9 | 1 | 14 | 4 | 6 | 11 | 5 | 13 | 10 | 2 | 15 |
| 13 | 4 | 12 | 9 | 11 | 6 | 5 | 14 | 7 | 3 | 13 | 16 | 15 | 8 | 1 | 2 | 10 |
| 14 | 14 | 10 | 12 | 3 | 2 | 4 | 9 | 11 | 7 | 8 | 5 | 16 | 15 | 6 | 13 | 1 |
| 15 | 12 | 14 | 6 | 3 | 9 | 13 | 5 | 15 | 8 | 1 | 11 | 4 | 10 | 7 | 2 | 16 |
| 16 | 16 | 4 | 9 | 1 | 12 | 3 | 7 | 13 | 11 | 8 | 10 | 2 | 6 | 15 | 14 | 5 |
| 17 | 8 | 3 | 10 | 7 | 14 | 11 | 13 | 6 | 5 | 1 | 9 | 4 | 2 | 12 | 15 | 16 |
| 18 | 7 | 12 | 1 | 5 | 16 | 4 | 11 | 8 | 2 | 9 | 10 | 13 | 15 | 3 | 6 | 14 |
| 19 | 2 | 12 | 14 | 11 | 3 | 4 | 10 | 15 | 16 | 5 | 6 | 8 | 7 | 1 | 9 | 13 |
| 20 | 3 | 1 | 8 | 2 | 11 | 6 | 4 | 9 | 10 | 12 | 14 | 5 | 15 | 13 | 16 | 7 |
| 21 | 8 | 14 | 2 | 1 | 10 | 11 | 6 | 9 | 13 | 4 | 7 | 5 | 15 | 3 | 16 | 12 |
| 22 | 15 | 6 | 13 | 12 | 8 | 14 | 4 | 16 | 10 | 1 | 9 | 5 | 11 | 2 | 7 | 3 |
| 23 | 10 | 6 | 2 | 14 | 8 | 9 | 11 | 5 | 7 | 4 | 1 | 3 | 15 | 12 | 13 | 16 |
| 24 | 15 | 14 | 13 | 2 | 8 | 1 | 10 | 11 | 16 | 3 | 12 | 4 | 7 | 5 | 6 | 9 |
| 25 | 9 | 1 | 6 | 8 | 12 | 3 | 5 | 14 | 16 | 7 | 10 | 11 | 15 | 13 | 2 | 4 |
| 26 | 9 | 11 | 10 | 7 | 1 | 15 | 8 | 16 | 2 | 6 | 5 | 3 | 12 | 14 | 13 | 4 |
| 27 | 8 | 5 | 13 | 10 | 2 | 1 | 7 | 4 | 11 | 9 | 16 | 3 | 6 | 12 | 15 | 14 |
| 28 | 10 | 12 | 8 | 6 | 5 | 3 | 14 | 7 | 1 | 16 | 2 | 9 | 11 | 15 | 13 | 4 |
| 29 | 1 | 13 | 4 | 3 | 11 | 2 | 6 | 10 | 12 | 14 | 8 | 5 | 9 | 16 | 15 | 7 |
| 30 | 8 | 2 | 5 | 10 | 4 | 9 | 16 | 6 | 3 | 14 | 12 | 13 | 1 | 15 | 7 | 11 |
| 31 | 9 | 15 | 16 | 12 | 8 | 7 | 14 | 6 | 3 | 13 | 10 | 5 | 1 | 11 | 4 | 2 |
| 32 | 8 | 1 | 15 | 14 | 16 | 7 | 11 | 12 | 13 | 6 | 10 | 5 | 4 | 9 | 2 | 3 |
| 33 | 7 | 2 | 16 | 15 | 4 | 9 | 10 | 5 | 11 | 3 | 12 | 13 | 6 | 14 | 1 | 8 |
| 34 | 16 | 15 | 10 | 7 | 4 | 11 | 1 | 8 | 3 | 12 | 5 | 6 | 13 | 2 | 14 | 9 |
| 35 | 6 | 16 | 10 | 5 | 12 | 7 | 9 | 4 | 1 | 2 | 15 | 14 | 13 | 11 | 8 | 3 |
| 36 | 1 | 5 | 15 | 9 | 13 | 6 | 16 | 7 | 4 | 8 | 11 | 2 | 10 | 14 | 12 | 3 |
| 37 | 3 | 6 | 14 | 9 | 4 | 10 | 1 | 7 | 5 | 2 | 13 | 16 | 15 | 12 | 11 | 8 |
| 38 | 12 | 10 | 11 | 8 | 16 | 1 | 14 | 9 | 15 | 7 | 2 | 5 | 6 | 13 | 4 | 3 |
| 39 | 3 | 5 | 12 | 4 | 15 | 8 | 13 | 1 | 7 | 9 | 14 | 16 | 6 | 11 | 2 | 10 |
| 40 | 13 | 15 | 16 | 1 | 7 | 8 | 3 | 14 | 12 | 9 | 5 | 2 | 11 | 4 | 10 | 6 |
| 41 | 11 | 7 | 12 | 15 | 8 | 9 | 4 | 3 | 10 | 1 | 2 | 6 | 5 | 14 | 16 | 13 |
| 42 | 4 | 8 | 15 | 10 | 2 | 5 | 12 | 1 | 13 | 7 | 9 | 6 | 16 | 3 | 11 | 14 |
| 43 | 6 | 5 | 9 | 14 | 8 | 13 | 4 | 10 | 15 | 3 | 2 | 7 | 12 | 1 | 16 | 11 |
| 44 | 12 | 10 | 7 | 3 | 8 | 13 | 14 | 6 | 9 | 15 | 16 | 5 | 4 | 2 | 1 | 11 |
| 45 | 11 | 5 | 9 | 6 | 2 | 15 | 13 | 1 | 8 | 12 | 14 | 16 | 3 | 4 | 10 | 7 |
| 46 | 9 | 8 | 3 | 13 | 12 | 6 | 10 | 5 | 11 | 14 | 16 | 15 | 4 | 7 | 2 | 1 |
| 47 | 12 | 11 | 16 | 4 | 10 | 6 | 3 | 2 | 15 | 5 | 7 | 14 | 8 | 13 | 1 | 9 |
| 48 | 13 | 14 | 12 | 1 | 7 | 3 | 8 | 10 | 16 | 11 | 4 | 9 | 15 | 5 | 2 | 6 |
| 49 | 10 | 13 | 3 | 9 | 15 | 1 | 4 | 7 | 12 | 6 | 16 | 11 | 5 | 14 | 8 | 2 |
| 50 | 4 | 8 | 7 | 15 | 14 | 16 | 9 | 10 | 3 | 2 | 11 | 13 | 6 | 1 | 12 | 5 |

# Appendix C  Calculational Procedures

| | | | | | | | | | | | | | | | | |
|---|---|---|---|---|---|---|---|---|---|---|---|---|---|---|---|---|
| 51 | 10 | 2 | 14 | 6 | 13 | 9 | 7 | 16 | 8 | 15 | 4 | 12 | 5 | 11 | 3 | 1 |
| 52 | 3 | 10 | 1 | 7 | 8 | 2 | 9 | 15 | 4 | 6 | 5 | 11 | 13 | 12 | 14 | 16 |
| 53 | 16 | 6 | 12 | 15 | 2 | 3 | 1 | 14 | 10 | 9 | 13 | 8 | 4 | 5 | 6 | 11 |
| 54 | 5 | 11 | 12 | 1 | 10 | 15 | 2 | 4 | 8 | 6 | 9 | 14 | 7 | 16 | 3 | 13 |
| 55 | 11 | 12 | 7 | 8 | 14 | 15 | 13 | 1 | 9 | 3 | 5 | 4 | 6 | 16 | 2 | 10 |
| 56 | 13 | 5 | 2 | 16 | 12 | 14 | 1 | 8 | 10 | 11 | 9 | 15 | 3 | 4 | 6 | 7 |
| 57 | 7 | 6 | 11 | 4 | 13 | 10 | 1 | 3 | 14 | 8 | 5 | 2 | 9 | 16 | 15 | 12 |
| 58 | 16 | 3 | 6 | 8 | 5 | 12 | 7 | 9 | 14 | 2 | 15 | 1 | 11 | 4 | 10 | 13 |
| 59 | 16 | 14 | 8 | 5 | 13 | 2 | 6 | 3 | 7 | 4 | 11 | 15 | 1 | 9 | 12 | 10 |
| 60 | 11 | 12 | 10 | 6 | 8 | 5 | 4 | 7 | 3 | 14 | 13 | 15 | 9 | 1 | 16 | 2 |
| 61 | 14 | 4 | 15 | 13 | 1 | 11 | 3 | 9 | 8 | 16 | 10 | 2 | 12 | 5 | 6 | 7 |
| 62 | 1 | 15 | 14 | 16 | 12 | 6 | 3 | 13 | 4 | 5 | 7 | 8 | 9 | 10 | 11 | 2 |
| 63 | 10 | 5 | 13 | 2 | 15 | 4 | 3 | 7 | 9 | 8 | 12 | 6 | 1 | 11 | 14 | 16 |
| 64 | 14 | 1 | 9 | 8 | 5 | 12 | 2 | 4 | 11 | 10 | 3 | 7 | 16 | 15 | 13 | 6 |
| 65 | 10 | 11 | 15 | 9 | 13 | 1 | 7 | 14 | 16 | 5 | 8 | 6 | 3 | 2 | 12 | 4 |
| 66 | 10 | 12 | 1 | 13 | 6 | 2 | 14 | 11 | 3 | 15 | 5 | 8 | 16 | 4 | 7 | 9 |
| 67 | 9 | 13 | 15 | 10 | 14 | 7 | 16 | 2 | 4 | 11 | 12 | 3 | 1 | 5 | 6 | 8 |
| 68 | 9 | 14 | 2 | 10 | 11 | 15 | 4 | 1 | 16 | 7 | 6 | 13 | 8 | 3 | 12 | 5 |
| 69 | 5 | 4 | 12 | 13 | 3 | 14 | 11 | 16 | 2 | 15 | 7 | 9 | 6 | 1 | 8 | 10 |
| 70 | 13 | 8 | 6 | 10 | 11 | 7 | 16 | 15 | 3 | 14 | 4 | 12 | 1 | 9 | 2 | 5 |
| 71 | 1 | 16 | 8 | 12 | 9 | 13 | 15 | 7 | 10 | 5 | 2 | 11 | 14 | 3 | 4 | 6 |
| 72 | 4 | 1 | 6 | 15 | 13 | 7 | 2 | 3 | 12 | 8 | 10 | 16 | 9 | 5 | 14 | 11 |
| 73 | 7 | 13 | 9 | 12 | 14 | 5 | 6 | 2 | 15 | 11 | 16 | 8 | 10 | 3 | 1 | 4 |
| 74 | 13 | 16 | 15 | 3 | 7 | 14 | 9 | 6 | 5 | 8 | 12 | 1 | 4 | 10 | 2 | 11 |
| 75 | 13 | 7 | 6 | 9 | 3 | 12 | 11 | 5 | 14 | 16 | 15 | 8 | 4 | 1 | 10 | 2 |
| 76 | 12 | 10 | 16 | 9 | 3 | 4 | 14 | 2 | 15 | 13 | 6 | 8 | 11 | 5 | 1 | 7 |
| 77 | 1 | 11 | 9 | 5 | 8 | 6 | 7 | 2 | 10 | 12 | 3 | 14 | 13 | 4 | 15 | 16 |
| 78 | 1 | 10 | 16 | 8 | 7 | 4 | 13 | 6 | 2 | 9 | 15 | 3 | 5 | 14 | 11 | 12 |
| 79 | 16 | 12 | 6 | 15 | 1 | 14 | 9 | 13 | 11 | 10 | 4 | 7 | 3 | 8 | 5 | 2 |
| 80 | 2 | 12 | 7 | 14 | 16 | 5 | 8 | 13 | 1 | 15 | 11 | 3 | 10 | 9 | 6 | 4 |
| 81 | 12 | 15 | 9 | 2 | 4 | 10 | 5 | 6 | 14 | 13 | 8 | 3 | 7 | 1 | 16 | 11 |
| 82 | 10 | 8 | 11 | 16 | 3 | 15 | 2 | 4 | 13 | 9 | 1 | 12 | 14 | 6 | 5 | 7 |
| 83 | 2 | 15 | 6 | 4 | 14 | 8 | 7 | 16 | 11 | 9 | 13 | 12 | 10 | 5 | 1 | 3 |
| 84 | 3 | 9 | 15 | 16 | 8 | 11 | 5 | 4 | 7 | 12 | 2 | 6 | 1 | 13 | 14 | 10 |
| 85 | 12 | 7 | 9 | 4 | 3 | 16 | 8 | 2 | 15 | 6 | 11 | 10 | 1 | 14 | 5 | 13 |
| 86 | 16 | 13 | 4 | 8 | 9 | 3 | 14 | 15 | 12 | 10 | 1 | 2 | 11 | 6 | 5 | 7 |
| 87 | 4 | 13 | 9 | 7 | 3 | 2 | 11 | 8 | 12 | 5 | 15 | 6 | 16 | 14 | 1 | 10 |
| 88 | 14 | 10 | 2 | 8 | 12 | 3 | 7 | 9 | 6 | 1 | 5 | 15 | 4 | 13 | 16 | 11 |
| 89 | 15 | 4 | 9 | 8 | 2 | 10 | 16 | 5 | 3 | 11 | 1 | 7 | 13 | 14 | 6 | 12 |
| 90 | 3 | 16 | 8 | 15 | 5 | 14 | 11 | 2 | 12 | 6 | 1 | 13 | 4 | 10 | 9 | 7 |
| 91 | 6 | 2 | 16 | 5 | 9 | 14 | 10 | 12 | 13 | 4 | 7 | 1 | 8 | 11 | 3 | 15 |
| 92 | 6 | 15 | 11 | 12 | 2 | 3 | 16 | 5 | 14 | 13 | 8 | 9 | 4 | 7 | 10 | 1 |
| 93 | 12 | 7 | 4 | 10 | 11 | 15 | 6 | 14 | 8 | 9 | 16 | 3 | 5 | 13 | 1 | 2 |
| 94 | 10 | 12 | 15 | 6 | 14 | 1 | 2 | 16 | 5 | 3 | 9 | 13 | 11 | 7 | 8 | 4 |
| 95 | 5 | 16 | 13 | 1 | 4 | 12 | 15 | 14 | 7 | 3 | 11 | 8 | 2 | 6 | 10 | 9 |
| 96 | 16 | 13 | 2 | 15 | 11 | 3 | 12 | 4 | 9 | 5 | 6 | 1 | 7 | 8 | 14 | 10 |
| 97 | 1 | 3 | 9 | 4 | 7 | 14 | 13 | 10 | 5 | 15 | 2 | 8 | 16 | 6 | 11 | 12 |
| 98 | 2 | 4 | 15 | 12 | 1 | 7 | 16 | 13 | 14 | 3 | 8 | 6 | 5 | 11 | 10 | 9 |
| 99 | 10 | 14 | 8 | 4 | 7 | 15 | 11 | 9 | 2 | 1 | 6 | 5 | 3 | 12 | 13 | 16 |
| 100 | 15 | 4 | 7 | 1 | 9 | 16 | 3 | 5 | 13 | 8 | 12 | 6 | 10 | 11 | 14 | 2 |

**Table C.3.** Critical Values of $r$[a]

| df | p | | | |
|---|---|---|---|---|
|  | .10 | .05 | .01 | .001 |
| 6 | .622 | .707 | .834 | .925 |
| 7 | .582 | .666 | .798 | .898 |
| 8 | .549 | .632 | .765 | .872 |
| 9 | .521 | .602 | .735 | .847 |
| 10 | .497 | .576 | .708 | .823 |
| 11 | .476 | .553 | .684 | .801 |
| 12 | .458 | .532 | .661 | .780 |
| 13 | .441 | .514 | .641 | .760 |
| 14 | .426 | .497 | .623 | .742 |
| 15 | .412 | .482 | .606 | .725 |
| 16 | .400 | .468 | .590 | .708 |
| 17 | .389 | .456 | .575 | .693 |
| 18 | .378 | .444 | .561 | .679 |
| 19 | .369 | .433 | .549 | .665 |
| 20 | .360 | .423 | .537 | .652 |
| 25 | .323 | .381 | .487 | .597 |
| 30 | .296 | .349 | .449 | .554 |
| 35 | .275 | .325 | .418 | .519 |
| 40 | .257 | .304 | .393 | .490 |
| 45 | .243 | .288 | .372 | .465 |
| 50 | .231 | .273 | .354 | .443 |
| 60 | .211 | .250 | .325 | .408 |
| 70 | .195 | .232 | .302 | .380 |
| 80 | .183 | .217 | .283 | .357 |
| 90 | .173 | .205 | .267 | .338 |
| 100 | .164 | .195 | .254 | .321 |

[a] Taken from Table VII of R. A. Fisher and F. Yates, *Statistical tables for biological, agricultural and medical research*, published by Longman Group Ltd., London (previously published by Oliver and Boyd, Edinburgh). Used by permission of the authors and publishers.

### Evaluating the Statistical Significance of $\chi^2$

To use Table C.4 to evaluate the relationship between two variables as measured by $\chi^2$ you must have a value of $\chi^2$ and know the degrees of freedom. The degrees of freedom are equal to $(R - 1) \times (C - 1)$, where $R$ = the number of rows in your contingency table and $C$ = the number of columns.

Look down the *df* column until you reach a number equal to your degrees of freedom. (If you have more than 30, you can't use this table.) Then compare your calculated $\chi^2$ with the numbers in that row. If your $\chi^2$ is bigger than a number in that row, then the relation between your two variables is significant at the level shown at the top of that column.

# Appendix C  Calculational Procedures

*Example.* Suppose we had a $\chi^2 = 30.0$ which had been calculated from a contingency table with 4 rows and 7 columns. The degrees of freedom would be equal to $(4 - 1) \times (7 - 1) = 3 \times 6 = 18$, so we would look at the row where $df = 18$. In that row we see that 30.0 is bigger than 28.9, but smaller than 34.8. Therefore, we know that the $\chi^2$ is significant at the .05 level, but not at the .01 level. It would be correct to write that $\chi^2(18) = 30.0$, $p < .05$.

If we had a $\chi^2 = 14.9$ with $df = 8$ the relationship between the two variables would not be significant, $p > .10$.

Table C.4. Critical Values of $\chi^2$ [a]

| df | p | | | |
|---|---|---|---|---|
|    | .10 | .05 | .01 | .001 |
| 1  | 2.7  | 3.8  | 6.6  | 10.8 |
| 2  | 4.6  | 6.0  | 9.2  | 13.8 |
| 3  | 6.3  | 7.8  | 11.3 | 16.3 |
| 4  | 7.8  | 9.5  | 13.3 | 18.5 |
| 5  | 9.2  | 11.1 | 15.1 | 20.5 |
| 6  | 10.6 | 12.6 | 16.8 | 22.5 |
| 7  | 12.0 | 14.1 | 18.5 | 24.3 |
| 8  | 13.4 | 15.5 | 20.1 | 26.1 |
| 9  | 14.7 | 16.9 | 21.7 | 27.9 |
| 10 | 16.0 | 18.3 | 23.2 | 29.6 |
| 11 | 17.3 | 19.7 | 24.7 | 31.3 |
| 12 | 18.5 | 21.0 | 26.2 | 32.9 |
| 13 | 19.8 | 22.4 | 27.7 | 34.5 |
| 14 | 21.1 | 23.7 | 29.1 | 36.1 |
| 15 | 22.3 | 25.0 | 30.6 | 37.7 |
| 16 | 23.5 | 26.3 | 32.0 | 39.3 |
| 17 | 24.8 | 27.6 | 33.4 | 40.8 |
| 18 | 26.0 | 28.9 | 34.8 | 42.3 |
| 19 | 27.2 | 30.1 | 36.2 | 43.8 |
| 20 | 28.4 | 31.4 | 37.6 | 45.3 |
| 21 | 29.6 | 32.7 | 38.9 | 46.8 |
| 22 | 30.8 | 33.9 | 40.3 | 48.3 |
| 23 | 32.0 | 35.2 | 41.6 | 49.7 |
| 24 | 33.2 | 36.4 | 43.0 | 51.2 |
| 25 | 34.4 | 37.7 | 44.3 | 52.6 |
| 26 | 35.6 | 38.9 | 45.6 | 54.0 |
| 27 | 36.7 | 40.1 | 47.0 | 55.5 |
| 28 | 37.9 | 41.3 | 48.3 | 56.9 |
| 29 | 39.1 | 42.6 | 49.6 | 58.3 |
| 30 | 40.3 | 43.8 | 50.9 | 59.7 |

[a] Taken from Table IV of R. A. Fisher and F. Yates, *Statistical tables for biological, agricultural and medical research*, published by Longman Group Ltd., London (previously published by Oliver and Boyd, Edinburgh). Used by permission of the authors and publishers.

**Table C.5.** Critical Values of $t$[a]
(Two-Tailed)

| df | \.10 | \.05 | \.01 | \.001 |
|---|---|---|---|---|
| | | $p$ | | |
| 6 | 1.94 | 2.45 | 3.71 | 5.96 |
| 7 | 1.90 | 2.37 | 3.50 | 5.41 |
| 8 | 1.86 | 2.31 | 3.36 | 5.04 |
| 9 | 1.83 | 2.26 | 3.25 | 4.78 |
| 10 | 1.81 | 2.23 | 3.17 | 4.59 |
| 11 | 1.80 | 2.20 | 3.11 | 4.44 |
| 12 | 1.78 | 2.18 | 3.06 | 4.32 |
| 13 | 1.77 | 2.16 | 3.01 | 4.22 |
| 14 | 1.76 | 2.14 | 2.98 | 4.14 |
| 15 | 1.75 | 2.13 | 2.95 | 4.07 |
| 16 | 1.75 | 2.12 | 2.92 | 4.02 |
| 17 | 1.74 | 2.11 | 2.90 | 3.97 |
| 18 | 1.73 | 2.10 | 2.88 | 3.92 |
| 19 | 1.73 | 2.09 | 2.86 | 3.88 |
| 20 | 1.72 | 2.09 | 2.84 | 3.85 |
| 21 | 1.72 | 2.08 | 2.83 | 3.82 |
| 22 | 1.72 | 2.07 | 2.82 | 3.79 |
| 23 | 1.71 | 2.07 | 2.81 | 3.77 |
| 24 | 1.71 | 2.06 | 2.80 | 3.75 |
| 25 | 1.71 | 2.06 | 2.79 | 3.73 |
| 30 | 1.70 | 2.04 | 2.75 | 3.65 |
| 40 | 1.68 | 2.02 | 2.70 | 3.55 |
| 60 | 1.67 | 2.00 | 2.66 | 3.46 |
| 120 | 1.66 | 1.98 | 2.62 | 3.37 |
| $\infty$ | 1.64 | 1.96 | 2.58 | 3.29 |

[a] Taken from Table III of R. A. Fisher and F. Yates, *Statistical tables for biological, agricultural and medical research*, published by Longman Group Ltd., London (previously published by Oliver and Boyd, Edinburgh). Used by permission of the authors and publishers.

### Evaluating the Statistical Significance of $t$

To use Table C.5 to evaluate the statistical significance of the difference between two means you must have a value of $t$ and know the degrees of freedom. See the instructions for the two kinds of $t$ tests to determine the degrees of freedom.

Look down the *df* column until you reach a number equal to your degrees of freedom; if your number isn't there, use the next lowest number. Then compare your calculated $t$ with the numbers in that row. If your $t$ is bigger than a number in that row (regardless of sign), then the difference you're testing is significant at the level shown at the top of that column.

# Appendix C  Calculational Procedures

*Example.* Suppose we had a $t = 2.10$ with 26 degrees of freedom. We look at the row where $df = 25$, since 26 isn't listed. In that row we see that 2.10 is bigger than 2.06 but isn't bigger than 2.79. Therefore, we know that the difference between the means is significant at the .05 level but not at the .01 level. It would be correct to write that $t(26) = 2.10$, $p < .05$.

If we had a $t = 1.90$ with $df = 10$, the difference between the means would not be significant, $p > .05$. It also would be correct to write $.10 > p > .05$.

## Evaluating the Statistical Significance of F

To use Table C.6 to evaluate the statistical significance of differences between means (including main effects and interactions), you must have a value of $F$ and know two kinds of degrees of freedom: the degrees of freedom for the term you're trying to evaluate and the degrees of freedom for the error term you used in calculating $F$. See the instructions for the different analyses of variance to determine the degrees of freedom.

Using the degrees of freedom for your error term, look down that column until you reach a number equal to your degrees of freedom; if your number isn't there, use the next lowest number. Then use the degrees of freedom for the effect to be evaluated. Go across that row until you reach the column corresponding to your degrees of freedom or the next lowest number.

Then look at the tabled values where your row and column intersect. Compare your calculated $F$ with the tabled values. If your $F$ is bigger than a tabled value, then the effect you're testing is significant at the level of $p$ at the left end of that row.

*Example.* Suppose we found an $F = 3.50$. The degrees of freedom corresponding to that effect are 3; the degrees of freedom in the error term are 25. We find the 24 $df$ in the error term column and use those rows, because 25 isn't listed. We go across until we reach the column corresponding to 3 $df$ in the effect to be evaluated. The tabled values there are 2.33, 3.01, 4.72, and 7.55. We see that 3.50 is bigger than 3.01 but not bigger than 4.72. Therefore, we know that this effect is significant at the .05 level but not at the .01 level. It would be correct to write that $F(3, 25) = 3.50$, $p < .05$.

If we had an $F = 2.50$ with 4 and 18 degrees of freedom, it would not be significant, $p > .05$. It would also be correct to write $.10 > p > .05$.

**Table C.6.** Critical Values of $F$[a]

| Degrees of freedom in error term | $p$ | Degrees of Freedom in Term of Interest ||||||||| |
| --- | --- | --- | --- | --- | --- | --- | --- | --- | --- | --- | --- |
| | | 1 | 2 | 3 | 4 | 5 | 6 | 8 | 12 | 24 | $\infty$ |
| 6 | .10 | 3.78 | 3.46 | 3.29 | 3.18 | 3.11 | 3.05 | 2.98 | 2.90 | 2.82 | 2.72 |
| | .05 | 5.99 | 5.14 | 4.76 | 4.53 | 4.39 | 4.28 | 4.15 | 4.00 | 3.84 | 3.67 |
| | .01 | 13.74 | 10.92 | 9.78 | 9.15 | 8.75 | 8.47 | 8.10 | 7.72 | 7.31 | 6.88 |
| | .001 | 35.51 | 27.00 | 23.70 | 21.90 | 20.81 | 20.03 | 19.03 | 17.99 | 16.89 | 15.75 |
| 7 | .10 | 3.59 | 3.26 | 3.07 | 2.96 | 2.88 | 2.83 | 2.75 | 2.67 | 2.58 | 2.47 |
| | .05 | 5.59 | 4.74 | 4.35 | 4.12 | 3.97 | 3.87 | 3.73 | 3.57 | 3.41 | 3.23 |
| | .01 | 12.25 | 9.55 | 8.45 | 7.85 | 7.46 | 7.19 | 6.84 | 6.47 | 6.07 | 5.65 |
| | .001 | 29.22 | 21.69 | 18.77 | 17.19 | 16.21 | 15.52 | 14.63 | 13.71 | 12.73 | 11.69 |
| 8 | .10 | 3.46 | 3.11 | 2.92 | 2.81 | 2.73 | 2.67 | 2.59 | 2.50 | 2.40 | 2.29 |
| | .05 | 5.32 | 4.46 | 4.07 | 3.84 | 3.69 | 3.58 | 3.44 | 3.28 | 3.12 | 2.93 |
| | .01 | 11.26 | 8.65 | 7.59 | 7.01 | 6.63 | 6.37 | 6.03 | 5.67 | 5.28 | 4.86 |
| | .001 | 25.42 | 18.49 | 15.83 | 14.39 | 13.49 | 12.86 | 12.04 | 11.19 | 10.30 | 9.34 |
| 9 | .10 | 3.36 | 3.01 | 2.81 | 2.69 | 2.61 | 2.55 | 2.47 | 2.38 | 2.28 | 2.16 |
| | .05 | 5.12 | 4.26 | 3.86 | 3.63 | 3.48 | 3.37 | 3.23 | 3.07 | 2.90 | 2.71 |
| | .01 | 10.56 | 8.02 | 6.99 | 6.42 | 6.06 | 5.80 | 5.47 | 5.11 | 4.73 | 4.31 |
| | .001 | 22.86 | 16.39 | 13.90 | 12.56 | 11.71 | 11.13 | 10.37 | 9.57 | 8.72 | 7.81 |
| 10 | .10 | 3.28 | 2.92 | 2.73 | 2.61 | 2.52 | 2.46 | 2.38 | 2.28 | 2.18 | 2.06 |
| | .05 | 4.96 | 4.10 | 3.71 | 3.48 | 3.33 | 3.22 | 3.07 | 2.91 | 2.74 | 2.54 |
| | .01 | 10.04 | 7.56 | 6.55 | 5.99 | 5.64 | 5.39 | 5.06 | 4.71 | 4.33 | 3.91 |
| | .001 | 21.04 | 14.91 | 12.55 | 11.28 | 10.48 | 9.92 | 9.20 | 8.45 | 7.64 | 6.76 |
| 11 | .10 | 3.23 | 2.86 | 2.66 | 2.54 | 2.45 | 2.39 | 2.30 | 2.21 | 2.10 | 1.97 |
| | .05 | 4.84 | 3.98 | 3.59 | 3.36 | 3.20 | 3.09 | 2.95 | 2.79 | 2.61 | 2.40 |
| | .01 | 9.65 | 7.20 | 6.22 | 5.67 | 5.32 | 5.07 | 4.74 | 4.40 | 4.02 | 3.60 |
| | .001 | 19.69 | 13.81 | 11.56 | 10.35 | 9.58 | 9.05 | 8.35 | 7.63 | 6.85 | 6.00 |
| 12 | .10 | 3.18 | 2.81 | 2.61 | 2.48 | 2.39 | 2.33 | 2.24 | 2.15 | 2.04 | 1.90 |
| | .05 | 4.75 | 3.88 | 3.49 | 3.26 | 3.11 | 3.00 | 2.85 | 2.69 | 2.50 | 2.30 |
| | .01 | 9.33 | 6.93 | 5.95 | 5.41 | 5.06 | 4.82 | 4.50 | 4.16 | 3.78 | 3.36 |
| | .001 | 18.64 | 12.97 | 10.80 | 9.63 | 8.89 | 8.38 | 7.71 | 7.00 | 6.25 | 5.42 |
| 13 | .10 | 3.14 | 2.76 | 2.56 | 2.43 | 2.35 | 2.28 | 2.20 | 2.10 | 1.98 | 1.85 |
| | .05 | 4.67 | 3.80 | 3.41 | 3.18 | 3.02 | 2.92 | 2.77 | 2.60 | 2.42 | 2.21 |
| | .01 | 9.07 | 6.70 | 5.74 | 5.20 | 4.86 | 4.62 | 4.30 | 3.96 | 3.59 | 3.16 |
| | .001 | 17.81 | 12.31 | 10.21 | 9.07 | 8.35 | 7.86 | 7.21 | 6.52 | 5.78 | 4.97 |
| 14 | .10 | 3.10 | 2.73 | 2.52 | 2.39 | 2.31 | 2.24 | 2.15 | 2.05 | 1.94 | 1.80 |
| | .05 | 4.60 | 3.74 | 3.34 | 3.11 | 2.96 | 2.85 | 2.70 | 2.53 | 2.35 | 2.13 |
| | .01 | 8.86 | 6.51 | 5.56 | 5.03 | 4.69 | 4.46 | 4.14 | 3.80 | 3.43 | 3.00 |
| | .001 | 17.14 | 11.78 | 9.73 | 8.62 | 7.92 | 7.43 | 6.80 | 6.13 | 5.41 | 4.60 |
| 15 | .10 | 3.07 | 2.70 | 2.49 | 2.36 | 2.27 | 2.21 | 2.12 | 2.02 | 1.90 | 1.76 |
| | .05 | 4.54 | 3.68 | 3.29 | 3.06 | 2.90 | 2.79 | 2.64 | 2.48 | 2.29 | 2.07 |
| | .01 | 8.68 | 6.36 | 5.42 | 4.89 | 4.56 | 4.32 | 4.00 | 3.67 | 3.29 | 2.87 |
| | .001 | 16.59 | 11.34 | 9.34 | 8.25 | 7.57 | 7.09 | 6.47 | 5.81 | 5.10 | 4.31 |

[a] Taken from Table V of R. A. Fisher and F. Yates, *Statistical tables for biological, agricultural and medical research,* published by Longman Group Ltd., London (previously published by Oliver and Boyd, Edinburgh). Used by permission of the authors and publishers.

# Appendix C  Calculational Procedures

**Table C.6.** (Continued)

| Degrees of freedom in error term | p | \multicolumn{10}{c}{Degrees of Freedom in Term of Interest} |
|---|---|---|---|---|---|---|---|---|---|---|---|
| | | 1 | 2 | 3 | 4 | 5 | 6 | 8 | 12 | 24 | ∞ |
| 16 | .10 | 3.05 | 2.67 | 2.46 | 2.33 | 2.24 | 2.18 | 2.09 | 1.99 | 1.87 | 1.72 |
| | .05 | 4.49 | 3.63 | 3.24 | 3.01 | 2.85 | 2.74 | 2.59 | 2.42 | 2.24 | 2.01 |
| | .01 | 8.53 | 6.23 | 5.29 | 4.77 | 4.44 | 4.20 | 3.89 | 3.85 | 3.18 | 2.75 |
| | .001 | 16.12 | 10.97 | 9.00 | 7.94 | 7.27 | 6.81 | 6.19 | 5.55 | 4.85 | 4.06 |
| 17 | .10 | 3.03 | 2.64 | 2.44 | 2.31 | 2.22 | 2.15 | 2.06 | 1.96 | 1.84 | 1.69 |
| | .05 | 4.45 | 3.59 | 3.20 | 2.96 | 2.81 | 2.70 | 2.55 | 2.38 | 2.19 | 1.96 |
| | .01 | 8.40 | 6.11 | 5.18 | 4.67 | 4.34 | 4.10 | 3.79 | 3.45 | 3.08 | 2.65 |
| | .001 | 15.72 | 10.66 | 8.73 | 7.68 | 7.02 | 6.56 | 5.96 | 5.32 | 4.63 | 3.85 |
| 18 | .10 | 3.01 | 2.62 | 2.42 | 2.29 | 2.20 | 2.13 | 2.04 | 1.93 | 1.81 | 1.66 |
| | .05 | 4.41 | 3.55 | 3.16 | 2.93 | 2.77 | 2.66 | 2.51 | 2.34 | 2.15 | 1.92 |
| | .01 | 8.28 | 6.01 | 5.09 | 4.58 | 4.25 | 4.01 | 3.71 | 3.37 | 3.00 | 2.57 |
| | .001 | 15.38 | 10.39 | 8.49 | 7.46 | 6.81 | 6.35 | 5.76 | 5.13 | 4.45 | 3.67 |
| 19 | .10 | 2.99 | 2.61 | 2.40 | 2.27 | 2.18 | 2.11 | 2.02 | 1.91 | 1.79 | 1.63 |
| | .05 | 4.38 | 3.52 | 3.13 | 2.90 | 2.74 | 2.63 | 2.48 | 2.31 | 2.11 | 1.88 |
| | .01 | 8.18 | 5.93 | 5.01 | 4.50 | 4.17 | 3.94 | 3.63 | 3.30 | 2.92 | 2.49 |
| | .001 | 15.08 | 10.16 | 8.28 | 7.26 | 6.61 | 6.18 | 5.59 | 4.97 | 4.29 | 3.52 |
| 20 | .10 | 2.97 | 2.59 | 2.38 | 2.25 | 2.16 | 2.09 | 2.00 | 1.89 | 1.77 | 1.61 |
| | .05 | 4.35 | 3.49 | 3.10 | 2.87 | 2.71 | 2.60 | 2.45 | 2.28 | 2.08 | 1.84 |
| | .01 | 8.10 | 5.85 | 4.94 | 4.43 | 4.10 | 3.87 | 3.56 | 3.23 | 2.86 | 2.42 |
| | .001 | 14.82 | 9.95 | 8.10 | 7.10 | 6.46 | 6.02 | 5.44 | 4.82 | 4.15 | 3.38 |
| 22 | .10 | 2.95 | 2.56 | 2.35 | 2.22 | 2.13 | 2.06 | 1.97 | 1.86 | 1.73 | 1.57 |
| | .05 | 4.30 | 3.44 | 3.05 | 2.82 | 2.66 | 2.55 | 2.40 | 2.23 | 2.03 | 1.78 |
| | .01 | 7.94 | 5.72 | 4.82 | 4.31 | 3.99 | 3.76 | 3.45 | 3.12 | 2.75 | 2.31 |
| | .001 | 14.38 | 9.61 | 7.80 | 6.81 | 6.19 | 5.76 | 5.19 | 4.58 | 3.92 | 3.15 |
| 24 | .10 | 2.93 | 2.54 | 2.33 | 2.19 | 2.10 | 2.04 | 1.94 | 1.83 | 1.70 | 1.53 |
| | .05 | 4.26 | 3.40 | 3.01 | 2.78 | 2.62 | 2.51 | 2.36 | 2.18 | 1.98 | 1.73 |
| | .01 | 7.82 | 5.61 | 4.72 | 4.22 | 3.90 | 3.67 | 3.36 | 3.03 | 2.66 | 2.21 |
| | .001 | 14.03 | 9.34 | 7.55 | 6.59 | 5.98 | 5.55 | 4.99 | 4.39 | 3.74 | 2.97 |
| 26 | .10 | 2.91 | 2.52 | 2.31 | 2.17 | 2.08 | 2.01 | 1.92 | 1.81 | 1.68 | 1.50 |
| | .05 | 4.22 | 3.37 | 2.98 | 2.74 | 2.59 | 2.47 | 2.32 | 2.15 | 1.95 | 1.69 |
| | .01 | 7.72 | 5.53 | 4.64 | 4.14 | 3.82 | 3.59 | 3.29 | 2.96 | 2.58 | 2.13 |
| | .001 | 13.74 | 9.12 | 7.36 | 6.41 | 5.80 | 5.38 | 4.83 | 4.24 | 3.59 | 2.82 |
| 28 | .10 | 2.89 | 2.50 | 2.29 | 2.16 | 2.06 | 2.00 | 1.90 | 1.79 | 1.66 | 1.48 |
| | .05 | 4.20 | 3.34 | 2.95 | 2.71 | 2.56 | 2.44 | 2.29 | 2.12 | 1.91 | 1.65 |
| | .01 | 7.64 | 5.45 | 4.57 | 4.07 | 3.75 | 3.53 | 3.23 | 2.90 | 2.52 | 2.06 |
| | .001 | 13.50 | 8.93 | 7.19 | 6.25 | 5.66 | 5.24 | 4.69 | 4.11 | 3.46 | 2.70 |
| 30 | .10 | 2.88 | 2.49 | 2.28 | 2.14 | 2.05 | 1.98 | 1.88 | 1.77 | 1.64 | 1.46 |
| | .05 | 4.17 | 3.32 | 2.92 | 2.69 | 2.53 | 2.42 | 2.27 | 2.09 | 1.89 | 1.62 |
| | .01 | 7.56 | 5.39 | 4.51 | 4.02 | 3.70 | 3.47 | 3.17 | 2.84 | 2.47 | 2.01 |
| | .001 | 13.29 | 8.77 | 7.05 | 6.12 | 5.53 | 5.12 | 4.58 | 4.00 | 3.36 | 2.59 |

**Table C.6.** (Continued)

| Degrees of freedom in error term | p | \multicolumn{10}{c}{Degrees of Freedom in Term of Interest} |||||||||| 
|---|---|---|---|---|---|---|---|---|---|---|---|
| | | 1 | 2 | 3 | 4 | 5 | 6 | 8 | 12 | 24 | ∞ |
| 40  | .10  | 2.84  | 2.44 | 2.23 | 2.09 | 2.00 | 1.93 | 1.83 | 1.71 | 1.57 | 1.38 |
|     | .05  | 4.08  | 3.23 | 2.84 | 2.61 | 2.45 | 2.34 | 2.18 | 2.00 | 1.79 | 1.51 |
|     | .01  | 7.31  | 5.18 | 4.31 | 3.83 | 3.51 | 3.29 | 2.99 | 2.66 | 2.29 | 1.80 |
|     | .001 | 12.61 | 8.25 | 6.60 | 5.70 | 5.13 | 4.73 | 4.21 | 3.64 | 3.01 | 2.23 |
| 60  | .10  | 2.79  | 2.39 | 2.18 | 2.04 | 1.95 | 1.87 | 1.77 | 1.66 | 1.51 | 1.29 |
|     | .05  | 4.00  | 3.15 | 2.76 | 2.52 | 2.37 | 2.25 | 2.10 | 1.92 | 1.70 | 1.39 |
|     | .01  | 7.08  | 4.98 | 4.13 | 3.65 | 3.34 | 3.12 | 2.82 | 2.50 | 2.12 | 1.60 |
|     | .001 | 11.97 | 7.76 | 6.17 | 5.31 | 4.76 | 4.37 | 3.87 | 3.31 | 2.69 | 1.90 |
| 120 | .10  | 2.75  | 2.35 | 2.13 | 1.99 | 1.90 | 1.82 | 1.72 | 1.60 | 1.45 | 1.19 |
|     | .05  | 3.92  | 3.07 | 2.68 | 2.45 | 2.29 | 2.17 | 2.02 | 1.83 | 1.61 | 1.25 |
|     | .01  | 6.85  | 4.79 | 3.95 | 3.48 | 3.17 | 2.96 | 2.66 | 2.34 | 1.95 | 1.38 |
|     | .001 | 11.38 | 7.31 | 5.79 | 4.95 | 4.42 | 4.04 | 3.55 | 3.02 | 2.40 | 1.54 |
| ∞   | .10  | 2.71  | 2.30 | 2.08 | 1.94 | 1.85 | 1.77 | 1.67 | 1.55 | 1.38 | 1.00 |
|     | .05  | 3.84  | 2.99 | 2.60 | 2.37 | 2.21 | 2.09 | 1.94 | 1.75 | 1.52 | 1.00 |
|     | .01  | 6.64  | 4.60 | 3.78 | 3.32 | 3.02 | 2.80 | 2.51 | 2.18 | 1.79 | 1.00 |
|     | .001 | 10.83 | 6.91 | 5.42 | 4.62 | 4.10 | 3.74 | 3.27 | 2.74 | 2.13 | 1.00 |

**Appendix D
Sample Report
Typed in APA
Format**

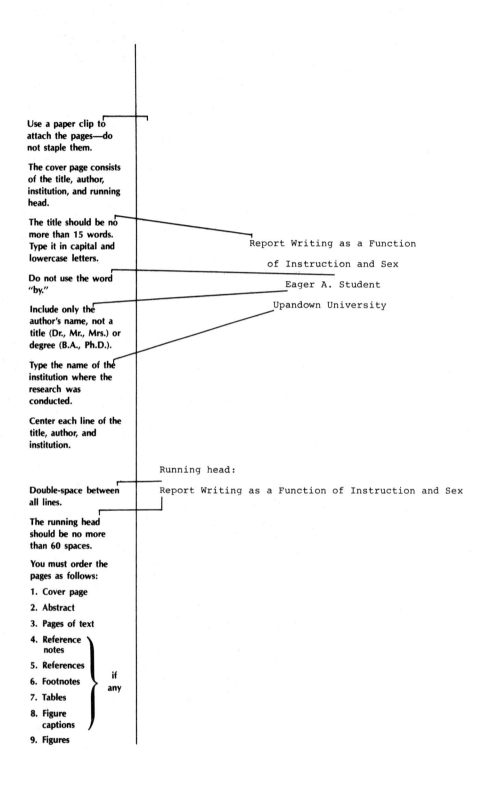

Report Writing
1

Abstract

The present study provided corroboration of previous observations that instruction in research writing facilitates such writing and that female students benefit more from instruction than male students. An undergraduate experimental psychology class consisting of 14 males and 14 females was required to write a report either after instruction in the organization and content of the report or without such instruction. The instructed students received a mean grade of 91 on the report, while the noninstructed students received a mean grade of 80. The increase in grades with instructions was greater for females than for males. Overall there was no difference between males and females.

---

Begin the abstract on a new page.

Type the page identification in the upper right corner.

The abstract is always page 1.

Do not include any title or name other than the word "Abstract."

Do not indent—type the abstract in block style.

Remember to double-space throughout.

Allow margins of 1–1½ in. or 2½–4 cm at the top, bottom, right, and left of every page.

The abstract should run 100 to 175 words.

Report Writing

2

Report Writing as a Function

of Instruction and Sex

Teachers often express interest in methods for improving the research writing of their students (cf. Whine & Complain, 1978). However, methods for improving such writing have been studied little. One careful observer (Newman, Note 1) suggested that one of the most important areas of instruction was a description of the major sections of the research report (introduction, method, and so on). Along this line Freeblemutzer (1932) reported that students instructed in the format of research reports obtained higher grades than students not so informed.

Butternut and Cheese (1916) stated their beliefs succinctly: "Telling students more about the reports to be written leads to better written reports" (p. 416). It has also been reported (Potrzebie, Hurley, & Burley, 1880) that female students benefited more from instructions about research writing than did male students. The conclusions of Freeblemutzer, Butternut and Cheese and Potrzebie et al. were based on their personal observations while teaching rather than upon experimental evidence. The only existing experimental report (Dum & Dee, 1955) hopelessly confounded the effects of instructions with those of time of year and previous instructional techniques.

*Annotations:*
- Begin the text on a new page.
- Repeat the complete title; be sure to center it.
- Don't include any name.
- Use the author-date method to cite references in the text.
- Cite notes as "Notes," rather than dates.
- If the authors' names are part of the text, cite only the year of publication in parentheses.
- Note that "and" is used outside parentheses and & is used inside parentheses.
- Note the comma before the ampersand.
- If a reference has more than two authors, cite all names for the first reference. From then on, use the name of the first author followed by "et al."
- If the authors' names are not part of the text, include both the authors' names and the date in parentheses.

The present study attempted to provide experimental evidence about these questions. The effect of instruction on the grades assigned for research reports were studied for male and female students. On the basis of the reports mentioned it was expected that instruction would increase assigned grades and that females would benefit more from instruction.

*Use the past tense to describe your study.*

## Method

*Center and underline all main headings.*

### Subjects

The subjects were students in an undergraduate class in experimental psychology. There were 14 males and 14 females. Within each sex the students were assigned to one of the two instruction conditions by a block-randomization procedure, resulting in seven blocks with two students in each block. Thus there were seven males and seven females in each of the two instruction groups.

*Do not use Ss, Es, or Os.*

### Procedure

After assembling at the usual class hour, all students were lectured on the experiment with which their reports were to deal. They were not required to conduct the experiment themselves. The major conditions of a hypothetical experiment were verbally presented, and mimeographed sets of data from that experiment were distributed.

After answering questions pertinent to the experiment, the instructor distributed sets of rules concerning the

*Flush side headings to the left margin. Underline them and then double-space.*

details of report writing to all students. Such details included, for example, instructions to underline certain words, information about page numbering, margin widths, and so forth. The students were told that the rules were to be read outside of class and could be referred to in writing up the experiment.

Then all the students were told they would be given additional materials and that 30 min. would be allowed for reading these materials. For instructed students the additional materials were information about the major divisions of the report, telling what each division should include. For noninstructed students the additional material was a report of another hypothetical experiment. After 30 min. passed, all students were told to work individually on their reports.

*Avoid abbreviations.*

## Results

*Explain the meaning of your numbers.*

One grader who was unaware of the nature of the experiment graded all reports, therefore guarding against effects due to differential grading. A predetermined number of points had been allocated to various aspects of the reports such that the total grade was the sum of the scores on the separate grading categories and could vary from 0 to 100. Fractional scores were not assigned.

Means of report grades for each of the four groups of the experiment are shown in Figure 1. Examination of the

*Be sure you mention each figure and table in the text.*

figure shows that the means for the various groups were

------------------------
Insert Fig. 1 about here
------------------------

ordered as expected. For both male and female students the mean grades of the instructed students were higher than for the noninstructed students. An analysis of variance showed that the main effect of instruction was significant, $F(1, 24) = 9.50$, $p < .01$. Further, differences between the instruction conditions were greater for females than for males. The analysis of variance showed this interaction to be significant, $F(1, 24) = 4.83$, $p < .05$. Finally, the mean grades of females were higher than for males, but this effect failed to reach significance, $F(1, 24) = 2.03$, $p > .05$.

As an afterthought, at the end of the course, the students were given a questionnaire asking what they found most difficult about writing research papers. The results of that questionnaire are shown in Table 1. As can be

------------------------
Insert Table 1 about here
------------------------

seen, learning what to put into each of the several divisions of the paper gave relatively little trouble to the students who received the extra instruction but gave a great deal of trouble to the others.

## Discussion

The finding that instruction in research report writing raised the grades assigned was consistent with the observations of Freeblemutzer (1932) and Butternut and Cheese (1916). A number of explanations of this result are possible: The information itself contained in the additional instructions may have produced the effect. Or the instructions could have increased the motivation of the students in the instructed condition. Or perhaps knowing that they were expected to perform differently caused these students to perform better. The present experiment did not provide tests of alternative interpretations of the results.

Similarly, the finding that instructions had a more profound effect for females than for males was consistent with the report of Potrzebie et al. (1880). Again, the present experiment did not provide a choice among alternative explanations. Whatever the reason, females benefited more from additional instructions than males.

No significant difference was found between the overall performance of males and females. This result is difficult to evaluate since it is due in part to the particular males and females employed.

It is apparent from the results of the questionnaire that special instruction about the divisions of research papers causes students to feel less troubled by that factor.

Report Writing

7

Reference Note

1. <u>Personal opinions about teaching</u>. Lecture from a course in personal hygiene, taught in Fall 1962 at Frogstown University, Angels Camp, California.

**Begin the reference notes on a new page.**

**Notice that the title is underlined.**

Report Writing

8

References

Butternut, I. M., & Cheese, P. U.  Obvious observations made while teaching experimental psychology.  In B. B. Bread (Ed.), Obvious observations in teaching. Milpitas, Calif.: Artichoke Press, 1916.  Pp. 400-417.

Dum, T., & Dee, D. T.  Effects of instructions on report writing. Journal of Inadequate Studies, 1955, 316, 1024-1099.

Freeblemutzer, Z.  Evaluation of a class of experimental psychology students with concern for their research writing. Journal of Archaic Psychology, 1932, 2, 121-179.

Potrzebie, P. P., Hurley, Q. Q., & Burley, R. R.  Observations of the scientific ability of a class of psychology students at the University of Southwestern Borneo. Journal of Educational Speculations, 1880, 7, 347-456.

Whine, C. B., & Complain, A. D.  Comments on the state of modern day student mentality.  Rock, N.Y.: Hardplace Press, 1978.

---

**Annotations:**

- Begin the references on a new page.
- Arrange references in alphabetical order by author's last name.
- Invert the order of the authors' names, that is, last names first followed by the initials.
- Allow one space between people's initials.
- Insert a comma between the names of multiple authors.
- Insert two spaces after the last author's initials.
- Use a period after book or article titles, followed by two spaces.
- Type the year of publication, the volume number, and the pages of the article after a journal title. Separate this information by commas, and end with a period.
- Capitalize the first letter of all major words in journal titles.
- Underline journal titles, book titles, and volume numbers.
- Capitalize the first letter of the first word only in article, chapter, or book titles.
- Type the publisher and city of publication after a book title. Insert a colon between the place of publication and the book publisher.
- Reference everything you cited specifically in the text; do not reference anything you don't cite.

# Appendix D  Sample Report Typed in APA Format

Report Writing

9

Table 1

Number of Students Citing Each Point as Very Difficult

| Point | Group | |
|---|---|---|
| | Instructed[a] | Noninstructed[b] |
| Getting material into proper division | 2 | 13 |
| Typing | 12 | 14 |
| Getting the punctuation right | 3 | 2 |
| Understanding the report | 11 | 12 |

[a] $\underline{N} = 12$.
[b] $\underline{N} = 14$.

Type each table on a new page.

Center the word "Table" and the arabic number over the table.

Provide a complete and explanatory title.

Mark rules very lightly in pencil or not at all.

Center column headings over the columns. Capitalize the first letter of the first word of all headings.

Align numbers on the decimal point.

Type your table in such a way that straight rules could be easily drawn dividing all the rows and columns—but don't draw them. Allow as much space as you can between columns.

Indent 5 spaces to the table footnotes.

Report Writing

10

Figure Caption

Figure 1. Mean score on research reports as a function of instruction and sex.

**Begin the figure captions on a new page.**

**Indent 5 spaces.**

**Capitalize only the first letter of the first word of a figure caption.**

**You would list any additional figure captions on this page.**

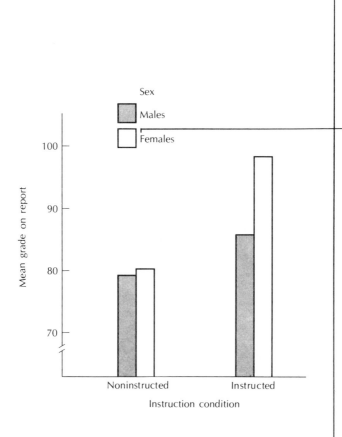

Each figure requires a separate page.

Figures should be carefully and precisely drawn—use them only to help illustrate complex interactions.

For publication, typing is inappropriate; you must use other letter techniques.

Be sure you include all necessary labels.

# Glossary

**Analysis of variance:** An inferential statistical technique for comparing means.

**Association between variables:** A patterned relationship between variables. Demonstration of an association does not necessarily imply that a causal relationship exists. See also positive and negative associations, and linear, nonlinear, and nonmonotonic relationships.

**Baseline level:** The level of the dependent variable prior to the occurrence of the independent variable in a before-after design.

**Before-after design:** An experimental design in which the dependent variable is measured before and again after the independent variable occurs.

**Before-after variable:** The independent variable in a before-after design.

**Between-subjects design:** An experimental design in which the independent variables are between-subjects variables.

**Between-subjects variable:** An independent variable having different subjects at different levels; each subject is measured only once on the dependent variable. Compare with within-subject variable.

**Block randomization:** A technique of ordering repeated levels of an independent variable, arranging the levels in blocks, then randomizing the order of levels within the blocks. Compare with counterbalancing and complete randomization.

**Case study:** A kind of scientific observation typically gathering a wide range of information about a single individual.

**Causal relationship:** An association between variables with the property that changes in one variable cause changes in the other.

**Complete randomization:** A technique of ordering repeated levels of an independent variable, arranging the repetitions of the levels in a completely random order. Compare with counterbalancing and block randomization.

**Confounded variable:** A variable combined with an independent variable in such a way that the levels of the two variables overlap, or coincide. Whenever this occurs, the effects of the two variables on a third cannot be separated. See extraneous-variable confounds, variable-connected confound, practice or fatigue effects, novelty effect, subject bias, and experimenter bias.

**Control group:** A group of subjects receiving a control treatment.

**Control techniques:** Techniques for eliminating confounded variables. Either hold the confounded variable constant, or distribute its levels equally across the levels of the independent variable.

**Control treatment:** A level of the independent variable added to control variable-connected confounds; often a placebo treatment, sometimes no treatment at all.

**Correlation between variables:** Same as association between variables. See also correlation coefficient.

**Correlation coefficient:** A descriptive statistic that describes the amount of agreement between two sets of scores.

**Correlational study:** Usage varies. Usually, any study in which it is not possible to conclude that a causal relationship exists. Can also mean a pseudo-experiment, an observational study, or any study in which a correlation coefficient is used to describe the data.

**Counterbalancing:** A technique of ordering repeated levels of an independent variable, arranging the levels so that each appears equally often in each position. Compare with complete randomization and block randomization.

**Criterion variable:** A variable that can be predicted by another variable, especially when no causal relationship can be said to exist between the variables.

**Cross-sectional variable:** Another name for a between-subject variable, most commonly used with time-related nonmanipulated variables, such as age or year in school. Compare with longitudinal variable.

**Demand characteristics:** Characteristics of an experiment that enable subjects to guess its purpose and/or the expectations of the experimenter. See also subject bias, experimenter bias, and reactive measure.

**Dependent variable:** In this text, a variable in an experiment which is measured after the level of the independent variable is manipulated or selected. (Some instructors use this term only when the independent variable is manipulated.) See also pseudo-dependent variable and criterion variable.

**Descriptive statistics:** Statistics used to summarize and describe collections of data.

**Double-blind study:** An experiment in which both the subject and the experimenter are unaware of the level of the independent variable occurring. Used to control subject bias and experimenter bias. See also placebo.

**Experiment:** In this text, a kind of scientific observation typically gathering a narrow range of information, designed to make it easier to understand the relationships among a small number of variables. (Some instructors use the word experiment as this text has defined true experiment.) See also pseudo-experiment, correlational study, and observational study.

**Experimental design:** A summary of the pattern of combinations of variables in an experiment. Important factors include the number and nature of the independent and dependent variables, the number of levels of each independent variable, and whether they varied between subjects, within subjects, or as before-after variables, and how they were combined.

**Experimental hypothesis:** A prediction, sometimes derived from a theory, to be tested in an experiment. An experimental hypothesis predicts an association among two or more variables.

**Experimenter bias:** A source of confounding arising when the experimenter's attitude or behavior is related to the level of the independent variable occurring.

**Extraneous-variable confounds:** Confounds with variables unrelated to the variables in the experiment.

**Factor analysis:** A descriptive statistical procedure for expressing the degree to which a large number of items or events tend to vary together.

**Factorial design:** An experimental design in which each level of each independent variable occurs with each level of every other independent variable.

**Fatigue effects:** Decline in performance over trials or repetitions caused by fatigue. See practice or fatigue effects.

**Field study:** A study performed in a real-life situation rather than in a laboratory or other artificial situation.

**Frequency distribution:** A technique for summarizing data showing the number of times each level of a variable occurred.

**Hawthorne effect:** See novelty effect.

**Hypothesis:** A relatively informal theory, or hunch, summarizing a few observations and attempting to explain them. See also experimental hypothesis and null hypothesis.

**Hypothesis-testing statistics:** See inferential statistics.

**Independent variable:** In this text, a variable in an experiment that is either manipulated or on the basis of which events are selected to be observed. (Some instructors use this term only when the variable is manipulated.) See also pseudo-independent variable and predictor variable.

**Inferential statistics:** Statistics used to estimate the probability that events have been caused by random variation, i.e., the probability that the null hypothesis is true. See significant and nonsignificant differences.

**Informed consent:** Consent from subjects to participate in an experiment after they have been informed of all features of the experiment that might influence their willingness to participate.

**Interaction:** An association among variables such that the association between some independent and dependent variables changes depending on another independent variable. See also two-way and three-way interactions.

**Interval scale of measurement:** A scale of measurement in which the values are related by a quantitative dimension with equal intervals between successive scale values. The zero point may be arbitrary. Contrast with nominal, ordinal, and ratio scales of measurement.

**Inverse relationship:** See negative association.

**Law:** A well-accepted, well-supported theory, summarizing many observations.

**Level of a variable:** A value of a variable. With variables measured on an ordinal, interval, or ratio scale of measurement each level is usually represented by a number. With variables measured on a nominal scale each level is indicated with just a label.

**Line of best fit:** See regression line.

**Linear relationship:** An association between variables that can be summarized by a straight line. See positive and negative associations.

**Longitudinal variable:** Another name for a within-subject variable, most commonly used with time-related nonmanipulated variables, such as age or year in school. Compare with cross-sectional variable.

**Main effect:** A statistical concept, expressing the association between one independent and one dependent variable. See also interaction.

**Manipulated variable:** An independent variable directly controlled by the experimenter. When variables are manipulated it is sometimes possible to conclude that a causal relationship exists.

**Matching:** A control technique consisting of distributing the levels of possibly confounded variables equally across the levels of the independent variable.

**Mean:** A descriptive statistic representing the "average" or central tendency of a set of scores, appropriate for use with variables having interval or ratio scales of measurement.

**Measurement:** Assigning a value, often a numerical value, in a meaningful way. See scale of measurement.

**Mixed design:** Usage varies. Either an experimental design in which some independent variables are between-subject and others are within-subject, or an experimental design with some manipulated and some nonmanipulated variables.

**Negative association:** An association between variables in which high scores on one variable go with low scores on the other.

**Negative correlation:** See negative association and correlation coefficient.

**Nominal scale of measurement:** A scale of measurement in which the values are not necessarily related by a quantitative dimension, but may represent only qualitatively different categories. Contrast with ordinal and interval scales of measurement, and ratio scales.

**Nominal variable:** A variable that is measured only on a nominal scale of measurement.

**Nonequivalent groups:** A possible source of confounding arising when subjects are assigned to different levels of a between-subject variable in any manner other than random assignment.

**Nonlinear relationship:** An association between variables that cannot be summarized as a straight line. See also nonmonotonic relationship.

**Nonmanipulated variable:** An independent variable in which the experimenter either accepts the levels that occur naturally or actively selects some naturally occurring levels. When no variables are manipulated it is usually impossible to conclude that a causal relationship exists.

**Nonmonotonic relationship:** An association between variables that changes direction; often a U-shaped relationship. Contrast with linear relationship.

**Nonparametric statistics:** Statistics that may be used with variables having nominal or ordinal scales of measurement as well as with variables having interval or ratio scales. Contrast with parametric statistics.

**Nonreactive measure:** A measurement taken in such a way that the fact of measurement does not influence the subject's behavior. Often, the subject is unaware of being observed. Sometimes subjects are not observed at all, but only inanimate objects reflecting previous behavior.

**Nonsignificant difference:** A difference between means or frequencies shown by inferential statistics to have a relatively high (often, greater than .05) probability of having been caused by random variations, i.e., a relatively high probability that the null hypothesis is true.

**Nonsignificant relationship:** An association between variables shown by inferential statistics to have a relatively high (often, greater than .05) probability of having been caused by random variations.

**Novelty effect:** Changes in behavior caused by a change in routine, especially a change caused by experimental measurements being taken. A possible source of confounding, especially in before-after designs.

# Glossary

**Null hypothesis:** A statement that there is no real association between variables, or that any demonstrated association was caused by random variations. See also significant and nonsignificant relationships.

**Observational study:** Usage varies. Usually, a field study in which no variables are manipulated. Can also be any field study, a pseudo-experiment, or a study using nonreactive measures.

**Operational definition:** A way of defining a concept, especially an experimental variable, by telling what concrete operations must be performed to demonstrate or measure the concept.

**Order effects:** Changes in behavior caused by performing events in a particular order. Sometimes used in a general sense as this text uses practice or fatigue effects.

**Ordinal scale of measurement:** A scale of measurement in which values are related by a quantitative dimension, but in which there are not necessarily equal intervals between successive scale values. Contrast with nominal, interval, and ratio scales of measurement.

**Parametric statistics:** Statistics that should be used only with variables assumed to have interval or ratio scales of measurement.

**Placebo:** A substance or treatment indistinguishable by the subjects from a real treatment, but lacking the key component of the real treatment. Used to control subject bias and variable-connected confounds. See also single-blind and double-blind studies.

**Population:** A group of individuals or events. See also random sample.

**Position effects:** Changes in behavior caused by the position in a series of trials at which an event occurs. See practice or fatigue effects.

**Positive association:** An association between variables in which high scores on one variable go with high scores on the other, and low scores go with low scores.

**Positive correlation:** See positive association and correlation coefficient.

**Post hoc:** After the fact. Post hoc theories are created after an experiment is performed, and are less trustworthy than those tested by an experiment. Post hoc inferential statistics test the significance of relationships or differences not predicted by an experimental hypothesis, but recognized only after the experiment is completed.

**Practice effects:** Improvement in performance over trials or repetitions caused by practice. See practice or fatigue effects.

**Practice or fatigue effects:** In this text, a general term meaning changes in behavior over trials or repetitions. A possible source of confounding arising when within-subject variables or before-after variables are used. See also practice effects, fatigue effects, order effects, position effects, time effects, and repeated-measures effects.

**Predictor variable:** A variable that can be used to predict another variable (a criterion), especially when no causal relationship can be said to exist between the variables.

**Pseudo-dependent variable:** The dependent variable in a pseudo-experiment.

**Pseudo-experiment:** An experiment in which the independent variable is nonmanipulated, but in which the experimenter actively selects naturally occurring levels of the independent variable to observe. It is not possible to conclude that causal relationships exist on the basis of pseudo-experiments. See also true experiment and observational study.

**Pseudo-independent variable:** The independent variable in a pseudo-experiment.

**Quasi-experiment:** See pseudo-experiment.

**Random:** Chance events, distributed by the formal rules of probability. Not the same as haphazard or accidental. See random assignment and random sample.

**Random assignment:** A technique of assigning subjects to levels of a between-subject variable in such a way that every subject has an equal chance of being assigned to any level. Can be used only with manipulated variables.

**Random sample:** A group of subjects or events selected from a population using procedures to ensure that every member of the population had an equal chance of being selected.

**Range:** A descriptive statistic used to estimate the amount of variability in any set of scores with an ordinal, interval, or ratio scale of measurement.

**Ratio scale of measurement:** A scale of measurement in which values are related by a quantitative dimension with equal intervals between scale values and a value corresponding to absolute zero. Contrast with nominal, ordinal, and interval scales of measurement.

**Reactive measure:** A measurement taken in such a way that it causes the subject to behave differently than if no measurement were being taken. See also nonreactive measure.

**Regression line:** A straight line that summarizes a linear relation between two variables.

**Repeated-measures design:** An experimental design including a repeated-measures variable.

**Repeated-measures effect:** Changes in performance caused by the fact that more than one measurement is made on each subject in a within-subject design. See practice or fatigue effects.

**Repeated-measures variable:** A within-subject variable on which each subject is measured more than once on each level.

**Scale of measurement:** A set of numbers or labels assigned to subjects or events to designate the amount of some characteristic possessed by the subjects or events, or the category to which they belong. See nominal, ordinal, interval, and ratio scales of measurement, and measurement.

**Scatter graph:** A technique for displaying the association between two variables by representing the paired observations of the two variables as points at the appropriate coordinates of a graph.

**Significant difference:** A difference between two means or frequencies shown by inferential statistics to have a relatively low (often, less than .05) probability of having been caused by random variations, i.e., a relatively low probability that the null hypothesis is true.

**Significant relationship:** An association between variables shown by inferential statistics to have a relatively low (often less than .05) probability of having been caused by random variations.

**Single-blind study:** An experiment in which the subjects are unaware of the level of the independent variable that is occurring. Used to control subject bias. See also placebo.

**Standard deviation:** A descriptive statistic used to estimate the amount of variability around the mean of a set of scores. Its square is the variance.

**Statistical significance:** See inferential statistics, significant and nonsignificant differences.

**Subject bias:** A source of confounding arising when subjects influence their responses according to their beliefs about the purpose of the experiment or the expectations of the experimenter.

**Subject variable:** A variable that is an integral part of the personality or habitual behavior of experimental subjects.

**Subjects:** The individuals observed in an experiment.

**Survey:** A kind of scientific observation typically gathering an organized but wide-range set of observations about many individuals.

**Theory:** A summary of scientific observations that attempts to explain the observations, and that can be used to provide predictions for the future.

**Three-way interaction:** A statistical concept, expressing an association among three independent and one dependent variable, such that the two-way interaction between two of the independent variables changes, depending on the level of the third independent variable.

**Time effects:** Changes in performance caused by the time at which an event occurs, especially in a series of trials. See practice or fatigue effects.

**True experiment:** An experiment in which the independent variable is manipulated. See also pseudo-experiment and observational study.

**Two-way interaction:** A statistical concept, expressing an association among two independent and one dependent variable, such that the association between one independent variable and the dependent variable changes, depending on the level of the other independent variable.

**Unobtrusive measure:** See nonreactive measure.

**Variability:** The spread of a set of scores around the "average" or central tendency. Commonly estimated by the range, standard deviation, or variance.

**Variable:** The events or behaviors with which an experiment is concerned. Variables take different values, or levels; the levels can be measured on some scale of measurement.

**Variable-connected confound:** Confounds with variables intimately connected with the independent variable.

**Variance:** A descriptive statistic used to estimate the amount of variability around the mean of a set of scores. Its square root is the standard deviation.

**Within-subject variable:** An independent variable having the same subjects at all levels; each subject is measured several times on the dependent variable. Compare with between-subjects variable.

**Within-subjects design:** An experimental design in which the independent variables of interest are within-subject variables.

# References

Alper, T. G., & Korchin, S. J. Memory for socially relevant material. *Journal of Abnormal and Social Psychology,* 1952, *47,* 25–37.

Athanasiou, R., Oppel, W., Michelson, L., Unger, T., & Yager, M. Psychiatric sequelae to term birth and induced early and late abortion: A longitudinal study. *Family Planning Perspectives,* 1973, *5,* 227–231.

Bain, J. A. *Thought control in everyday life.* New York: Funk & Wagnalls, 1928.

Belluci, J. E. Microcounseling and imitation learning: A behavioral approach to counselor education. *Counselor Education and Supervision,* 1972, *12,* 88–97.

Bernstein, B. Language and social class. *The British Journal of Sociology,* 1960, *11,* 271–276.

Bernstein, B. Social class and linguistic development: A theory of social learning. In A. H. Halsey, J. Floud, & C. A. Anderson (Eds.), *Education, economy, and society.* Glencoe: The Free Press, 1961.

Blum, R. H., & Associates. *Students and drugs: College and high school observations.* San Francisco: Jossey-Bass, 1969.

Bracken, M. B., Grossman, G., Hachamovitch, M., Sussman, D., & Schrier, D. Abortion counseling: An experimental study of three techniques. *American Journal of Obstetrics and Gynecology,* 1973, *117,* 10–20.

Brigham, J. C., & Cook, S. W. The influence of attitude on the recall of controversial material: A failure to confirm. *Journal of Experimental Social Psychology,* 1969, *5,* 240–243.

Bruning, J. L., & Kintz, B. L. *Computational handbook of statistics.* Glenview, Ill.: Scott, Foresman, 1972.

Christian Science Publishing Company. *A century of Christian Science healing.* Boston: Christian Science Publishing Co., 1966.

Cohen, R. J., & Smith, F. J. Socially reinforced obsessing: Etiology of a disorder in a Christian Scientist. *Journal of Consulting and Clinical Psychology,* 1976, *44,* 142–144.

Costin, F., Greenough, W. T., & Menges, R. J. Student rating of college teaching: Reliability, validity, and usefulness. *Review of Educational Research,* 1971, *41,* 511–535.

Cross, H. J., & Davis, G. L. College students' adjustment and frequency of marijuana use. *Journal of Counseling Psychology,* 1972, *19,* 65–67.

David, H. P. Psychological studies in abortion. In J. T. Fawcett (Ed.), *Psychological perspectives on population.* New York: Basic Books, 1973.

Deutsch, M., Katz, I., & Jensen, A. R. *Social class, race, and psychological development.* New York: Holt, Rinehart & Winston, 1968.

Dugan, M., & Sheridan, C. Effects of instructed imagery on temperature of hands. *Perceptual and Motor Skills,* 1976, *42,* 14.

Dutta, S., & Kanungo, R. N. Retention of affective material: A further verification of the intensity hypothesis. *Journal of Personality and Social Psychology,* 1967, *5,* 476–481.

Eddy, M. B. *Science and health with key to the scriptures.* Boston: Trustees under the will of Mary Baker Eddy, 1875.

Edwards, A. L. *Experimental design in psychological research.* New York: Holt, Rinehart & Winston, 1972.

Edwards, A. L. Political frames of reference as a factor influencing recognition. *Journal of Abnormal and Social Psychology,* 1941, *36,* 34–61.

Elliot, D. H. Characteristics and relationships of various criteria of college and university teaching. *Purdue University Studies in Higher Education,* 1950, *70,* 5–61.

Frey, P. W. Student instructional ratings and faculty performance. Paper presented at the Annual Meeting of the American Educational Research Association, New Orleans, February–March, 1973.

Games, P. A. Multiple comparisons of means. *American Educational Research Journal,* 1971, *8,* 531–565.

Gessner, P. K. Evaluation of instruction. *Science,* 1973, *180,* 566–569.

Gottlieb, A., Gleser, G., & Gottschalk, L. Verbal and physiological responses to hypnotic suggestion of attitudes. *Psychosomatic Medicine,* 1967, *29,* 172–183.

Gottschalk, L. A study of conditioned vasomotor response in ten human subjects. *Psychosomatic Medicine,* 1946, *8,* 16–27.

Greenwald, H. *Direct decision therapy.* San Diego: Edits, 1973.

Harris, M. D., O'Hare, D., Pakter, J., & Nelson, F. G. Legal abortion 1970–1971—the New York City experience. *American Journal of Public Health,* 1973, *63,* 409–418.

Hays, W. L. *Statistics for the social sciences.* New York: Holt, Rinehart & Winston, 1973.

Herr, E. L., & Cramer, S. H. *Vocational guidance and career development in the schools: Toward a systems approach.* New York: Houghton Mifflin, 1972.

Herr, E. L., Horan, J. J., & Baker, S. B. Clarifying the counseling mystique. *American Vocational Journal,* 1973, *48,* 66–72.

Hogan, R., Mankin, D., Conway, J., & Fox, S. Personality correlates of undergraduate marijuana use. *Journal of Consulting and Clinical Psychology,* 1970, *35,* 58–63.

Ivey, A. E. *Microcounseling innovations in interviewing training.* Springfield, Ill.: Charles C Thomas, 1971.

Jones, E. E., & Aneshanel, J. The learning and utilization of contravaluant material. *Journal of Abnormal and Social Psychology,* 1956, *53,* 27–33.

Jones, E. E., & Kohler, R. The effects of plausibility on the learning of controversial statements. *Journal of Abnormal and Social Psychology,* 1958, *57,* 315–320.

Kaltreider, N. B. Emotional patterns related to delay in decision to legal abortion. *California Medicine, Western Journal of Medicine,* 1973, *118,* 23–27.
Katchadourian, H. A., & Lunde, D. T. *Fundamentals of human sexuality.* New York: Holt, Rinehart & Winston, 1972.
Keeler, M. H. Marijuana-induced hallucinations. *Diseases of the Nervous System,* 1968, *29,* 314–315.
Keniston, K. Heads and seekers. Drugs on campus, counter-cultures and American society. *American Scholar,* 1968/1969, *69,* 97–112.
Keppel, G. *Design and analysis: A researcher's handbook.* Englewood Cliffs, N.J.: Prentice-Hall, 1973.
Krumboltz, J. D., & Thoresen, C. E. (Eds.) *Behavioral counseling cases and techniques.* New York: Holt, Rinehart & Winston, 1969.
Lazarus, A. A. *Behavior therapy and beyond.* New York: McGraw-Hill, 1971.
Levine, J. M., & Murphy, G. Learning and forgetting controversial material. *Journal of Abnormal and Social Psychology,* 1943, *38,* 507–517.
Linton, M., & Gallo, P. S. *The practical statistician: Simplified handbook of statistics.* Monterey, Ca.: Brooks/Cole, 1977.
Loftus, E. F., & Palmer, J. C. Reconstruction of automobile destruction: An example of the interaction between language and memory. *Journal of Verbal Learning and Verbal Behavior,* 1974, *13,* 585–589.
Loftus, E. F., & Zanni, G. Eyewitness testimony: The influence of the wording of a question. *Bulletin of the Psychonomic Society,* 1975, *5,* 86–88.
Lutkus, A. D. The effect of "imaging" on mirror image drawing. *Bulletin of the Psychonomic Society,* 1975, *5,* 389–390.
Martin, C. D. Psychological problems of abortion for the unwed teenage girl. *Genetic Psychology Monographs,* 1973, *88,* 23–110.
Maslach, C., Marshall, G., & Zimbardo, P. Hypnotic control of peripheral skin temperature. *Psychophysiology,* 1972, *9,* 600–605.
McDaniel, W. F., & Vestal, L. C. Issue relevance and source credibility as a determinant of retention. *Bulletin of the Psychonomic Society,* 1975, *5,* 481–482.
Melamed, L. Therapeutic abortion in a midwestern city. *Psychological Reports,* 1975, *37,* 1143–1146.
Morsh, J. E., Burgess, G. G., & Smith, P. N. Student achievement as a measure of instructor effectiveness. *Journal of Educational Psychology,* 1956, *47,* 79–88.
Murstein, B. I. Introductory on Philip Goldberg's review of sentence completion methods in personality assessment. In B. I. Murstein (Ed.), *Handbook of projective techniques.* New York: Basic Books, 1965.
Myers, J. L. *Fundamentals of experimental design.* Boston: Allyn & Bacon, 1972.
Nowlis, H. *Drugs on the college campus.* Garden City, N.Y.: Doubleday (Anchor Books), 1969.
Osofsky, J. D., & Osofsky, H. J. The psychological reaction of patients to legalized abortion. *American Journal of Orthopsychiatry,* 1972, *42,* 48–60.
Pakter, J., Harris, D., & Nelson, F. Surveillance of the abortion program in New York City: Preliminary report. *Clinical Obstetrics and Gynecology,* 1971, *14,* 267–299.
Peck, R. F., & Veldman, D. J. Personal characteristics associated with effective teaching. Paper presented at the Annual Meeting of the American Educational Research Association, New Orleans, February–March, 1973.
Podmore, F. *From Mesmer to Christian Science: A short history of mental healing.* New Hyde Park, N.Y.: University Books, 1963.

Reiss, A. J. *Occupations and social status.* New York: Free Press, 1961.
Robbins, E., Robbins, L., Frosch, W., & Stern, M. College student drug use. *American Journal of Psychiatry,* 1970, *126,* 1743–1750.
Rodin, M., & Rodin, B. Student evaluations of teachers. *Science,* 1972, *177,* 1164–1166.
Rotter, J. B., & Rafferty, J. *Manual for the Rotter Incomplete Sentences Blank, College Form.* New York: Psychological Corporation, 1950.
Ryder, N. B. Contraceptive failure in the United States. *Family Planning Perspectives,* 1973, *5,* 133–142.
Schutz, S. R., & Keislar, E. R. Young children's immediate memory of word classes in relation to social class. *Journal of Verbal Learning and Verbal Behavior,* 1972, *11,* 13–17.
Siegel, S. *Non-parametric statistics.* New York: McGraw-Hill, 1956.
Smith, E. M. A follow-up study of women who request abortion. *American Journal of Orthopsychiatry,* 1973, *43,* 547–585.
Smith, S. S., & Jamieson, B. D. Effects of attitude and ego involvement on the learning and retention of controversial material. *Journal of Personality and Social Psychology,* 1972, *22,* 303–310.
Sokolov, Y. *Perception and the conditioned reflex.* New York: Macmillan, 1963.
Starch, D. A demonstration of the trial and error method of learning. *Psychological Bulletin,* 1910, *7,* 20–23.
Sullivan, A. M. A structured individualized approach to the teaching of introductory psychology. *Programmed Learning,* 1969, *6,* 231–242.
Sullivan, A. M., & Skanes, G. R. Validity of student evaluation of teaching and the characteristics of successful instructors. *Journal of Educational Psychology,* 1974, *66,* 584–590.
Supreme Court Reporter, 1973, 3: Rules of Evidence for United States Courts and Magistrates.
Taft, R. Selective recall and memory distortion of favorable and unfavorable material. *Journal of Abnormal and Social Psychology,* 1954, *49,* 23–28.
Taub, E. Some methodological issues in the training of self-regulation of skin temperature. *Proceedings of the Biofeedback Research Society, Sixth Annual Meeting.* Monterey, Ca., 1975.
Wallace, W. G., Horan, J. J., Baker, S. B., & Hudson, G. R. Incremental effects of modeling and performance feedback in teaching decision-making counseling. *Journal of Counseling Psychology,* 1975, *22,* 570–572.
Waly, P., & Cook, S. W. Attitude as a determinant of learning and memory: A failure to confirm. *Journal of Personality and Social Psychology,* 1966, *4,* 280–288.
Winer, B. J. *Statistical principles in experimental design.* New York: McGraw-Hill, 1971.
Woodworth, R. S., & Schlosberg, N. *Experimental psychology.* New York: Holt, Rinehart & Winston, 1954.
Yeatts, L. M., & Brantley, J. C. Improving a cerebral palsied child's typing with operant techniques. *Perceptual and Motor Skills,* 1976, *42,* 197–198.
Zimmerman, C., & Bauer, R. A. The effect of an audience upon what is remembered. *Public Opinion Quarterly,* 1956, *20,* 238–248.

# Index

Boldface numbers refer to definitions in the glossary.

abbreviations, 285
abortion survey. *See* Melamed survey.
abscissa. *See* horizontal axis.
abstracts, 20
  how to read, 269–271
  how to write, 281–282
  and other incomplete descriptions, 273–275
*Abstracts, Psychological,* 22, 293
American Psychological Association
  code of ethics, 301–303
  *Ethical principles in the conduct of research with human participants,* 61, 151
  format for report, 357–369
  *Publication manual,* 276
  *Rules regarding animals,* 61, 151
analysis of variance (F ratio), 213, 222, **371**.
  *See also* interactions; main effects.
  deciding to use, 221–224
  determining statistical significance of, 353–356

analysis of variance *(cont.)*
  one-way between-groups, 321–324
  one-way within-subjects, 325–329
  table for, 353–356
  two-way between-groups, 329–334
  two-way mixed, 334–341
animal subjects, 61, 151
*Annual Review of Psychology,* 22
anonymity, 103, 152
appendices, 282, 286
applied research, 8. *See also* practical implications.
assignment of subjects. *See also* nonequivalent groups.
  compared to measurement, 39–45
  compared to selection, 44–45
  to matched groups, 141
  to mixed and unmixed treatment groups, 117, 119, 124, 140–141
  random, 122–123, 142
association between variables, 32–35, 37, 39, **371**
  approaches in expressing, 157

association between variables, *(cont.)*
  distinguished from causation, 39–51
  interpretations of, 194–195
  patterns of, 34
  problems about, 40, 41, 44–45, 48–49
  seeing in your data, 204–209, 221
  trustworthiness of. *See* inferential statistics; trustworthiness of results.
attrition of subjects, 125, 142–143
average. *See* central tendency; mean.

bar graphs
  drawing, 217–220
  interpreting, 173–175, 180, 182, 184, 187–188, 193–194
baseline, 88, 158, 159, **371**
before-after design, 86–89, **371**
  results section using, 158
before-after variable, 86–89, **371**
  deciding to use, 100–102
  minimizing confounds with, 134–139, 147–148, 154
  operationally defining, 103–108
  possible confounds with, 112–132
between-groups design. *See* between-subjects design.
between-subjects design, **371**
  inferential statistics for, 315–318, 321–324, 329–334
between-subjects variable, 81–84, **371**
  deciding to use, 100–102
  minimizing confounds with, 134–143, 154
  operationally defining, 103–108
  possible confounds with, 112–125
  problems about, 82–83, 84
  recognizing, 86, 89–90
bias, experimenter. *See* experimenter bias.
bias, subject. *See* subject bias.
biserial, 163
  deciding to use, 214
block randomization, 127, 142, 146, **371**
Bruning and Kintz (statistics book), 213, 223

C (contingency coefficient), 214
calculator, 305
card file, 274–275
case study, 8, 9, 17, **371**. *See also* Cohen and Smith case study.

causal relationship between variables, **371**. *See also* control techniques.
  distinguished from associations, 39–51
  drawing conclusions about, 87, 101–102, 111–112, 119–122, 131, 147–148, 227, 230
  problems about, 40, 41, 44–45, 48–49
  as suggested by article titles, 267–268
central tendency, 170
  calculating measures of, 306–307
  selecting measure of, 215–216
cerebral palsy study. *See* Yeatts and Brantley typing study.
chance. *See* inferential statistics, random.
chi-square ($X^2$), 165, 213
  calculating, 312–315
  deciding to use, 221–224
  determining statistical significance of, 350–351
  table of, 350–351
Christian Scientist case study. *See* Cohen and Smith case study.
coefficient of concordance (K), 163
Cohen and Smith case study, 9–13
complete randomization, 127, 146, **372**
conclusions. *See also* causal relationship between variables; control techniques.
  drawing your own, 255–259
  evaluating, 228–233, 273–275
  problems about, 230–233, 258–259, 262–263
confidentiality, 103, 152
confounded variables, 112, 151, 154, 228, 260–261, **372**. *See also* experimenter bias; extraneous-variable confounds; nonequivalent groups; novelty effect; practice or fatigue effects; subject bias; variable-connected confounds.
contingency coefficient (C), 214
control group, 121, 135, 136, 148, **372**
  problems about, 138–139
control techniques, 112, **372**
  for experimenter bias, 115, 135
  for extraneous-variable confounds, 112–113, 140–143, 147, 154
  for nonequivalent groups, 122–123, 140–143
  for novelty effects, 130–131, 147–148

# Index

control techniques, *(cont.)*
  for practice or fatigue effects, 126–127, 143–147
  for subject bias, 114–115, 134–135
  for variable-connected confounds, 119–122, 136–137, 147
control treatment, 121, 135, 136, 148, **372**
  problems about, 138–139
correlational approach, 160–165, 197
  calculating statistics for, 310–315
  deciding to use, 200
  inferential statistics for, 221–223
  organization of data in, 203, 205–206
correlational studies, 43–44, **372**
correlation between variables. *See* association between variables.
correlation coefficient, 34–35, 162–165, **372**
  calculating Pearson $r$, 310–312
  determining statistical significance of, 221–223, 349–350
  list of, 163
  as related to scatter graphs and regression lines, 161–164
  selecting, 214–215
counselling study. *See* Wallace et al. teaching method study.
counterbalancing, 126–127, 145, 146, **372**
cover page, 283, 358
criterion, 81
criterion variable, 43–44, **372**
Cross and Davis marijuana study
  abstract, 269
  discussion, 233–234
  introduction, 29–30
  method, 70–71
  reference, 29
  results, 170–171
cross-sectional variable, 84, **372**. *See also* between-subjects variable.
  deciding to use, 100–101
  problems about, 84
cumulative responses, 196

data
  collecting, 154
  display of, 159, 160, 205–209, 216–220
  organization of, 200–204
  recording, 153

data, *(cont.)*
  summarizing, 213–216
debriefing, 152
deception, 62, 152
definition, operational. *See* operational definition.
degrees of freedom. *See* analysis of variance; chi-square; Pearson $r$; $t$ test.
demand characteristics, 114, **372**. *See also* subject bias.
dependent variable, 42–51, **372**. *See also* experimental designs.
  choosing your own, 53–56, 59
  in descriptions of results, 157, 159
  interpretation of, 195–197
  operationally defining, 107
  problems about, 43, 44–45, 48–49
descriptive statistics, 159, **372**
  calculating, 306–312
  selecting, 214–216
design, experimental. *See* experimental designs.
$df$ (degrees of freedom). *See* analysis of variance; chi-square; Pearson $r$; $t$ test.
differences-between-frequencies approach, 167–170, 197–198
  calculating statistics for, 306–307, 312–315
  deciding to use, 200, 215–216
  organization of data in, 201–202
  selecting inferential statistics for, 221–224
differences-between-levels approaches. *See* differences-between-frequencies approach; differences-between-means approach.
differences-between-means approach, 170, 198
  calculating statistics for, 307–309, 315–345
  deciding to use, 200, 215–216
  organization of data in, 201–202
  selecting inferential statistics for, 221–224
difference scores, 195–196
directed observations, 17–18
discrete data, 211
discussion sections, 20
  combined with results sections, 248–249
  how to read, 227–230, 233–240, 246–255
  how to write, 280–281
display techniques, 216–220
double-blind studies, 115, **372**

drop-out, 125, 142–143
Dugan and Sheridan hand temperature experiment, 272–273
Duncan's multiple range test, 222
Dunnett's test, 172

E (experimenter), 129, 361
ecological validity. See experimental situation; generalization.
Edwards (statistics book), 223
environmental variables, 42, 60, 100–101
et al. (et alia), 37, 360
ethical considerations, 301–303
  for animals, 103, 151
  coercion, 97
  comfort of human subjects, 151
  confidentiality, 103, 152
  deception, 152
  explanation of experiment, 152
  informed consent, 93–96, 151
  in manipulating a variable, 61–63
Ethical principles in the conduct of research with human participants, 61, 151
experiment, 18, **372**
experimental designs, 17–18, 81, **372**
  before-after, 86–89, **371**
  between-subjects. See between-subjects variable.
  correlational, 43–44, **372**
  descriptions of, 89–90, 92, 148–150
  factorial, 47–51, 89, **373**
  mixed, 89, 90, **374**
  problems about, 48–49, 55–56, 61
  repeated-measures, 127, **376**
  selecting your own, 53–56, 59–64, 100–108, 142, 147–148
  within-subject. See within-subjects design.
experimental hypothesis, 7, 31, 37, 39, 51, 64–65, **373**
experimental reports. See reports of experiments.
experimental situation
  generalization from, 77, 80, 86, 94
  selecting your own, 93–96
experimenter bias, 73–74, 103, 108, 115, **373**
  control techniques for, 115, 135
  problems about, 116–119, 129
ex post facto studies. See pseudo-experiments.

external validity. See generalization.
extraneous-variable confounds, 112–113, **373**
  control techniques for, 112–113, 140–143, 147, 154
  problems about, 113–114, 116–119, 123–124

F. See analysis of variance.
factor analysis, 197, **373**
factorial designs, 47–51, 89, 149–150, **373**. See also interactions, main effects.
  inferential statistics for, 221–224
  problems about, 48–49, 82–84
fatigue effects, **373**. See also practice or fatigue effects.
field studies, 77, **373**
  deciding to perform, 93–96
figures
  deciding which kind to use, 217–218
  drawing, 218–220
  interpreting, 159, 160–163, 173–175, 181–184, 186–189, 193–196
  in reports of experiments, 280, 295, 363, 368–369
flow charts
  activities after completing an experiment, 291
  deciding how to display results, 218
  deciding which correlation coefficient to use, 214
  deciding which inferential statistics to use, 223–224
  deciding which measures of central tendency and variability to use, 215
  determining scale of measurement of a variable, 212
  how to read, 212–213
F ratio. See analysis of variance.
frequencies, 167, 171
  translating to percentages or proportions, 197–198, 205, 207
frequency distributions, **373**
  drawing, 206–209
  interpreting, 173–175

generalization, 69
  beyond operational definitions, 78
  from the experimental situation, 77, 80, 86, 94

# Index

generalization *(cont.)*
  problems about, 75–76, 77–78, 79
  from samples to larger groups, 74–75, 80, 85, 89, 97–98
groups
  assigning subjects to. *See* assignment of subjects.
  control. *See* control group.
  mixed and unmixed treatment, 117, 119, 124, 140–141
  nonequivalent. *See* nonequivalent groups.
  random. *See* random assignment.

hand temperature experiment. *See* Dugan and Sheridan hand temperature experiment.
Hawthorne effect, 130. *See also* novelty effect.
headings, 284, 361
horizontal axis, 159, 218–220
hypothesis, **373**. *See also* experimental hypothesis; null hypothesis; scientific hypothesis.
hypothesis-testing statistics. *See* inferential statistics.

ideas for your own experiment, 26–27, 57–58
independent variable, 42–51, **373**. *See also* before-after variable; between-subjects variable; experimental designs; within-subjects variable.
  choosing your own, 53–56, 59
  in descriptions of results, 157, 159
  operationally defining, 105–106
  problems about, 43, 44–45, 48–49
inferential statistics, **373**
  calculating, 312–356
  compared to estimates of variability, 175–177
  interpreting, 165–167, 197–198, 221–222
  post hoc, 172
  as related to conclusions, 229
  reporting, 280, 286, 363
  selecting, 221–224
informed consent, 93–96, 151–152, **373**
instructions, 152
interactions, 51, 181–195, **373**
  in correlational approach, 254
  problems about, 182–185, 188–190, 193–194

interactions, *(cont.)*
  testing statistical significance of, 222
interval scale of measurement, 211, **373**
introduction sections, 19
  how to read, 29–31, 35–39, 45–47, 49–51
  how to write, 277–278
inverse relationship between variables, 34

jargon, 19
journal articles
  difficulty in reading, 19
  how to find, 20–25, 57
  sections of, 19–20
journals, 20–21
  books listing, 293
  list of titles and Library of Congress call numbers, 293–300

$K$ (coefficient of concordance), 163
Kendall's tau ($\tau$), 163
Keppel (statistics book), 223

laboratory experiments
  deciding to perform, 93–96
  generalizing from, 77, 80, 86, 94
law, 6, **373**
leading-question study. *See* Loftus and Zanni question-wording study.
levels of a variable, 31, **373**
  in between-subjects and within-subjects variables, 82
  in descriptions of experimental designs, 89–90, 92
  discrete, 211
  measuring, 209–211
  number of, 105–107
  operationally defining, 105–108
  order of, 144
levels of significance, 166
linear relationship, 34, 161, **373**
line of best fit. *See* regression lines.
line graphs
  drawing, 217–220
  interpreting, 159, 181–182, 183, 184, 186–187, 189, 194, 195
Linton and Gallo (statistics book), 213
Loftus and Zanni question-wording study
  abstract, 270

Loftus and Zanni *(cont.)*
    controls for confounding in, 126–127
    discussion, 239–240
    method, 85
    reference, 85
    results, 168–169
longitudinal variable, 84, **373**. *See also*
        within-subjects variable.
    deciding to use, 100–101
    problems about, 84
Lutkus mirror-image drawing study
    controls for confounding in, 122–123
    method, 80
    reference, 80
    results, 178

M. *See* mean.
McDaniel and Vestal source credibility study
    abstract, 270
    introduction, 46
    method, 118–119
    reference, 45
    results, 185–186
main effects, 177–179, 194–195, **374**
    problems about, 179–180, 182–185,
        188–190
    testing statistical significance of, 222
manipulated variables, 39, 42–45, 47–51, **374**
    deciding to use, 60–63
    minimizing confounds with, 134–148, 154
    operationally defining, 105–106
    possible sources of confounding with,
        112–119, 121–132
    problems about, 43, 44–45, 48–49, 60–61,
        124
Mann-Whitney U, 169, 213
    deciding to use, 221–224
marijuana study. *See* Cross and Davis
        marijuana study.
matching, 119, 122, 136–137, 141, 143, **374**
mean, 71, 170, 213, **374**
    calculating, 307–308
    deciding to use, 215
measurement, 31, 209–211, **374**
    scale of. *See* scales of measurement.
measurement of subjects
    compared to assignment, 39–45
    compared to selection, 63

measures
    nonreactive, 77, 96
    reactive, 77, 94–95
    repeated. *See* repeated measures.
    unobtrusive, 77, 94, 96
median, 213
    calculating, 308
    deciding to use, 215
Melamed survey, 14–17
method sections, 20
    how to read, 69–71, 79–81, 84–85, 88–89,
        90–92, 116–119, 122–123, 125–131
    how to write, 279–280
minimal statistics approach, 158, 197
    deciding to use, 200
    organization of data in, 201
mirror-image study. *See* Lutkus mirror-image
        drawing study.
mixed design, 89, 90, **374**
mixed treatment groups, 117, 119, 124,
        140–141
mode, 213
    calculating, 306
    deciding to use, 215
MS (mean square). *See* analysis of variance.
mu ($\mu$). *See* mean.
Myers (statistic book), 223

n, N, 170
naturalistic observation. *See* field studies;
        nonmanipulated variables; nonreactive
        measure.
negative associations, 34, **374**
negative correlations, 34–35, 161–164
Newman-Keuls test, 172, 222
    calculating, 341–345
    deciding to use, 221–224
nominal scale of measurement, 210, **374**
nominal variables, 31, **374**
    in the correlational approach, 203
    operationally defining, 105–106
    in your experiment, 58–59
nonequivalent groups, 122, 142–143, **374**
    after random assignment, 261
    control techniques for avoiding, 122–123,
        140–143
    problems about, 123–125

nonlinear relationships between variables, 34, **374**
nonmanipulated variables, 39, 42–45, 47–51, **374**
 deciding to use, 60–63
 minimizing confounds with, 134–136, 140–148, 154
 operationally defining, 103–108
 possible sources of confounding with, 112–120, 121–132
 problems about, 43, 44–45, 48–49, 124
 reasons for lack of causal conclusions from, 119–120
nonmonotonic relationships between variables, 34, 161, **374**
nonparametric statistics, 213, **374**
nonreactive measure, 77, 96, **374**
nonsignificant difference, **374**. *See also* inferential statistics.
nonsignificant relationship, 165–167, 221–222, 260, **374**. *See also* inferential statistics.
normal distribution, 212
 problems using, 174
novelty effect, 130–132, **374**
 control techniques for, 130–131, 147–148
 problems about, 131–132
null hypothesis, 65, 166, **375**
number of subjects, 99–100, 148–150
numbers, 285

observation, 4–8, 13, 17
 directed, 17–18
 problems about, 5
observational strategies, 8–18. *See also* experimental designs.
observational studies, 43–44, **375**
operational definition, 71–74, **375**
 creating your own, 103–108
 generalizing beyond, 78
 post hoc, 235
 problems about, 73, 79, 103–104
 recognizing, 80–81, 86, 89
 as related to conclusions, 230
oral reports, 20, 287, 290
order effects, 125, **375**. *See also* practice or fatigue effects.
 as related to conclusions, 261–262

order, treatment, 125–130, 144–147
ordinal scale of measurement, 210, **375**
ordinate. *See* vertical axis.

$p$, 165–166
page numbers, 284, 359
parametric statistics, 213, **375**
*PASAR*, 23
Pearson $r$, 213
 calculating, 310–312
 deciding to use, 214
 determining statistical significance of, 345, 350
percentages, 167, 205, 207
permutations, 85, 126
 random, table of, 346–347
phi coefficient ($\phi$), 163
 deciding to use, 214
pilot study, 154
placebo, 115, 135, **375**
point biserial, 163
 deciding to use, 214
population, 74–75, 98–99, **375**
position effects, 125, **375**. *See also* practice or fatigue effects.
post hoc, **375**
post hoc comparisons, 171–172
 in combined results and discussion sections, 248–249
 tests for statistical significance of, 222
post hoc operational definition, 235
post hoc relationships, 65
post hoc theories, 264
practical implications, 236–237, 265. *See also* applied research.
practice articles
 questions about, 24, 51–52, 92, 132–133, 198, 255–256, 275
 selecting, 23
practice effects, **375**
practice or fatigue effects, 125, **375**
 control techniques for, 126–127, 143–147
 problems about, 127–130
 as related to conclusions, 261–262
prediction, 6–8, 17–18, 32–35, 60, 69, 74–79
 compared to causation, 39–51
 as influenced by interactions, 195
 from regression lines, 161–162

prediction (cont.)
  as related to conclusions, 227
  testing, 245–246, 264–265
  from theories, 7, 55, 245–246, 264–265
predictor variable, 43–44, **375**
privacy, invasion of, 95–96
probability. See inferential statistics.
proportions, 167, 171, 205, 207
pseudo-dependent variable, 43–44, **375**
pseudo-experiments, 43–44, **375**
pseudo-independent variable, 43–44, **375**
*Psychological Abstracts,* 22, 293
*Psychological Bulletin,* 22, 294
*PsycInfo,* 23
*Publication manual,* 276
publication style, 276–286, 290, 357–369
punctuation, 283
pure research, 8. See also theoretical implications.

q (Studentized range statistic), table of, 343
quasi-experiment. See before-after design; correlational studies; pseudo-experiments.
question-wording study. See Loftus and Zanni question-wording study.

r. See Pearson r; correlation coefficient.
random, 75, **375**
random assignment, 122–123, 142, **376**
random blocks, 142, 348–349
random groups. See between-subjects design; random assignment.
random numbers, 98, 346–347
random permutations, 142, 348–349
random samples, 74–75, 80, 98–99, **376**
range, 173, **376**
  calculating, 307
  deciding to use, 215
ratio scale of measurement, 211, **376**
raw data. See data.
reactive measure, 77, 94–95, **376**
reference notes, 281, 286, 365
references
  interpreting, 21
  locating, 20–25, 27
  reporting, 281, 284, 286, 360, 365, 366
  using to make a point, 278–279

regression lines, 34, 161–162, **376**
  problem about, 162
relationship between variables. See association between variables.
reliability, 70, 72
reliability of results. See trustworthiness of results.
repeated-measures designs, 127, **376**
repeated-measures effects, **376**. See also practice or fatigue effects.
repeated-measures variables, 127, **376**
  deciding to use, 101–102, 144–146
replication, 27, 289
reports of experiments, 19
  difficulty in reading, 19
  incomplete, 273–275
  locating, 20–25, 27
  with no division of sections, 271–273
  oral, 20, 287, 290
  publication style as different from class style, 290
  reading whole, 271
  sections of, 19–20
  typed sample, 357–369
  typing instructions, 282–286, 357–369
  writing, 276–282
results sections, 20
  combined with discussion sections, 248–249
  how to read, 158–159, 163–165, 167–172, 175–179, 185–187, 190–191, 197–198
  how to write, 280
rho ($\rho$). See Spearman's rho.
role-playing, 63, 96
*Rules regarding animals,* 61, 151
running head, 283, 358

S (subject), 129, 361
s, S. See standard deviation.
$s^2$, $S^2$. See variance.
sample articles. See practice articles; typing a report.
samples
  generalizing from, 74–75, 80, 85, 89, 97–98
  random, 74, 98–99
  selecting your own, 96–99, 140–142

# Index

scales of measurement, 209–212, **376**
scatter graphs, **376**
  drawing, 205–206, 217–220
  interpreting, 160–164
Scheffe's test, 172, 222
Schutz and Keislar word-class study
  abstract, 270
  introduction, 49–50
  method, 128–129
  reference, 49
  results, 190–191
  Study II, plus discussion, 246–249
science, 3–8
*Scientific citations*, 23
scientific hypothesis, 6
SD. *See* standard deviation.
selection of subjects
  compared to assignment, 44
  compared to measurement, 63
  in your experiment, 96–99, 140–142
Siegel (statistics book), 224
sigma ($\sigma$). *See* standard deviation.
sigma$^2$ ($\sigma^2$). *See* variance.
significant difference, **376**. *See also* inferential statistics.
significant relationship, 165–167, 221–222, **376**. *See also* inferential statistics.
single-blind studies, 115, **376**
source credibility study. *See* McDaniel and Vestal source credibility study.
spacing, 283, 363
Spearman's rho ($\rho$), 163
  deciding to use, 214
*S*s (subjects), 129, 361
*SS* (sum of squares). *See* analysis of variance.
standard deviation, 173–174, 213, **376**
  calculating, 308–309
  deciding to use, 215
  problems about, 174
statistical significance. *See* inferential statistics.
statistics. *See* descriptive statistics; inferential statistics.
statistics books
  "cookbook" types, 213
  general, 223
  nonparametric statistics, 224
stratification, 99. *See also* matching.

subject attrition, 125, 142–143
subject bias, 114–115, **376**. *See also* demand characteristics; reactive measure.
  control techniques for, 114–115, 134–135
  problems about, 116–119, 123, 129
subject drop out, 125, 142–143
subjects, **376**
  assignment of. *See* assignment of subjects.
  between- . *See* between-subjects design, between-subjects variable.
  consent of. *See* informed consent.
  generalization from samples of, 74–75, 80, 85, 89, 97–98
  number needed for experiment, 99–100, 148–150
  protection of rights of. *See* ethical considerations.
  selecting your own, 96–99, 140–142
  within- . *See* within-subjects design; within subjects variable.
subject selection. *See* selection of subjects.
Subjects × Treatment design. *See* within-subjects design.
subject variables, 42, 60, 100, 140–142, **376**
Sullivan and Skanes teacher evaluation study
  introduction, 36
  method, 90–91
  reference, 35
  results and discussion, 164, 249–254
summary, 20. *See also* abstracts.
survey, 8, 13, 17, **376**
  Melamed, 14–17

*t*. *See t* test.
tables,
  of critical values of $\chi^2$, 350–351
  of critical values of $F$, 353–356
  of critical values of $r$, 345, 350
  of critical values of Student's $q$ statistic, 343
  of critical values of $t$, 352–353
  interpreting, 159, 179, 183, 185, 190–192
  for organizing raw data, 201–204
  of proportions or percentages, 205, 208
  of random blocks, 142, 348–349
  of random numbers, 98, 346–347
  of random permutations, 348–349

tables *(cont.)*
  in reports of experiments, 216–217, 280, 285, 363, 367
tau (τ). *See* Kendall's tau.
teacher evaluation study. *See* Sullivan and Skanes teacher evaluation study.
teaching method study. *See* Wallace et al. teaching method study.
tetrachoric, 214
theoretical implications, 241. *See also* pure research.
theories, 6–8, **377**
  developing your own, 244–245, 264
  evaluating, 242–243
  post hoc, 264
  prediction from, 245–246
  purposes, 241
  testing, 245–246, 264–265
  in your experiment, 26, 54–55, 58, 60–61, 94–95, 97–98
three-way interactions, 190–195, **377**
  problems about, 193–194
time effects, 125, **377**. *See also* practice or fatigue effects.
titles of articles
  how to read, 267–268
  how to write, 282
  problems about, 268
  in typed report, 283, 286
titles of figures, 220
titles of journals, 293–300
titles of tables, 217
treatment order, 125–130, 144–147
Treatment × Subjects design. *See* within-subjects design.
triple interaction. *See* three-way interaction.
true experiments, 43–44, **377**
trustworthiness of results
  as estimated by inferential statistics. *See* inferential statistics.
  as estimated by variability, 172–175
  influenced by post hoc comparisons, 172
  as influenced by replication, 289
  as related to conclusions, 229, 260
*t* test, 170
  calculating for independent measures, 315–318
  calculating for related measures, 318–321

*t* test *(cont.)*
  deciding to use, 221–224
  determining the statistical significance of, 352–353
Tukey test, 222
two-way interactions, 181–195, **377**
  problems about, 182–185, 188–190
typing a report, 282–286, 357–369
typing study. *See* Yeatts and Brantley typing study.

*U*. *See* Mann-Whitney *U*.
unmixed treatment groups, 117, 119, 124, 140–141
unobtrusive measures, 77, 94, 96

variability, 172–177, **377**
  around a regression line, 162–164
  calculating measures of, 307–309
  reducing, 260
  selecting a measure of, 215–216
variable-connected confounds, 119, **377**
  control techniques for, 119–122, 136–137, 147
  with manipulated variables, 121–122
  with nonmanipulated variables, 119–121
  problems about, 124, 129, 138–139
variables, 31–32, 47, **377**. *See also* association between variables; before-after variable; between-subjects variable; criterion variable; cross-sectional variable; dependent variable; environmental variables; independent variable; longitudinal variable; manipulated variables; nominal variables; nonmanipulated variables; predictor variable; pseudo-dependent variable; pseudo-independent variable; subject variables; within-subjects variable.
  problems about, 32, 33
  selecting your own, 53–56
variance, 173–174, **377**
  calculating, 308–309
  deciding to use, 215
  problems about, 174
vertical axis, 159, 218–220

Wallace et al. teaching method study
   discussion, 237–238
   introduction, 38
   method, 116–117
   reference, 37
   results, 176
Wilcoxon test, 169, 213
   deciding to use, 221–224
Winer (statistics book), 223
within-subjects design, **377**
   inferential statistics for, 318–321; 325–329, 334–341
within-subjects variable, 81–84, **377**
   deciding to use, 100–102
   minimizing confounds with, 134–139, 143–147, 154
   operationally defining, 103–108

within-subjects variable *(cont.)*
   possible confounds with, 112–121, 125–130
   problems about, 82–83, 84
   recognizing, 86, 89–90
word-class study. *See* Schutz and Keislar word-class study.

$\bar{X}$. *See* mean.

Yeatts and Brantley typing study
   abstract, 269
   discussion, 228
   method, 88
   reference, 88
   results, 158